5th Symposium on Large TPCS for Low Energy Rare Event Detection and Workshop on Neutrinos from Supernovae 2010

AA002803

Journal of Physics: Conference Series Volume 309

Paris, France
14-17 December 2010

ISBN: 978-1-61839-129-2
ISSN: 1742-6588

Printed from e-media with permission by:

Curran Associates, Inc.
57 Morehouse Lane
Red Hook, NY 12571

Some format issues inherent in the e-media version may also appear in this print version.

Printed by Curran Associates, Inc. (2011)

For permission requests, please contact the Institute of Physics at the address below.

Institute of Physics
Dirac House, Temple Back
Bristol BS1 6BE UK

Phone: 44 1 17 929 7481
Fax: 44 1 17 920 0979

techtracking@iop.org

Additional copies of this publication are available from:

Curran Associates, Inc.
57 Morehouse Lane
Red Hook, NY 12571 USA
Phone: 845-758-0400
Fax: 845-758-2634
Email: curran@proceedings.com
Web: www.proceedings.com

Papers

PREFACE

The Fifth International Symposium on large TPCs for low-energy rare-event detection was held at the auditorium of the Astroparticle and Cosmology (APC) Laboratory in Paris, on 14–17 December 2010. As for all previous meetings, always held in Paris in 2008, 2006, 2004 and 2002, it brought together a significant community of physicists involved in rare event searches and/or development of time projection chambers (TPCs). As a novelty this year, the meeting was extended with two half-day sessions on Supernova physics. These proceedings also include the contributions corresponding to the supernova sessions.

The purpose of the meeting was to present and discuss the status of current experiments or projects involving the use of TPCs to search for rare events, like low-energy neutrinos, double beta decay, dark matter or axion experiments, as well as to discuss new results and ideas in the framework of the last developments of Micro Pattern Gaseous Detectors (MPGD), and how these are being – or could be – applied to these searches. As in previous meetings in this series, the format included an informal program with some recent highlighted results, rather than exhaustive reviews, with time for discussion and interaction. The symposium, the fifth of the series, is becoming consolidated as a regular meeting place for the synergic interplay between the fields of rare events and TPC development.

The meeting started with a moving tribute by Ioannis Giomataris to the memory of George Charpak, who recently passed away. We then moved on to the usual topics like the status of some low-energy neutrino physics and double beta decay experiments, dark matter experiments with directional detectors, axion searches, or development results. A relevant subject this time was the electroluminescence in Xe TPCs, covered by several speakers. Every time the conference program is enriched with original slightly off-topic contributions that trigger the curiosity and stimulate further thought. As mentioned before, this time we enjoyed a number of contributions on supernova physics, both theoretical and experimental.

This volume contains the proceedings of the conference. The articles correspond to 31 of the talks given in the conference and they all represent high quality work in the above mentioned fields.

The successful organization of the Symposium was made possible thanks to the contribution of the members of the Organizing Committee and International Advisory Committee. We want to thank in particular D Nygren and D Sinclair for their help with the program, as well as to all the session chairmen, J Zinn-Justin, J L Vuilleumier, J Busto, D Atti'e, G Chardin, N Spooner, D Vignaud, I Giomataris, K Scholberg, H T Wong, W Fulgione and D Sinclair, for their contribution to the smooth running of the workshop. The symposium was free of fees, which was made possible thanks to the financial support from CEA/IRFU, CNRS/IN2P3, University of Zaragoza, European Research Council (through grant ERC-2009-StG-240054) and ETH-Zurich, that are all gratefully acknowledged.

Finally we want to thank the speakers for the high quality of their talks and all participants for coming to Paris and actively contributing to the meeting.

Latest results and prospects of the CERN Axion Solar Telescope

I G Irastorza[6,22], S Aune[1], K Barth[2], A Belov[3], S Borghi[2,23], H Bräuninger[4], G Cantatore[5], J M Carmona[6], S A Cetin[7], J I Collar[8], T Dafni[6], M Davenport[2], C Eleftheriadis[9], N Elias[2], C Ezer[7], G Fanourakis[10], E Ferrer-Ribas[1], P Friedrich[4], J Galán[6], A Gardikiotis[11], E N Gazis[12], T Geralis[10], I Giomataris[1], S Gninenko[3], H Gómez[6],E Gruber[13],T Guthörl[13],R Hartmann[14], F Haug[2], M D Hasinoff[15], D H H Hoffmann[16], F J Iguaz[6,24], J Jacoby[17], K Jakovčić[18], M Karuza[5], K Königsmann[13], R Kotthaus[19], M Krčmar[18], M Kuster[4,16,25], B Lakić[18], J M Laurent[2], A Liolios[9], A Ljubičić[18], V Lozza[5], G Lutz[14,23], G Luzón[6], J Morales[6,26], T Niinikoski[2,27], A Nordt[4,16,28], T Papaevangelou[1], M J Pivovaroff[20], G Raffelt[19], T Rashba[21], H Riege[16], A Rodríguez[6], M Rosu[16], J Ruz[6,28], I Savvidis[9], P S Silva[2], S K Solanki[21], R Soufli[20], L Stewart[2], A Tomás[6], M Tsagri[11,28], K van Bibber[20,29], T Vafeiadis[2,9], J Villar[6], J K Vogel[13,20,30], S C Yildiz[7], K Zioutas[2,11]

[1] IRFU, Centre d'Études Nucléaires de Saclay (CEA-Saclay), Gif-sur-Yvette, France
[2] European Organization for Nuclear Research (CERN), Genève, Switzerland
[3] Institute for Nuclear Research (INR), Russian Academy of Sciences, Moscow, Russia
[4] Max-Planck-Institut für extraterrestrische Physik, Garching, Germany
[5] Instituto Nazionale di Fisica Nucleare (INFN), Sezione di Trieste and Università di Trieste, Trieste, Italy
[6] Instituto de Física Nuclear y Altas Energías, Universidad de Zaragoza, Zaragoza, Spain
[7] Dogus University, Istanbul, Turkey
[8] Enrico Fermi Institute and KICP, University of Chicago, Chicago, IL, USA
[9] Aristotle University of Thessaloniki, Thessaloniki, Greece
[10] National Center for Scientific Research "Demokritos", Athens, Greece
[11] Physics Department, University of Patras, Patras, Greece
[12] National Technical University of Athens, Athens, Greece
[13] Albert-Ludwigs-Universität Freiburg, Freiburg, Germany
[14] MPI Halbleiterlabor, München, Germany
[15] Department of Physics and Astronomy, University of British Columbia, Vancouver, Canada
[16] Technische Universität Darmstadt, IKP, Darmstadt, Germany

[22] Attending speaker.
[23] Present address: Department of Physics and Astronomy, University of Glasgow, Glasgow, UK
[24] Present address: IRFU, Centre d'Études Nucléaires de Saclay (CEA-Saclay), Gif-sur-Yvette, France
[25] Present address: European XFEL GmbH, Notkestrasse 85, 22607 Hamburg, Germany
[26] Deceased
[27] Present address: Excellence Cluster Universe, Technische Universität München, Garching, Germany
[28] Present address: European Organization for Nuclear Research (CERN), Genève, Switzerland
[29] Present address: Naval Postgraduate School, Monterey, CA, USA
[30] Present address: Lawrence Livermore National Laboratory, Livermore, CA, USA

[17] Johann Wolfgang Goethe-Universität, Institut für Angewandte Physik, Frankfurt am Main, Germany
[18] Rudjer Bošković Institute, Zagreb, Croatia
[19] Max-Planck-Institut für Physik, Munich, Germany
[20] Lawrence Livermore National Laboratory, Livermore, CA, USA
[21] Max-Planck-Institut für Sonnensystemforschung, Katlenburg-Lindau, Germany

E-mail: `Igor.Irastorza@cern.ch`

Abstract. The CERN Axion Solar Telescope (CAST) experiment searches for axions from the Sun converted into few keV photons via the inverse Primakoff effect in the high magnetic field of a superconducting Large Hadron Collider (LHC) decommissioned test magnet. After results obtained with vacuum in the magnet pipes (phase I of the experiment) as well as with ^{4}He the collaboration is now immersed in the data taking with ^{3}He, to be finished in 2011. The status of the experiment will be presented, including a preliminary exclusion plot of the first ^{3}He data. CAST is currently sensitive to realistic QCD axion models at the sub-eV scale, and with axion-photon couplings down to the $\sim 2 \times 10^{-10}$ GeV^{-1}, compatible with solar life limits. Future plans include revisiting vaccuum and ^{4}He configurations with improved sensitivity, as well as possible additional search for non-standard signals from chamaleons, paraphotons or other WISPs. For the longer term, we study the feasibility of an altogether improved version of the axion helioscope concept, with a jump in sensitivity of about one order of magnitude in $g_{a\gamma}$ beyond CAST.

1. Introduction

Axions are light pseudoscalar particles that arise in theories in which the Peccei-Quinn U(1) symmetry has been introduced to solve the strong CP problem [1]. They could have been produced in early stages of the Universe being attractive candidates to the cold Dark Matter (and in some particular scenarios to the hot Dark Matter) that could compose up to ~1/3 of the ingredients of the Universe.

The axion-photon coupling $g_{a\gamma}$, unlike other axion couplings[2], is automatically present in every QCD axion model, and allows for the conversion of axions into photons (and viceversa) in the presence of electromagnetic fields, being the base of almost every axion detection strategy.

A combination of astrophysical and nuclear physics constraints restricts the allowed range for axion mass and couplings[3, 4, 5]. Pure cosmological arguments lead to a conservative, relatively model-independent allowed mass range of $10^{-6}\text{eV} \lesssim m_a \lesssim 1$ eV, the upper limit being recently set [6], by requiring thermal production of axions to be compatible with last CMB data. The coupling $g_{a\gamma}$ can be constrained independently by astrophysical observations, namely $g_{a\gamma} \lesssim 10^{-9}$ GeV^{-1} based on the solar standard model and helioseismological observations [7], or the so-called *globular cluster* limit of $g_{a\gamma} \lesssim 10^{-10}$ GeV^{-1} [8, 9].

Although the PQ scenario just described is probably the best motivated theoretical foundation for an axion, a phenomenologically similar scenario of a scalar or pseudoscalar particle coupled to the photon arises also in other proposed extensions of the Standard Model[31]. This broader class of models are generically referred to as axion-like particles or, referring to a even wider class of particles, WISPs, weakly interacting slim particle. In these cases, the axion-like particle is just viewed as a (peudo)scalar of mass m_a and coupling to two photons $g_{a\gamma}$, where m_a and $g_{a\gamma}$ are now two independent parameters, without further reference to the underlying theory. Therefore, every experiment (or observational consideration) based only on the property of the axions coupling to the photons could also be generalized to any axion-like particle. In particular, axion helioscope experiments are also sensitive to axion-like particles. Limits are

[31] Examples of such extensions are extra dimensions (Kaluza-Klein) theories [10, 11, 12], quintessential dark energy models [13] or superstring theories [14] among others

expressed in a phenomenological ($g_{a\gamma} - m_a$) parameter space of the axion-like particles, with the area corresponding to the QCD axions singled out (yellow band in Fig. 1). However, one must be cautious and not forget that some of the limits mentioned before are set using specific properties of the QCD axions (other than the coupling to the photons) and therefore do not hold for any axion-like particle.

Axions could be produced at early stages of the Universe by the so-called misalignment (or realignment) effect[2], plus in some cases also from decay of primordial topological defects like axion strings or walls. These relic axions could be a substantial part of the cold dark matter (CDM) if $m_a \sim 10^{-6} - 10^{-3}$ eV, and could be detected by microwave cavities as originally proposed in[15]. Such technique is followed by experiments like the Axion Dark Matter Experiment (ADMX)[16, 17], which is actively scanning the low axion mass range mentioned above. Axions could also be copiously produced in the core of the stars by means of the Primakoff conversion of the blackbody photons in the fluctuating electric field of the plasma. Solar axions have energies of 1–10 keV, and could be detected by looking for x-rays of those energies coming from their conversion in a strong laboratory magnetic field. This *helioscope* concept[15, 18], was first experimentally applied in [19] and later on by the Tokyo helioscope [20], before CAST.

Finally, evidence for axions could also be produced altogether in the laboratory, by means of strong light sources combined with strong magnetic fields[21]. Although the sensitivity of laboratory experiments is still far from that of ADMX or CAST, they do not rely on any assumption on astrophysical sources of the axions. Like CAST, but unlike ADMX, they are also sensitive to axion-like particles.

2. Status of CAST and latest results

The CAST experiment is making use of a decommissioned LHC test magnet that provides a magnetic field of 9 Tesla along its two parallel pipes of 2×14.5 cm^2 area and 10 m length, increasing the corresponding axion-photon conversion probability by a factor 100 with respect to the previous best implementation of the helioscope concept[22]. The magnet is able to track the Sun by about 3 hours per day, half in the morning and half in the evening. At its two ends x-ray detectors are placed, at the "sunrise" side, a Micromegas detector[23] and a CCD[24], and at the "sunset" side two additional Micromegas detectors, installed in 2007 replacing the former TPC[25]. All of the detector setups are conceived following low background techniques (shielding, radiopure materials). The CCD is coupled to a focusing X-ray device (X-ray telescope)[24] that enhances its signal-to-bakcground ratio by two orders of magnitude. Both the CCD and the X-ray telescope are prototypes developed for X-ray astronomy.

The experiment already released its phase I results from data taken in 2003 and 2004 with vacuum in the magnet bores[22, 26]. No signal above background was observed, implying an upper limit to the axion-photon coupling $g_{a\gamma} < 8.8 \times 10^{-11}$ GeV^{-1} at 95% CL for the low mass (coherence) region $m_a \lesssim 0.02$ eV (Fig. 1). Since 2006 the experiment runs its second phase, which makes use of a buffer gas inside the magnet bores to recover the coherence of the conversion for specific axion masses matching the effective photon mass defined by the buffer gas density. The pressure of the gas is changed in discrete small steps to scan the parameter space above $m_a \sim 0.02$ eV. The ^{4}He Run taken in 2006[27], allowed to scan axion masses up to 0.39 eV, for axion-photon couplings down to about 2.2×10^{-10} GeV^{-1}, entering into the QCD axion model band, as shown in Fig. 1. Due to gas condensation, in order to go to higher pressures, the experiment switched to ^{3}He as buffer gas in 2007. The experiment is currently inmersed in the ^{3}He Run since beginning of 2008. It should last until mid 2011 and should allow us to explore up to 1.2 eV in axion mass approximately, overlapping with the CMB upper limit on the axion mass disussed above.

At the moment of writing this paper, the experiment has explored a region of axion masses up to about $m_a \sim 1$ eV. The first part of these data, taken in 2008, and covering approximately

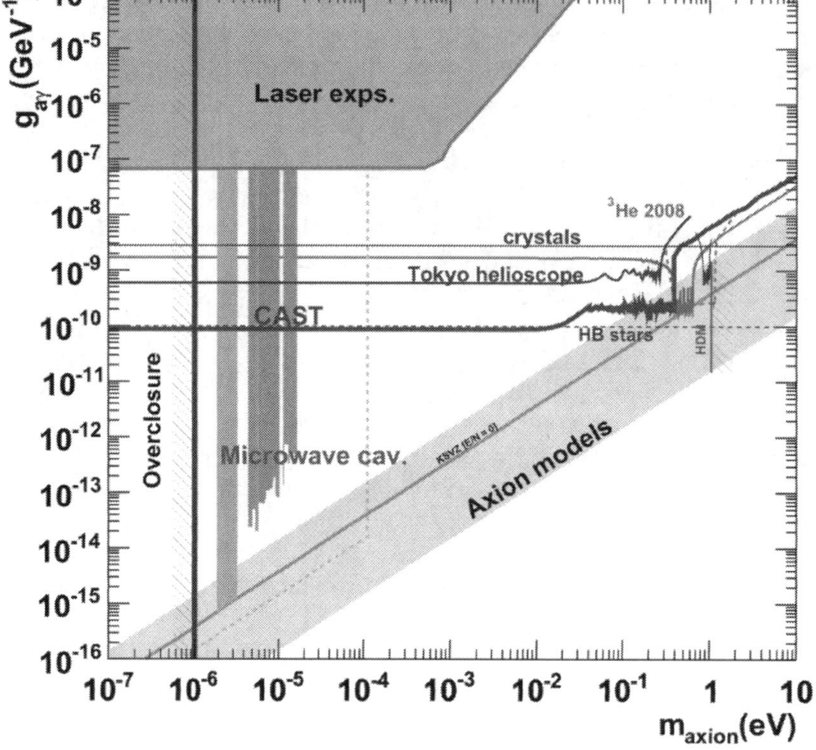

Figure 1. 95% CL exclusion line obtained from the previously published CAST data [22, 26] (blue line) and the preliminary analysis of the first 2008 ^{3}He data (red line, including only data from the 3 Micromegas detectors). Limits from other experiments or observations are included for comparison, such as the Tokyo helioscope and the cosmological upper limit on the axion mentioned in the text. The yellow area indicates the region of theoretical preference for axion models.

from 0.40 to 0.65 eV is almost fully analysed. They show no positive signal for axions and a preliminary exclusion plot (including for the moment only data from the 3 Micromegas detectors) is shown in Fig. 1. It extends previous CAST limit further into the QCD axion models band, and in particular now it traverses the benchmark model line (green line of Fig. 1), corresponding to a particular type of axions (KSVZ with E/N = 0 [28, 29]) for the mentioned axion masses. Everyday a new thin slice of untouched parameter space is being explored. Due to the sharp coherence effect, and to the fact that the parameter space to which we are sensitive now is populated by realistic QCD axion models and not excluded by previous experiments, a clear positive signal in CAST may appear at any moment.

The collaboration has performed by-product analysis of the data taken, to look for non-standard axion scenarios to which CAST would also be sensitive. Using data from the TPC in the CAST phase I, a search for 14 keV axions coming from M1 transitions in the Sun has been recently released[30]. In addition, data taken with a calorimeter during the phase I, were used to search for high energy (MeV) lines from high energy axion conversion[31]. Moreover, data has been taken with a visible detector coupled to one end of the CAST magnet, in search for axions with wavelengths in the "visible" range, possibly produced in the surface magnetic field

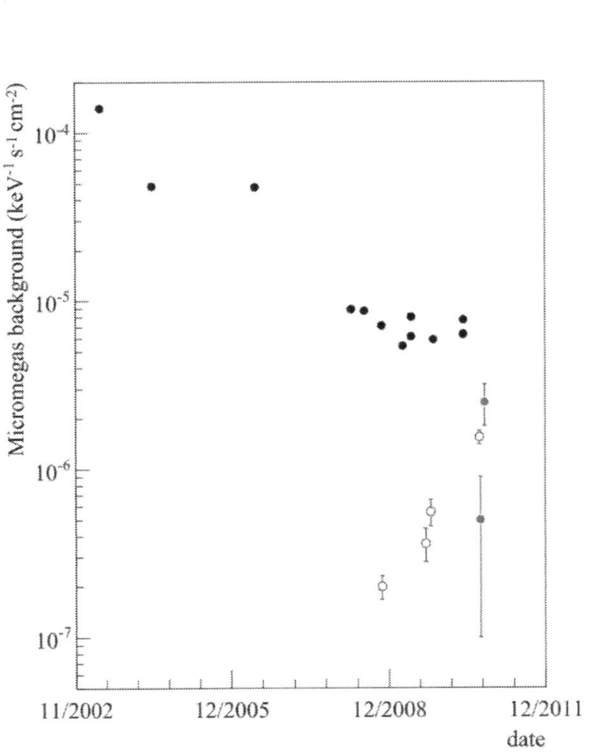

Figure 2. Background levels of Micromegas detectors over the years. Black points represents nominal values in CAST data taking campaigns. Open points correspond to ultralow background periods in CAST and the red points to data taken in special shielding conditions in the Canfranc underground laboratory.

of the Sun. A preliminary result was presented in [32], but more extended data taking is going on without interfering with the standard program of CAST thanks to a thin mirror foil that is transparent to x-rays but reflects visible photons into a dedicated setup.

Once the current ^{3}He phase finishes, the collaboration plans to revisit previous vacuum and ^{4}He configurations with improved sensitivity. This is possible thanks to the currently available detector's lower background levels. Micromegas detectors in CAST have gone through intense development over the years. By means of better shielding, cleaner detector materials, improved offline data treatment, and following the state-of-the-art *microbulk* fabrication technique, current Micromegas' backgrounds are now about a factor 20 lower than at the beginning of the experiment (see Fig. 2). A brief review of these efforts and prospects has been given by F. J. Iguaz at this conference.

3. Towards a new generation axion helioscope

For the longer term, a feasibility study is ongoing towards a new generation axion helioscope, with an improved sensitivity of at least one order of magnitude in $g_{a\gamma}$ beyond CAST. There is a important motivation to extend current CAST solar axion bounds further down to at least 10^{-11} or if possible to 10^{-12} GeV^{-1}. This region includes, at the high mass range, a large set of plausible QCD axion models, relevant as hot dark matter candidates. At lower axion masses, this region includes the ALP parameters invoked repeatedly for the emission mechanism of several classes of astrophysical phenomena[33, 34, 35, 36, 37, 38, 39]. This region is currently not excluded by any experimental results or astrophysical limits, and moreover, they are out of reach of foreseeable improvements of the experimental techniques for direct axion detection, with the exception of an enhanced axion helioscope.

Although the study is still ongoing, preliminary considerations point to the fact that an enhanced axion helioscope, based on pursuing innovations already introduced by CAST (specifically the use of x-ray optics to increase the signal-to-noise ratio and low background x-ray detectors), coupled with the use of a new, larger cross section, toroidal magnet, could achieve a sensitivity of 1 to 1.5 orders beyond current CAST limits, and therefore surpassing $g_{a\gamma} \sim 10^{-11}$ GeV^{-1}, for a wide axion mass range of up to ~0.25 eV. Such a tentatively accessible region is shown in Fig. 3, together with current limits and similar long-term prospects for the microwave cavity technique, resulting in largely complementary regions. Details on this preliminary results, including analysis of the three key elements of the new helioscope, magnet, optics and detectors, will be included in a publication in preparation.

4. Conclusions and prospects

The CAST experiment is currently finishing the ^{3}He phase of data taking. First preliminary results with ^{3}He have been presented, extending previous CAST limits further to higher masses (to ~0.65 eV) and deeper into the region favored by QCD axion models. After the ^{3}He phase, CAST plans to revisit previous configurations with improved detectors. Farther in the future, a feasibility study for a new generation axion helioscope is ongoing. Pushing innovations introduced by CAST, like the use of x-ray optics and of low background techniques, together with a specifically designed magnet, a substantial region, of at least one order of magnitude beyond current CAST limit, could be explored.

References

[1] Peccei R D and Quinn H R 1977 *Phys. Rev. Lett.* **38** 1440–1443
[2] Turner M S 1990 *Phys. Rept.* **197** 67–97
[3] Kolb E and Turner M 1990 *The Early Universe* ISBN 0-201-11603-0
[4] Burrows A, Ressell M T and Turner M S 1990 *Phys. Rev.* **D42** 3297–3309
[5] Engel J, Seckel D and Hayes A C 1990 *Phys. Rev. Lett.* **65** 960–963
[6] Hannestad S, Mirizzi A and Raffelt G 2005 *JCAP* **0507** 002 (*Preprint* hep-ph/0504059)
[7] Schlattl H, Weiss A and Raffelt G 1999 *Astropart. Phys.* **10** 353–359 (*Preprint* hep-ph/9807476)
[8] Raffelt G G 1999 *Ann. Rev. Nucl. Part. Sci.* **49** 163–216 (*Preprint* hep-ph/9903472)
[9] Raffelt G G 1986 *Phys. Rev.* **D33** 897
[10] Chang S, Tazawa S and Yamaguchi M 2000 *Phys. Rev.* **D61** 084005 (*Preprint* hep-ph/9908515)
[11] Dienes K R, Dudas E and Gherghetta T 2000 *Phys. Rev.* **D62** 105023 (*Preprint* hep-ph/9912455)
[12] Turok N 1996 *Phys. Rev. Lett.* **76** 1015–1018 (*Preprint* hep-ph/9511238)
[13] Carroll S M 1998 *Phys. Rev. Lett.* **81** 3067–3070 (*Preprint* astro-ph/9806099)
[14] Svrcek P and Witten E 2006 *JHEP* **06** 051 (*Preprint* hep-th/0605206)
[15] Sikivie P 1983 *Phys. Rev. Lett.* **51** 1415
[16] Asztalos S J *et al.* 2001 *Phys. Rev.* **D64** 092003
[17] Asztalos S J *et al.* 2004 *Phys. Rev.* **D69** 011101 (*Preprint* astro-ph/0310042)
[18] van Bibber K, McIntyre P M, Morris D E and Raffelt G G 1989 *Phys. Rev.* **D39** 2089
[19] Lazarus D M *et al.* 1992 *Phys. Rev. Lett.* **69** 2333–2336
[20] Moriyama S *et al.* 1998 *Phys. Lett.* **B434** 147 (*Preprint* hep-ex/9805026)
[21] Lindner A 2009 (*Preprint* 0910.1686)
[22] Zioutas K *et al.* (CAST) 2005 *Phys. Rev. Lett.* **94** 121301 (*Preprint* hep-ex/0411033)
[23] Abbon P *et al.* 2007 *New J. Phys.* **9** 170 (*Preprint* physics/0702190)
[24] Kuster M *et al.* 2007 *New J. Phys.* **9** 169 (*Preprint* physics/0702188)
[25] Autiero D *et al.* 2007 *New J. Phys.* **9** 171 (*Preprint* physics/0702189)
[26] Andriamonje S *et al.* (CAST) 2007 *JCAP* **0704** 010 (*Preprint* hep-ex/0702006)
[27] Arik E *et al.* (CAST) 2009 *JCAP* **0902** 008 (*Preprint* 0810.4482)
[28] Kim J E 1979 *Phys. Rev. Lett.* **43** 103
[29] Shifman M A, Vainshtein A I and Zakharov V I 1980 *Nucl. Phys.* **B166** 493
[30] Andriamonje S *et al.* (CAST) 2009 (*Preprint* 0906.4488)
[31] Andriamonje S *et al.* (CAST) 2009 (*Preprint* 0904.2103)
[32] Cantatore G *et al.* (CAST) 2008 (*Preprint* 0809.4581)
[33] Teshima M *et al.* (for the MAGIC) 2007 (*Preprint* 0709.1475)

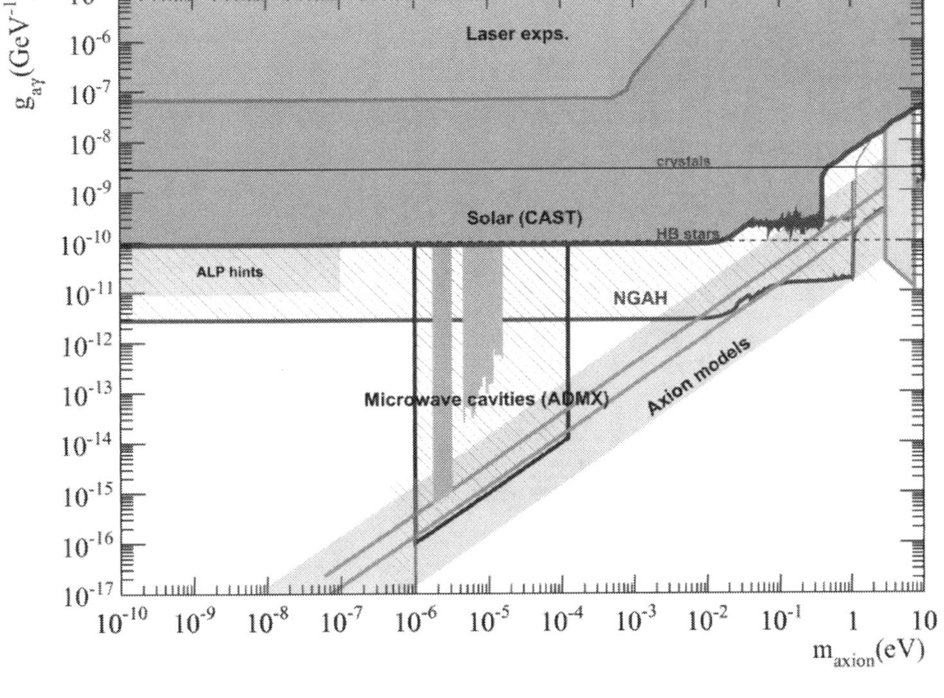

Figure 3. A comprehensive axion-like particle parameter space plot, including the regions tentatively explorable with a future new generation axion helioscope (dashed blue region) and with the microwave cavity technique (dashed brown region). Also shown is the region relevant to the astrophysical hints for ALPs (yellow rectangle in the low mass part of the plot) discussed in the text.

[34] De Angelis A, Mansutti O, Persic M and Roncadelli M 2008 (*Preprint* 0807.4246)
[35] Roncadelli M, De Angelis A and Mansutti O 2008 *AIP Conf. Proc.* **1018** 147–156 (*Preprint* 0902.0895)
[36] Gorbunov D S, Tinyakov P G, Tkachev I I and Troitsky S V 2004 *JETP Lett.* **80** 145–148 (*Preprint* astro-ph/0406654)
[37] Fairbairn M, Rashba T and Troitsky S V 2009 (*Preprint* 0901.4085)
[38] Albuquerque I F M and Chou A 2010 *JCAP* **1008** 016 (*Preprint* 1001.0972)
[39] Burrage C, Davis A C and Shaw D J 2009 *Phys. Rev. Lett.* **102** 201101 (*Preprint* 0902.2320)

Journal of Physics: Conference Series **309** (2011) 012002 doi:10.1088/1742-6596/309/1/012002

Ultralow background periods in CAST Micromegas detectors and tests in the Canfranc Underground Laboratory

F.J. Iguaz[1], T. Dafni[2], E. Ferrer-Ribas[1], J. Galán[2], J.A. García[2], A. Gardikiotis[3], I. Giomataris[1], I. Irastorza[2], J.P. Mols[1], T. Papaevangelou[1], A. Rodríguez[2], A. Tomás[2], T. Vafeiadis[4,5] and C. Yildiz[6]

[1] IRFU, Centre d'Études Nucléaires de Saclay (CEA-Saclay), Gif-sur-Yvette, France
[2] Laboratorio de Física Nuclear y Astropartículas, Universidad de Zaragoza, Zaragoza, Spain
[3] Physics Department, University of Patras, Patras, Greece
[4] European Organization for Nuclear Research (CERN), Genève, Switzerland
[5] Aristotle University of Thessaloniki, Greece
[6] Dogus University, Istanbul, Turkey

E-mail: `franciscojose.iguazgutierrez@cea.fr`

Abstract. Micromegas detectors have shown a low and stable background level in the CAST experiment due to their low radioactive materials, good energy resolution and spatial resolution. Since 2008, four ultralow background periods have been observed with the new *microbulk* detectors. During these periods, the background level reduced one order of magnitude, reaching a value around 10^{-7} s^{-1} keV^{-1} cm^{-2} between 2 and 7 keV. In all cases, the intensity of the three fluorescence lines (at 3, 6 and 8 keV) presented at the background spectrum reduced but the relative ratio remained stable. To clarify the origin of these periods and might determine the ultimate background level of these readouts, a CAST-like detector has been installed at the Canfranc Underground Laboratory. The actual level is compatible with the one measured in CAST. A new Micromegas detectors is being developped for working in CAST experiment during 2011, based on the conclusions of this work.

1. Introduction

The CERN Axion Solar Telescope (CAST) uses a prototype of a superconducting LHC dipole magnet to convert axions into detectable X-ray photons. Axions are pseudoscalar particles that appear in the Peccei-Quinn [1, 2] solution of the strog CP problem and are candidates to Dark Matter. Stars could produce axions via the Primakoff conversion of plasma photons [3]. The CAST experiment points to the Sun aiming to detect solar axions. These particles could couple to a virtual photon provided by the magnetic transverse field of 9 T, being transformed into a real photon that carries the energy and momentum of the original axion [4]. The magnet is able to follow the Sun twice a day during the sunset and sunrise for a total of about 3 hours per day. During the remaining time, i.e, when the CAST magnet is not aligned with the solar core, background data is taken.

Four X-ray detectors are installed at the ends of the 10 m long magnet in order to search for photons from Primakoff conversion. Two Micromegas detectors [6] follow the sunset. They have replaced the formerly used Time Projection Chamber [7] and show a better performance in terms of background discrimination than the previous detector. The sunrise is covered by another Micromegas detector and a combination of an X-ray mirror optics with a Charge Coupled Device (CCD) [8]. The mirror focuses the X-rays to a small spot on the CCD chip, improving the signal-to-background ratio and the sensitivity of the detector.

The experiment finished its first phase keeping the bores under vacuum by 2004 after having taken two years of data. As no significant signal above background was observed when following the Sun, an upper limit on the axion-to-photon coupling of $g_{a\gamma} < 8.8 \times 10^{-11}$ GeV^{-1} (95% CL) for axion masses $m_a < 0.02$ eV was set [5]. For extending the axion mass range, CAST was upgraded with a sophisticated gas system and novel cold windows for operation with helium gas at various pressures inside the magnetic field region. This gas would restore the coherent conversion probability from axions to photons for a narrow axion mass range.

During 2005 and 2006 the magnet was filled with ^{4}He gas and axion masses up to 0.39 eV were investigated by measuring a total of 160 presure steps between 0.08 mbar and 13.4 mbar. A typical upper limit on the axion-to-photon coupling of $g_{a\gamma} < 2.2 \times 10^{-10}$ GeV^{-1} (95% CL) for axion masses $m_a < 0.39$ eV was extracted [9]. Since the year 2008, CAST is taking data with ^{3}He in the magnet bore. At the moment of writing this proceeding, nearly 720 pressure steps have already been measured, equivalent to a gas pressure of 80 mbar and an axion mass of 0.93 eV. A preliminary exclusion plot for the ^{3}He data acquired during 2008 was generated [10], which includes data from three out of the four detectors and covers axion masses between 0.39 and 0.65 eV. The data of the remaining detector is being analyzed and the final combined limit is expected soon.

2. Micromegas detectors in CAST 2008-10

During the first phase of CAST and ^{4}He data-taking periods, a conventional Micromegas was used, covering one of the four magnet bore exits. Althought it was the only detector without any shielding, it showed the lowest background level of the three detectors: 5×10^{-5} s^{-1} keV^{-1} cm^{-2}. This fact motivated the installation of a shielding in 2008, consisting of a 25 mm thick archeological lead layer and an external 15-20 cm thick polyethylene layer. Between these two parts, there is a cadmium foil to absorb the neutrons thermalized in the polyethylene. A 5 mm-thick copper layer is placed inside the lead which works also as Faraday cage. The whole setup is flushed with nitrogen in order to remove radon.

The new shielding reduced the background level by a factor 3, reaching the level of 1×10^{-5} s^{-1} keV^{-1} cm^{-2} during the data-taking period between April and September of 2008. At the end of 2008, two of the Micromegas detectors were replaced by two new *microbulk* [13] type models. These detectors are built with very low-background materials: kapton and copper for the readout and plexiglas and aluminium for the chamber.

The data acquisition system registers the analogue signal induced in the mesh at each event with a 1 GHz FADC and the integrated charge on each strip using four Gassiplex cards. Several analysis parameters are extracted like the cluster size, multiplicity and width from the strips data and the risetime, width, amplitude and integral from the mesh pulses. An X-ray of less than 10 keV produces a primary ionization localized in a range less than 1 mm. The amplification of this charge gives a narrow pulse with a fixed risetime and a mean strip multiplicity which corresponds to around 5 mm. Cosmic muons and high energy gammas produce a spatially extended ionization, resulting to broader pulses and higher multiplicities.

Figure 1. Left: The Micromegas detector used in CAST. The active area is situated at the front part and is covered with a stainless steel window and a plexiglass piece. The strips are read by four Gassiplex cards situated at the rear part. Center: The circular lead shielding that surrounds the readout and the stainless steel tube that comes out from the magnet bore. Right: The Faraday cage and the external polyethylene shielding that covers the lead shielding. The electrical signals and the gas tubes are extracted from the copper box via feedthroughs.

The raw trigger rate of Micromegas detectors in CAST is around 1 Hz. Most of the background events like muons are rejected in the offline analysis with a discrimination algorithm, which keeps only events compatible with X-rays. As shown in figure 2 (left), the background level is reduced at least by one order of magnitude for all energies, while keeping an 80% of 6 keV X-rays from the iron source in the calibration runs. An example of the background level in CAST is shown in figure 2 (right). The remaining spectrum consists mainly in three peaks at 3, 6 and 8 keV, generated by the fluorescence lines of the Micromegas' copper and the stainless steel of the window and the tube coming from the magnet bore.

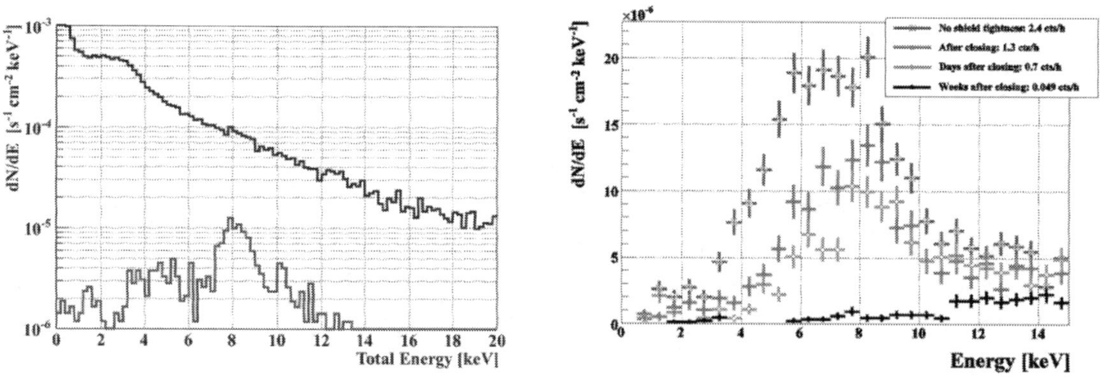

Figure 2. Left: Background energy spectrum of a CAST Micromegas detector shielded with 5 cm of lead. The black line considers all the events acquired and the blue one only those selected by the discrimination algorithms. Right: Evolution of the background energy spectra of CAST in 2008 after the application of the selection parameters. The four spectra correspond to data before the shield closing (blue line), just after (red line), one week later (green line) and two weeks later (black line).

3. The Ultralow background periods

The first ultralow background period was observed after the installation of the detector M11 in the sunrise line [12]. The detector took data during one month in stable conditions and calibrations were taken in random times each day to check the stability of the selection parameters. The background level of this detector was similar to previous ones in the same conditions but the level seemed to decrease with time. This reduction stopped on the 4th November and the background level returned to the initial values. This increase coincidenced in time with an interruption of the nitrogen flush inside the shielding. Later the background level decreased again and reached a value of 2×10^{-7} s^{-1} keV^{-1} cm^{-2}, which remained stable until the 27th of November, when a series of power cuts caused the end of the run.

The daily calibrations with the iron source showed that the detector response to the 6 keV X-rays remained unchanged, even if they were taken at random times. However, the parameters of background events changed along time, increasing the rejection power factor of the algorithms. This change was independent of the energy and as a consequence, the shape of the final background energy spectrum remained stable but it scaled down along time. This effect is shown in figure 2 (right). Note that no increase in the acquisition deadtime was observed and the raw trigger rate remained stable in 0.7 Hz during all days.

As shown in figure 3, three more ultralow background periods have been observed during the years 2009 and 2010. All of them occurred during September and October, which could indicate an environmental variation of background events. In all cases, the intensity of the three lines at 3, 6 and 8 keV drastically reduced one order of magnitude from the mean value of 6×10^{-6} s^{-1} keV^{-1} cm^{-2}. The environmental conditions (radon, humidity, temperature and air pressure) near the sunrise line are being monitored since October 2009. However, no clear correlation between radon and the ultralow background periods has been observed until now.

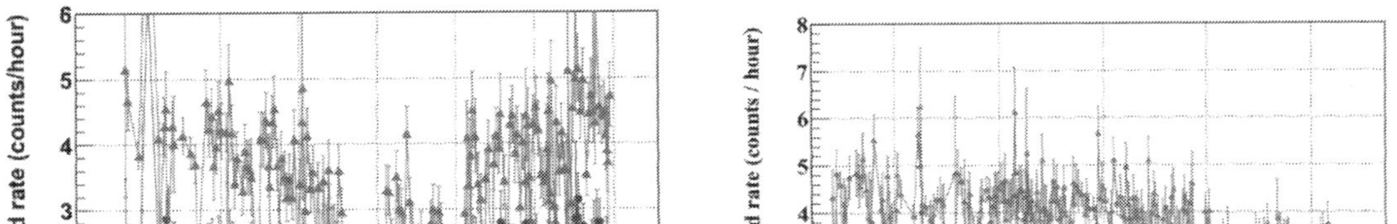

Figure 3. Background rate in the sunrise detector of CAST after the application of the selection parameters during 2009 (left) and 2010 (right) in the range of energies between 2 and 7 keV (blue circled line) and between 2 and 10 keV (red triangled line).

4. A Micromegas CAST-like detector at the LSC

A Micromegas CAST-like detector has been installed at the Canfranc Underground Laboratory with two main objectives: the determination of the ultimate background level of this kind of readouts and the clarification of the ultralow background periods observed at the CAST experiment. The detector tested is the M10, which took data in CAST only during some weeks.

Journal of Physics: Conference Series **309** (2011) 012002 doi:10.1088/1742-6596/309/1/012002

The readout has been characterized with an iron source in argon-isobutane mixtures of 2.3 and 5.0%, reaching gains up to 2×10^4 and showing an energy resolution of 13.5% FWHM at 5.9 keV in Ar-5%iC_4H_{10}.

As shown in figure 4 (left), the new setup reproduces the shielding of the sunrise detector. It consists in a 5 mm-thick copper layer, surrounded by a 25 mm-thick lead layer and all materials inside a Faraday copper box. This box is prepared for automatic calibrations with an iron source and is also gas-tight because all the gas and electrical connections are extracted via feedthroughs. These features assure a good radon extraction with a nitrogen flow of 30 l/h. The setup also includes a slow control which continuosly monitors the temperature, the gas pressure and the current of the detector.

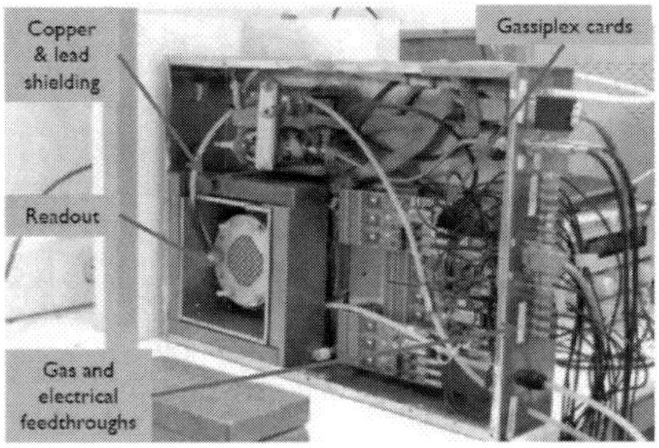

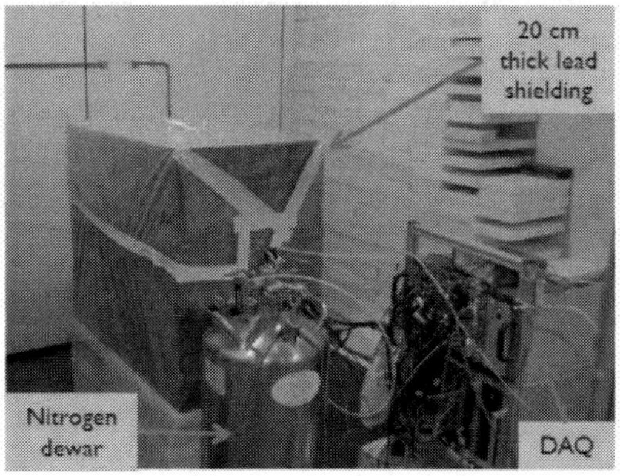

Figure 4. Left: A view of the setup installed in Canfranc. The M10 detector is shielded with a 5 mm-thick copper layer and a 25 mm-thick lead layer. The readout and the shielding are situated inside a copper Faraday cage and all the gas and electrical connections are extracted via feedthroughs. Right: The previous setup installed inside a 20 cm-thick lead castle, wrapped in a plastic bag at the Canfranc Underground Laboratory. A constant flux of nitrogen keeps the inside atmosphere clean of radon.

During the background tests in Canfranc, an external 20 cm-thick lead shielding (shown in figure 4, right) has been added around the setup to determinate the ultimate background level. This update has reduced the mean raw rate from 0.2 Hz to 0.01 Hz, but also the background rejection power due to the fewer number of external events detected. The preliminary result between 2 and 10 keV is $1.0 \pm 0.5 \times 10^{-6}$ s^{-1} keV^{-1} cm^{-2}, which is almost one order of magnitude better than in CAST. Other shielding configurations have been tested in order to check their effect in the background level and the final conclusions of this work are expected soon.

5. A new Micromegas detector for CAST

A new Micromegas detector is being designed with improved features and will be installed at the sunrise line in CAST in 2011. Two views of the actual design are shown in figure 5. Althought the final setup will also include the conclusions of Canfranc tests, the actual design supposes two main background sources in the actual detector: the radon which diffuses inside the shielding and the materials situated near the readout.

The lead shielding of the actual setup will be replaced by a lead gas-tight vessel, which will contain the argon-isobutane mixture and will conserve the shielding power. The inner vessel, composed of two pieces of stainless steel and plexiglass, will be replaced by a drift cage of

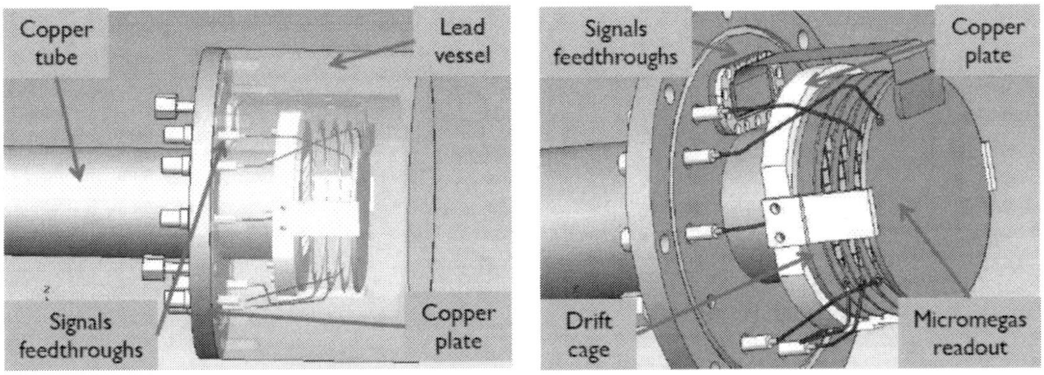

Figure 5. Two views of the new design of the sunrise Micromegas detector to be installed in CAST in 2011, designed by J.P. Mols. A detailed description of the setup is made in text.

copper and peek. In this way, a uniform drift field will be guaranteed and there will be less and cleaner material near the readout. Finally, strips will be digitized by the T2K electronics. This new acquisition system will make the detector a real TPC, which will increase the background reduction.

6. Conclusions and outlook

During 8 years of operation in CAST, Micromegas detectors have shown a low and stable background level due to their low radioactive materials, good energy resolution and spatial resolution. Since 2008, four ultralow background periods have been observed with the new *microbulk* detectors. During these periods, the background level reduced one order of magnitude, reaching a value around 10^{-7} s^{-1} keV^{-1} cm^{-2}. The energy spectrum after the application of the selection parameters shows three clear peaks at 3, 6 and 8 keV, caused by the fluorescence of near materials. The intensity variation of these lines could be correlated to variations of radon gas inside the shielding. Since October 2009, the radon is being monitored in CAST but no clear correlation with the ultralow background periods has been observed until now.

To clarify the origin of these periods and to determine the ultimate background level of these readouts, a CAST-like detector has been installed at the Canfranc Underground Laboratory. The actual level is compatible with the one measured in CAST. A new Micromegas detectors is being developped for working in CAST experiment during 2011, based on the conclusions of this work.

References

[1] R. Peccei, H. Quinn, *Phys. Rev. Letters* **38** (1977) 1440.
[2] S. Weinberg, *Phys. Rev. Letters* **40** (1978) 223; F. Wilczek, *Phys. Rev. Letters* **40** (1978) 279.
[3] H. Primakoff, *Phys. Rev. Letters* **81** (1951) 899; P. Sikivie, *Phys. Rev. Letters* **51** (1983) 1415 (Erratum ibid. **52** (1984) 695).
[4] K. van Bibber *et al.*, *Phys. Rev. Letters* **59** (1987) 759.
[5] K. Zioutas *et al.*, *Phys. Rev. Letters* **94** (2005) 121301; S. Andriamonje *et al.*, *JCAP* **4** (2007) 010.
[6] P. Abbon *et al.*, *NJP* **9** (2007) 170.
[7] D. Autiero *et al.*, *NJP* **9** (2007) 171.
[8] M. Kuster, *NJP* **9** (2007) 169.
[9] E. Arik *et al.*, *JCAP* **02** (2009) 008.
[10] F.J. Iguaz *et al.*, Proceedings of the conference IDM 2010, Montpellier (France).
[11] P. Baron *et al.*, *IEEE Nucl. Sci. Symp. Conf. Rec.* **55** (2008) 1744; D. Calvet *et al.*, *16th IEEE NPSS real time conference*, 2009, Beijing (China).

[12] T. Papaevangelou *et al.*, *J. Phys. Conf. Ser.* **179** (2009) 012015.
[13] T. Papaevangelou *et al.*, *JINST* **5** (2010) P02001.

Journal of Physics: Conference Series **309** (2011) 012003 doi:10.1088/1742-6596/309/1/012003

Search for double beta decay with the EXO-200 TPC and prospects for barium ion tagging in liquid xenon

Razvan Gornea on behalf of the EXO collaboration
LHEP, Albert Einstein Center for Fundamental Physics, Bern University, Switzerland
E-mail: gornea@gmail.com

Abstract. Neutrino oscillation experiments have shown that neutrinos have very small but non vanishing masses. These experiments are not able to determine neither the absolute mass scale of neutrinos nor whether they are Majorana particles. Neutrino-less double beta decay can only occur if the neutrinos are Majorana particles, a preferred scenario in most possible schemes leading to finite masses. Among several viable candidate isotopes, EXO has chosen Xe-136. The final state (i.e. the barium ion) can be tagged using optical spectroscopy. The efficient detection of the double beta decay daughter nucleus is a key step toward a background free measurement of such a rare process. An intermediate size detector (EXO-200) of 200 kg enriched xenon has been installed underground at WIPP (US). It is an ultra-low background detector with a design sensitivity of 6×10^{25} years for the half-life of neutrino-less double beta decay in Xe-136. A larger, ton-scale experiment is being designed with Ba ion tagging capability. We are presenting the status of the EXO-200 detector and review the R&D activities for a ton-scale EXO detector with barium ion tagging.

1. Introduction

The EXO collaboration aims for the development of a ton-scale, ultra low background TPC filled with xenon enriched in Xe-136 for the detection of neutrino-less double beta decay ($0\nu\beta\beta$) [1]. For the first phase called EXO-200, the collaboration has obtained 200 kg of 80% enriched xenon and has built a cryogenic liquid TPC equipped with both charge and scintillation light readout. Future detection schemes will involve identification of the final state (Ba^+ tagging) which enables complete rejection of the radioactive background (EXO-full).

Considerable efforts have been made to design and build the EXO-200 detector with very low radioactive background. All the materials and components have been screened. The design priority was minimizing the amount of the materials surrounding the TPC. EXO-200 will explore the quasi-degenerate neutrino mass scheme. The two-neutrino double beta decay ($2\nu\beta\beta$) has never been observed in Xe ($T_{1/2} > 10^{22}$ years) [2]. EXO-200 has the sensitivity to measure this channel if current theoretical predictions are accurate. Meanwhile, the collaboration is investigating various techniques for the transport and tagging of the barium ion. The expected performance of EXO-200 is given in table 1 along with the projections for the future phases with Ba^+ tagging.

The collaboration has set up an intensive screening program to determine the residual radioactive contamination of materials considered for the detector construction [5]. The K, U and Th concentrations of more than 350 materials have been measured. A Monte Carlo simulation that includes the geometry of the EXO-200 detector, the measured activities and

Table 1. Projected performance for EXO-200 and following phases. The detection efficiency is mostly dominated by the fiducial cut. For EXO-200 the background is dominated by the contribution of the residual activity. For future phases, the expected background is exclusively from the two-neutrino channel and is a function of the energy resolution. Calculations by [3, 4] used for nuclear matrix elements.

Detector name	Mass [ton]	Efficiency [%]	Run time [year]	$\Delta E(Q)$ [%]	Background [events]	$T_{1/2}$ limit [year]	$m_{\beta\beta}$ limit [meV]
EXO-200	0.2	70	2	1.6	40	6.4×10^{25}	130
EXO-full	1	70	5	1.6	0.5	2.0×10^{27}	24
EXO-full	10	70	10	1.0	0.7	4.1×10^{28}	5.3

event selection algorithms has been developed. It provided the prediction of the background for EXO-200. EXO has studied the energy resolution achievable in liquid xenon using a small test cell equipped with a UV sensitive PMT and a charge readout system [6]. Simultaneous measurements of light and charge provide better energy resolution than any single approach. The energy resolution at the Q value (2458 keV [7]) for the decay of Xe-136 is expected to be $\sigma = 1.6\%$.

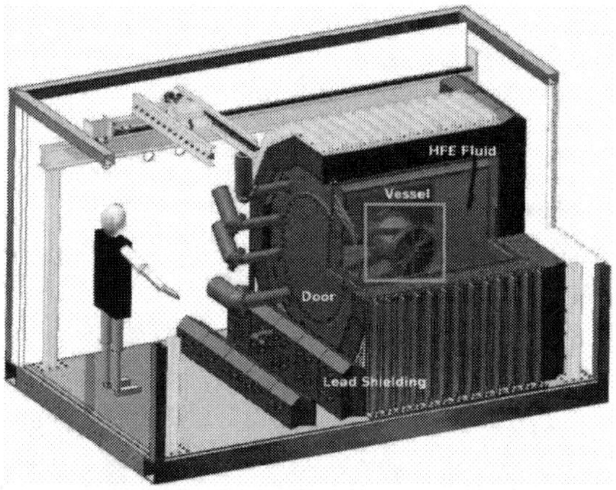

Figure 1. 3-D drawing of the cryostat.

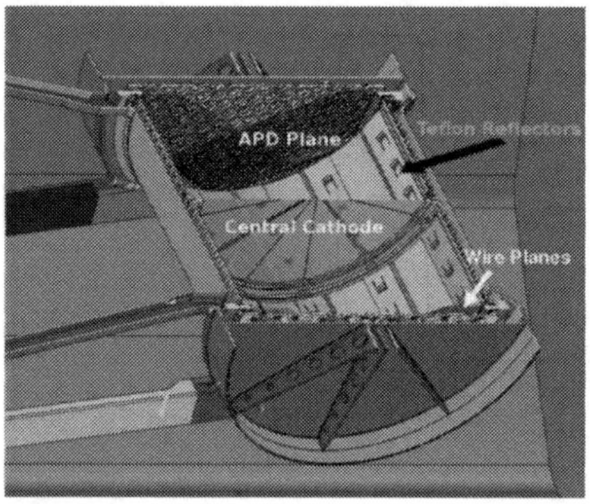

Figure 2. 3-D drawing of the TPC.

2. EXO-200 detector

The TPC has a cylindrical shape (40 cm diameter and 35 cm length) and is segmented in two zones by a central photo-etched cathode made of phosphor bronze. Both regions are equipped with induction and charge collection wire grids for ionization measurement and tracking, and with LAAPD (Large-Area Avalanche Photodiode) planes for the collection of scintillation light (peaked at 174 nm). Each LAAPD plane contains 250 UV sensors (with $QE > 1$ at 174 nm) which have an active diameter of 1.6 cm and are operated at a voltage around 1400 V for a gain of 150×. Lateral Teflon sheets serve as ultraviolet reflectors. Figure 2 shows a diagram and a 3-D drawing of the TPC. The chamber is made from ultra-low radioactivity copper by clean welding (i.e. e-beam and TIG welding) of 1.5 mm thick rings, performed in a controlled environment. The TPC is immersed in a refrigeration cryostat filled with 4.2 tons of high purity heat transfer fluid which serves as the inner gamma ray and neutron shield and as a thermal bath that maintains the chamber temperature uniform. The cryostat has a cylindrical shape (1.5 m diameter and 1.5 m length) and it is also made of ultra-low activity copper. A 3-D drawing is shown in figure 1.

The EXO-200 detector is installed at WIPP, in a salt mine at a depth of 665 m. Figure 3 shows pictures of the underground setup. A muon veto composed of plastic scintillator panels (20 units of 65×315 cm^2 and 11 units of 65×375 cm^2) covers the module that hosts the detector. Monte Carlo simulations have been used to optimize the configuration of these panels to obtain a high tagging efficiency (i.e. 99.7% resulting in the reduction of muon related background by a factor of 20×). The cryogenic system has been fully commissioned including the xenon and HFE handling. Data taking using natural xenon has started in November 2010. Operation with enriched xenon is expected for 2011.

Figure 3. EXO-200 underground experimental facility. 180° panoramic view inside the clean room (top) and outside in the drift (bottom).

3. Final state R&D

A ton-scale detector requires further radioactive background reduction that what can be provided by standard material screening techniques. To address this problem EXO envisioned the detection of the final state ^{136}Xe $\rightarrow^{136}$ Ba^{++} as an unique signature to the double beta decay. Precise ion location in the drift volume is important for an efficient final state detection. The reliable ion extraction and transport to the tagging apparatus is essential for a practical device. The Ba$^+$ ion can be tagged in a RF trap using resonant light scattering.

3.1. Barium tagging

The double beta decay of xenon produces Ba^{++} ions and, for the liquid phase, charge reduction to Ba$^+$ is expected. Laser based Ba$^+$ tagging has been accomplished in a RF trap using low pressure helium cooling ($P = 10^{-3}$ torr) [8, 9]. Figure 4 illustrates the excellent resolving power of this method for counting ions.

3.2. Ion collection probe

Several techniques are being investigated for ion collection. A promising avenue is the resonant ionization spectroscopy (RIS) which has been explored for pyrolytic graphite and silicon. In

Journal of Physics: Conference Series **309** (2011) 012003 doi:10.1088/1742-6596/309/1/012003

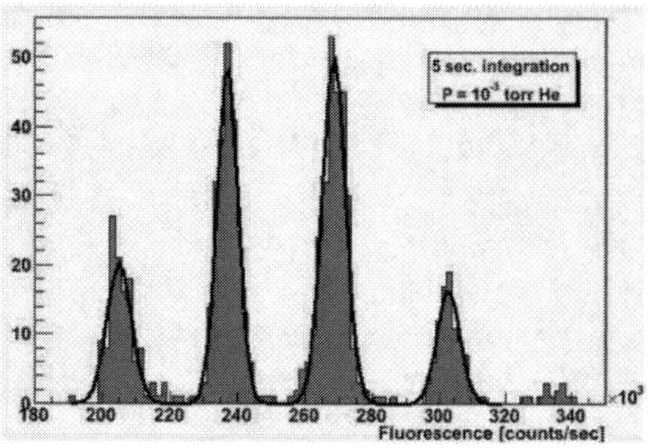

Figure 4. Histogram of the light scattered from barium ions (5 s time slices). The first peak is associated with the background and the following peaks correspond respectively to 1, 2 and 3 ions trapped.

this case, a desorption laser is used in combination with RIS lasers of specific frequency that selectively ionize the barium atom. The 553.5 nm and 389.7 nm lasers are tuned to barium transitions and are able to pump the atom to a highly excited state from which it decays to an ionized state. Figure 5 shows the time of flight spectrum of ions ejected by a desorption laser from a silicon surface with a thin barium layer produced by sputtering. When the RIS lasers are also employed, barium ions greatly dominate the spectrum. An efficiency of 0.1% has been reached and a significant improvement is expected for the next generation setup.

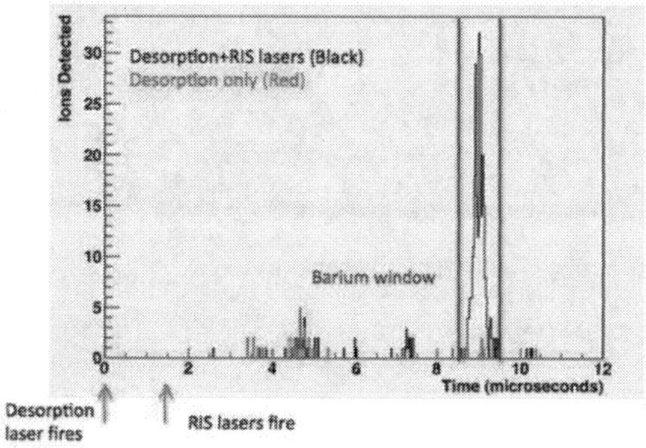

Figure 5. Time of flight spectrum for desorbed ions from a silicon surface with a thin layer of barium. The red curve illustrates the signal when only the desorption laser is employed. When RIS lasers, tuned to barium transitions, are also used then a large number of barium ions is observed as indicated by the black curve.

4. Conclusion

EXO has commissioned the EXO-200 detector and has started data taking. This first step will provide valuable data for future phases and hopefully interesting physics results. Various techniques for barium ion tagging are explored in preparation for the ton-scale detector.

References

[1] Akimov D et al. 2005 *Nucl. Phys.* B **138** 224-6
[2] R. Bernabei et al. 2002 *Phys. Lett.* B **546** 23-28
[3] Rodin V A et al. 2007 *Nucl. Phys.* A **793** 1-4 213-5
[4] Caurier E et al. 2008 *Phys. Rev. Lett.* **100** 052503
[5] Leonard D S et al. 2008 *Nucl. Instrum. Meth.* A **591** 490-509
[6] Conti E et al. 2003 *Phys. Rev.* B **68** 054201
[7] Redshaw M et al. 2007 *Phys. Rev. Lett.* **98** 053003
[8] Green M et al. 2007 *Phys. Rev.* A **76** 023404
[9] Flatt B et al. 2007 *Nucl. Instrum. Meth.* A **578** 399-408

R&D for high pressure gas TPC readout with segmented Micromegas

D. Franco on behalf of the EXO-Bern group.
Albert Einstein Center for Fundamental Physics, LHEP
Bern University, Sildestrasse 5, 3012 Bern (CH)

E-mail: domenico.franco@lhep.unibe.ch

Abstract. In the framework of the R&D program for the ton scale phase of the EXO experiment, the possibility of using a xenon gas TPC with segmented Micromegas readout plane is under investigation. A small gas TPC has been built in order to study the energy and spatial resolution with Micromegas. Tests with P10 and xenon gas are under way. The apparatus and some preliminary results are presented.

1. Introduction

In the last 15 years, data from neutrino oscillation experiments, using solar, atmospheric, accelerator and reactor generated neutrinos, clearly indicate that the neutrinos have non zero masses. These experiments however can only determine differences of squared masses. The absolute mass scale can be determined by studying the kinematics in the weak decays of Tritium [1] [2] or, if the neutrino is a Majorana particle, by searching for neutrinoless double beta decay.

The Enriched Xenon Observatory (EXO) [3] is an experiment searching for the double beta decay of ^{136}Xe. This isotope should decay through allowed double beta with the emission of two neutrinos to ^{136}Ba, though past experiments have only placed limits on the half-life [4]. Observation of the 2ν mode is interesting for nuclear theory, but does not provide information about neutrinos. In contrast, neutrinoless double beta decay implies a Majorana nature of the neutrino and the rate of this process is proportional to the square value of the effective neutrino mass [5]. The study of $0\nu\beta\beta$ may shed light on the absolute mass value, hierarchy and nature (Dirac-Majorana) of these amazing particles.

The first phase of EXO (EXO-200) is already running with a liquid TPC filled with 200 kg of pure xenon. Meanwhile a strong R&D program for a ton scale up-grade (EXO-full) of the system is pursued by the collaboration. Both liquid and gas option are under investigation.

A gas TPC implies a less compact apparatus, but electron tracks can well be visualized and identified, at pressures up to 5-10 bar, as demonstrated int the Gotthard TPC [6] and the MUNU TPC [7] [8]. Double beta decay events can be identified by using their topology: a candidate appears as a single track, with increased ionization at both ends. This may lead to a substantial background suppression. Moreover a gas apparatus allows the use novel sensors such as GEM or Micromegas for collecting charge and this should lead to a better spatial and energy resolution and an increased reliability.

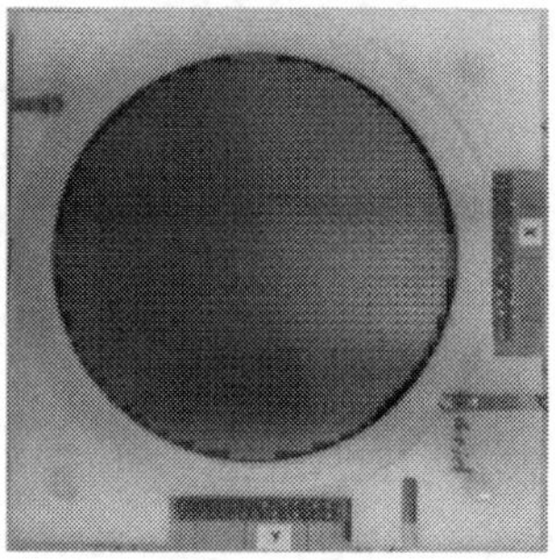

(a) Micromegas

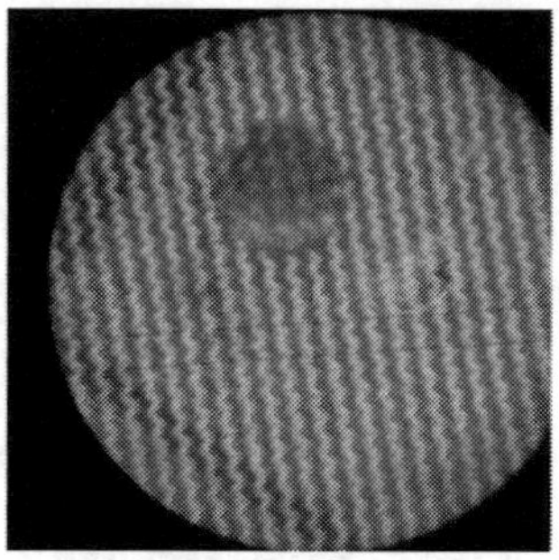

(b) Mesh

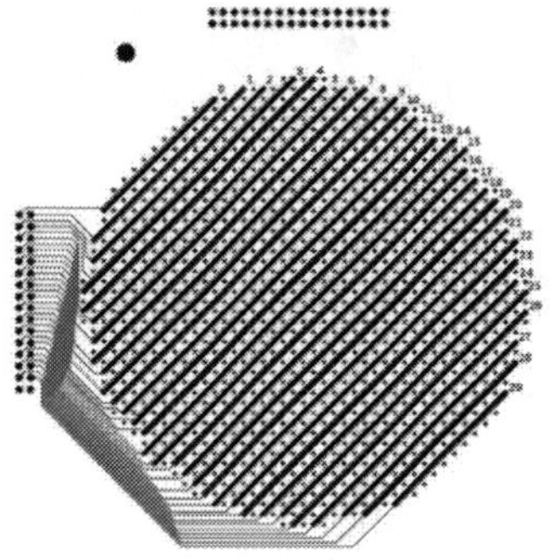

(c) X-strips layout

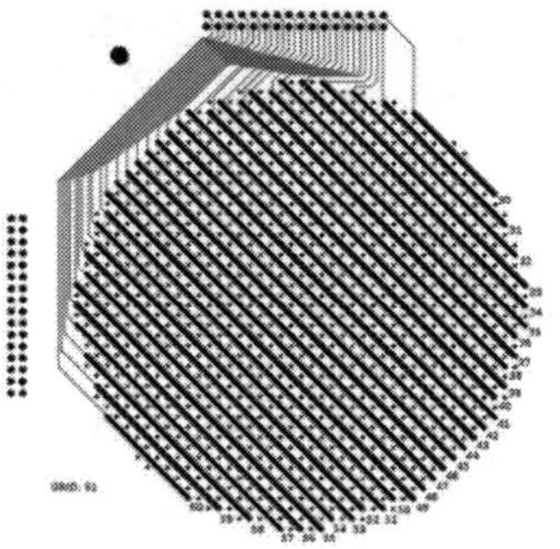

(d) Y-strips electric layout

Figure 1. (a) The 10 cm active area of the Bern Micromegas. Pixels are visible. (b) A detail of the grid. (c)-(d) The electrical layout of X and Y strips connection.

2. Bern mini-TPC setup

The Micromegas detector [9] is a fast gaseous detectors using a micro-mesh electrode separating the drift area from the amplification gap.

In order to study the adequacy of the Micromegas for double beta decay search, a mini-TPC apparatus (mTPC in the following) has been built at LHEP. Tests are performed using a Micromegas with a segmented anode plane built at CERN using photolithography by R. De Oliveira's group [10]. The anode plane consists in a matrix of 2.1x2.1 mm pixels on a PCB board connected diagonally in the back using connecting vias. These pixels form interlaced X and Y strips with a pitch of 3.113 mm. There are 31 X and 30 Y strips, for a total useful diameter of 10 cm. Photo-resist pillars of 500 μm diameter and 256 μm hight are placed off center on each pixel. They maintain the proper gap with the grid which is made of woven stainless steel wire of 18 μm diameter. A grounded guard ring surrounds the pixel matrix and provides a very uniform electric field on the edge of the detector.

The Micromegas is placed in a vessel at the end of 18 cm drift volume delimited by field shaping rings ad a cathode. Each strip is connected, using flat cables, to current preamplifier, developed at the LHEP institute, and sampled at 62.5 MHz by a 12 bit VME based Flash ADC. The grid is also sampled at the same speed.

The gas is constantly being recirculated through an Oxysorb filter and a cold trap, to remove oxygen, water and other contaminants. Moreover a Residual Gas Analyzer (RGA) allows to monitor the gas quality and its composition. A muon veto system is used to select or reject cosmic muons.

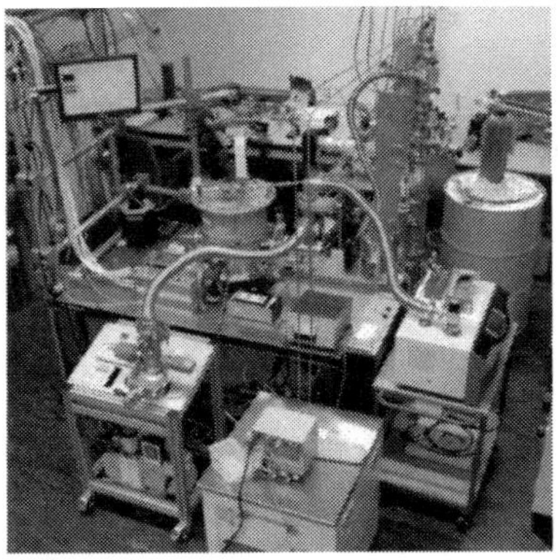

Figure 2. On the left the Bern mTPC is illustrated. The Micromegas is placed below the copper cage made by the field shaping rings. On the right the mTPC setup.

3. Muon tracks and gain factor

Preliminary tests have been performed filling the mTPC vessel with P10 (Ar(90 %)-CH_4(10 %)) at pressures ranging from 1 to 5 bar and looking at the ionization tracks generated by the cosmic muons. Data with a mixture of Xe (98%) and CF_4(2%) have been taken as well.

All the ADC channels have been calibrated. The raw current signals (fig. 3a) are integrated to get the charge (fig. 3b) then the noise level is evaluated (fig. 3c) and subtracted from the signal (fig. 3d). Finally the total charge deposited on each strip is estimated. The resulting data are organized in two bi-dimensional histograms that represent the projection of a track in the X-Z plane and the Y-Z plane where the Z axis is defined as the drift field direction. As shown in figure 4, very nice tracks in P10 up to 5 bar, and good tracks in xenon have been obtained. The analysis is still in progress, but one can readily state the the spatial resolution in X and Z is better than 2-3 mm.

To determine the gain factor of the Micromegas a software has been developed. The algorithm scans the bi-dimensional histograms and extracts the length of the tracks, the total collected charge, and then the dQ/dx. Simulations of the distribution for the muon dQ/dx in P10 and Xe-CF_4 have been performed with the HEED package [11]. Figure 5 shows one example of the simulated and experimental distribution of dQ/dx in which only tracks longer than 10 cm have been selected. The experimental distribution is fitted with a Landau function and the most probable value (MPV) is extracted from the fit. The ratio between the measured value and the simulated one gives the gain factor for a fixed amplification voltage and pressure. Data from Xe-CF_4 runs at the pressure of 1 and 2 bar have been analyzed.

4. Results

In figure 6, our results in Xe-CF_4 are compared with the previous one obtained with a 225μm amplification gap Micromegas and a non-segmented anode plane [12]. We were not able to reach the same amplification because discharges occur. The two experimental points correspond to a gain factor $G = 1674$ at 785V grid voltage and 1 bar of pressure and $G = 700$ at 1167V and 2 bars of pressure. They represent the highest achievable value for the gain before discharges developed between grid and pixels. The hypothesis is that the amplification gap is not large enough to allow the development of the shower and increasing the grid voltage induces the breakdown.

As the electron shower grows exponentially along the direction of the amplification field, a wider gap should lead to a better gain factor at fixed grid voltage. Two new Micromegas with gap of 320μm and 384μm respectively, are under construction at CERN and will be tested soon in Bern.

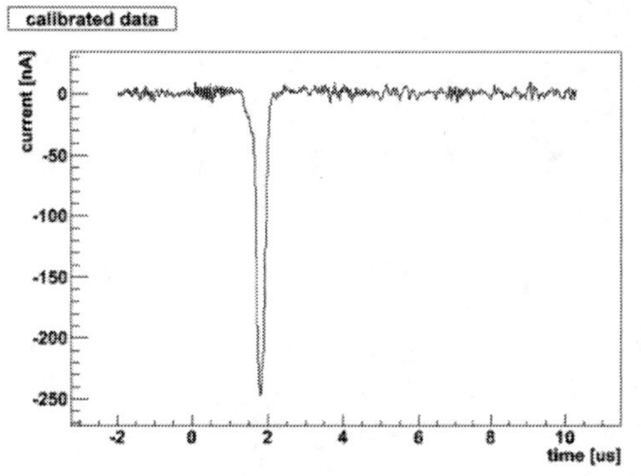

(a) calibrated current signal

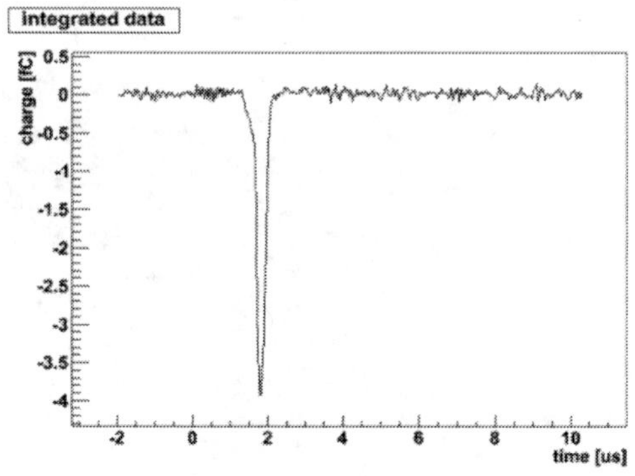

(b) Integrated signal

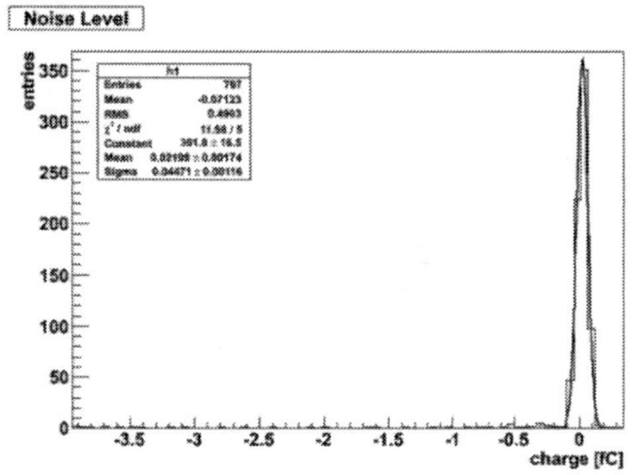

(c) Noise

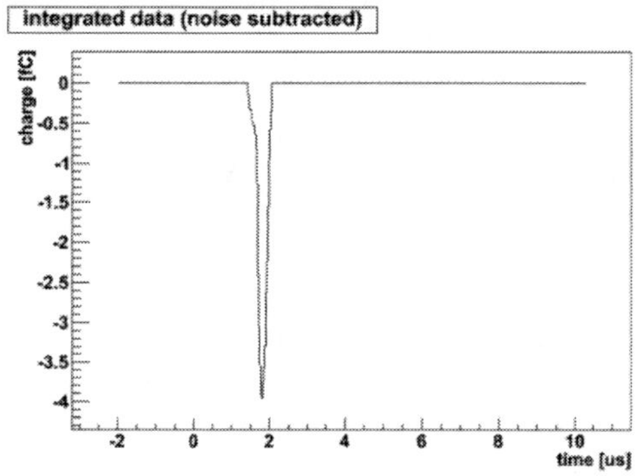

(d) Clean signal

Figure 3. The four histograms show the steps of the signal treatment. The average noise level is evaluated from a Gaussian fit.

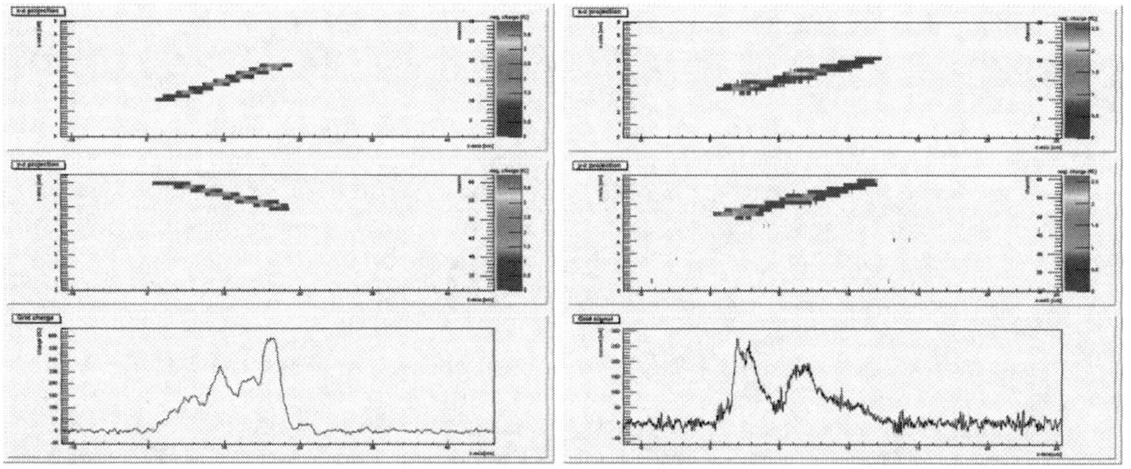

Figure 4. Left: track in P10 for a muon that cross the entire volume of the mTPC as it has been reconstructed by our software. Right: track in Xe-CF_4 at 2 bar. The Z distance is obtained multiplying the drift time and the electron drift velocity for the specific gas mixture, pressure and drift field.

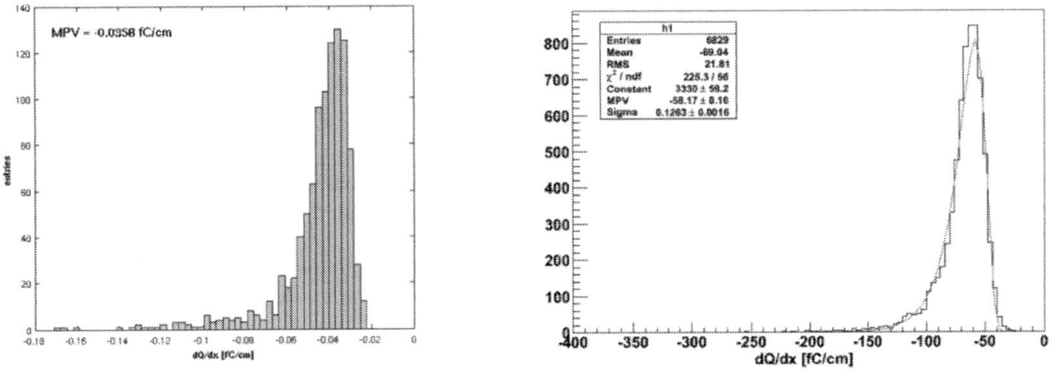

Figure 5. On the left the simulated dQ/dx distribution for muon in Xe-CF_4. On the right the measured distribution for one Xe-CF_4 run.

5. Prospectives

To complete the setup, a UV-light detector is under construction as well. It is well known that the spectrum of the scintillation light peaks around 174 nm in xenon. Scintillation and ionization are anti-correlated in Xe, the collection of the light as well as the charge, could provide a better energy resolution and a t_0 reference for the drift time.

We believe it is efficient in a large size TPC, to first catch the UV light in a light guide and then to convert it to visible using a wavelength shifter, finally detecting the visible light with sensors directly coupled on the guide.

The light guide must be relatively transparent to UV, which led us to chose polystyrene. Tetraphenylbutadiene (TPB) appears adequate as wavelength shifter [13] [14]. A large conversion ratio should be achieved with a 3 mm thick guide doped with 2% TPB per weight. The emission spectrum of TPB peaks around 440 nm which is optimal for photomultipliers and for solid state devices like Si photomultipliers (SiPM). The inner side of the TPC will be covered with TPB doped light guides attached to the field shaping rings. This scheme allows to collect UV photons over a large area in the TPC with a limited number of sensors. Si photomultipliers have been chosen as light sensors. These compact devices can be couple directly on the light guide inside the vessel and no light feed-troughs is needed to carry the light out of the vessel.

A bulk piece of scintillator consisting in TPB doped polystyrene has been produced by A. Olshevskiy and I. Nemchenok group at JINR Dubna. It has been already cut in slabs with dimensions 200x10x30 then polished. On one side they will be equipped with 3x3 mm SiPM which we have procured from JINR as well. These sensors are micro-pixel avalanche photo diodes operated in Geiger mode where each pixel acts as a photon counter and outputs a macroscopic current signal. Photons are counted by summing the signals arriving in coincidence from all pixels. The outputs of the SiPMs will be ganged together and the sum signal will be sampled at the same frequency as the TPC pixels.

Moreover for studying the energy resolution of the Micromegas at low energy level, the mTPC vessel will be provided with a mechanism that allows the insertion of a 5.9 KeV ^{55}Fe X-ray source.

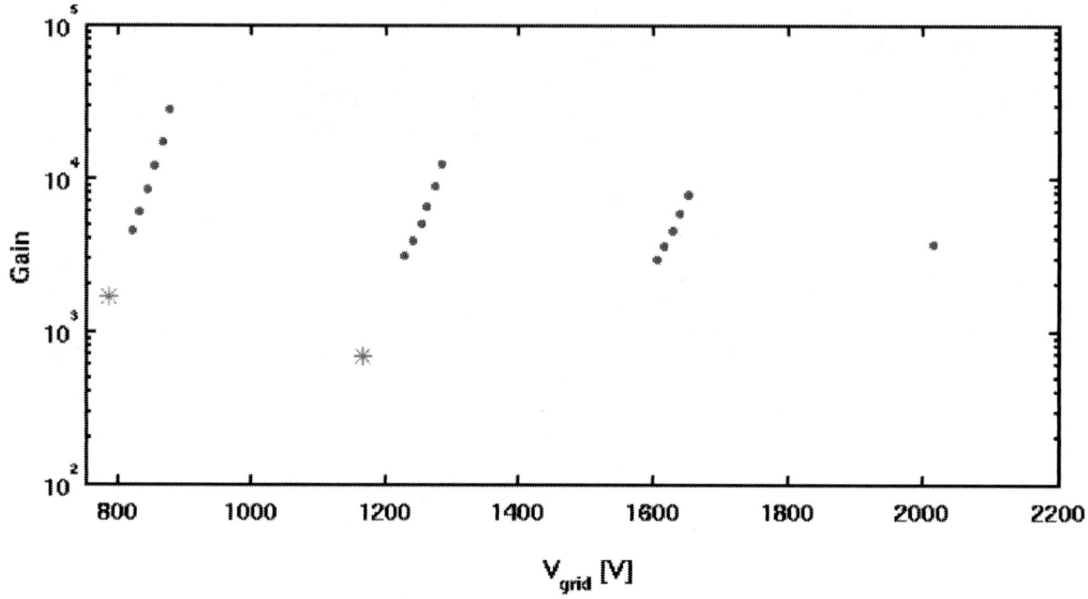

Figure 6. The Micromegas gain factor for Xe-CF_4 gas pressures ranging from 1 (starting from left) to 4 bars. Blue points are from ref [12], red points are our experimental results.

6. Conclusion

Xenon gas TPC with segmented Micromegas read-out plane could represent an interesting alternative for the EXO-full experiment. A strong R&D program is carried on by the Bern EXO group at LEHP using the mTPC setup. Encouraging result have been achieved with P10 up to 5 bars, and Xe-CF_4 up to 2 bars. New wider gap Micromegas, UV light system and radioactive source insertion mechanism will be available soon making the system complete and ready for detailed studies.

References

[1] V.M. Lobashev et al. 1999 *Phys. Lett. B* **460** 227-235.
[2] Ch. Kraus et al. 2005 *Eur. Phys. J. C* **40** 447-468.
[3] R. Gornea and the Exo collaboration 2010 *J. Phys.: Conf. Ser.* **259** 012039.
[4] R. Bernabei et al. 2002 *Phys. Lett. B* **546** 23-28.
[5] F.T. Avignone III et al. 2008 *Rew. Mod. Phys.* **80** 481-516.
[6] Luescher et al. 1998 *Phys. Lett. B* **434** 407-414.
[7] Z. Daraktchieva et al. 2005 (the MUNU collaboration) *Phys. Lett. B* **615** 153.
[8] Z. Daraktchieva et al. 2008 (the MUNU collaboration) *Phys. G: Nucl. Part. Phys.* **35** 125107.
[9] Y. Giomataris et al. 1996 *Nucl. Inst. Meth. A* **376** 96.
[10] R. De Oliveira, CERN, private comunication.
[11] I. Smirnov *HEED: Interactions of particles with gases* W5060, http://consult.cern.ch/writeup/heed/
[12] L.Ounalli et al. 2009 *JINST* **4** P01001.
[13] W. M. Burton and B.A. Powel 1973 *Appl. Opt* **12** 89.
[14] D. N. McKinsey et al. 1997*Nucl. Inst. Meth. B* **132** 351.

TPC2010 IOP Publishing
Journal of Physics: Conference Series **309** (2011) 012005 doi:10.1088/1742-6596/309/1/012005

Prospects for Barium Tagging in Gaseous Xenon

D Sinclair[1,2], E Rollin[1], J Smith[3], A Mommers[4], N Ackeran[5], B Aharmin[6], M Auger[7], PS Barbeau[8], C Benitez-Medina[9], M Breidenbach[5], A Burenkov[10], S Cook[5], A Coppens[1], T Daniels[11], R DeVoe[8], A Dobi[12], MJ Dolinski[8], K Donato[8], W Fairbank Jr.[9], J Farine[6], G Giroux[7], G Gornea[7], K Graham[1], G Gratta[8], M Green[8], C Hagemann[1], C Hall[12], K Hall[9], D Hallman[6], C Hargrove[1], S Herrin[5], LK Kaufman[13], DS Leonard[14], F LePort[8], D Mackay[5], R MacLennan[15], B Mong[9], M Montero Díez[8], AR Müller[8], R Neilson[8], E Niner[15], A Odian[5], K O'Sullivan[8], C Ouellet[1], A Piepke[15], A Pocar[11], CY Prescott[5], K Pushkin[15], PC Rowson[5], S Slutsky[12], V Stekhanov[10], K Twelker[8], N Voskanian[12], J-L Vuilleumier[7], U Wichoski[6], J Wodin[5], L Yang[5] and Y-R Yen[12]

[1]Dept. of Physics, Carleton University, Ottawa, ON, Canada
[2]TRIUMF, Vancouver, BC, Canada
[3]Dept. of Chemistry, Carleton University, Ottawa, ON, Canada
[4]Dept. of Chemistry, University of Ottawa, Ottawa, ON, Canada
[5]SLAC National Accelerator Laboratory, Menlo park CA, USA
[6]Dept. of Physics, Laurentian University, Sudbury, ON, Canada
[7]LHEP, Physikalisches Institute, University of Bern, Bern, Switzerland
[8]Dept. of Physics, Stanford University, Stanford, CA, USA
[9]Dept. of Physics, Colorado State University, Fort Collins, CO, USA
[10]Institute for Theoretical and Experimental Physics, Moscow, Russia
[11]Dept. of Physics, Univeersity of Massachusetts, Amherst, MA, USA
[12]Dept. of Physics, University of Maryland, College Park, MD, USA
[13]Dept. of Physics and IU Center for Exploration of Energy and Matter, Indiana University, Bloomington IN, USA
[14]Dept. of Physics, University of Seoul, Seoul, Korea
[15]Dept. of Physics and Astronomy, University of Alabama, Tuscaloosa, AL, USA

E-mail: Sinclair@physics.carleton.ca

Abstract. Tagging events with the coincident detection of a barium ion would greatly reduce the background for a neutrino-less double beta decay search in xenon. This paper describes progress towards realizing this goal. It outlines a source that can produce large quantities of Ba++ in gas, shows that this can be extracted to vacuum, and demonstrates a mechanism by which the Ba++ can be efficiently converted to Ba+ as required for laser identification.

1. Introduction

The concept of rejecting the backgrounds in the search for neutrino-less double beta decay by tagging events with the coincident detection of the barium daughter ion was first suggested by M.Moe [1]. This suggestion was developed into a conceptual design for a gaseous xenon detector by M. Danilov *et al.* [2] based on extending the design of the Gotthard detector [3]. At the time, single ions of Ba+ had been successfully detected using laser fluorescence for ions contained in RF quadrupole traps [4]. The concept presented by Danilov *et al.* introduced laser beams into a gas TPC and guided the beams by means of movable mirrors to the location of the possible barium ion. The fluorescence would be detected by an array of photomultiplier tubes placed around the field cage.

There are a number of potential difficulties with this scheme.

- The barium ion is created as Ba++ (it may be in a higher charge state but will rapidly find electrons to reach the doubly charged state) but the species that can be fluoresced is only the Ba+ ion. Ba++ cannot pick up an electron from a single Xe atom and it is not known if it will convert to Ba+ due to the formation of xenon clusters.
- If the barium ion is in high pressure gas, the transitions will be pressure broadened. This requires increasing the laser power by a factor of about 1000 if the fluorescence rate is to be maintained at the level used in traps. With this laser power in high pressure gas there will be a significant scattered background which will make the positive identification of a single barium ion challenging.
- It is highly desirable to detect the scintillation light from the ionizing event and if one uses normal gas gain detection devices for the energy measurement, it is likely that the quench gas used will quench the scintillation by de-exciting the Xe dimmers.

We have therefore been exploring an alternate concept for the detector. In this concept we operate a TPC in pure xenon. The scintillation light and ionization electrons are measured using an electroluminescence process as discussed by Graham [5]. When a candidate event is observed in the ionization signal, the barium ion will be guided towards a nozzle leading to a vacuum region, by shaping the electric drift field. The ion would be swept out of the chamber by the gas flow and then separated from the gas using an ion guide system. This type of system is commonly used for the production of secondary (possibly radioactive) beams for nuclear physics research, and by the bio-molecular community in their mass spectrometers. The ion would then be converted to Ba+ (if it does not naturally happen in xenon) and detected using the proven techniques of laser fluorescence.

In order to establish the feasibility of this concept several measurements must be carried out.

- We need to understand the ability of the electroluminescence detection to determine the decay energy and the event vertex.
- We need to measure the mobility of Ba++ in Xe.
- We need to understand any loss mechanisms of Ba++ either to Xe or to contaminants.
- We need to demonstrate that Ba++ ions can be extracted from the high pressure gas and determine the efficiency.
- We need to establish a feasible way to effectively convert Ba++ to Ba+.
- We need to develop and prove a concept for fluorescence of the Ba+ ions either by introducing them to a trap or by looking at the fluorescence in flight.

2. An Ion Source

The first step in addressing these issues has been to find a mechanism for conveniently producing Ba++ ions in high-pressure gas. Several sources were investigated.

Production of barium ions using a spark discharge with barium on the electrodes is a technique that can produce barium ions. Figure 1 shows such a discharge with a laser tuned to the valance S1/2 to P1/2 transition passing through the arc. The fluorescence of the ions is obvious. This source can be used over a small pressure range to study the energy shift and broadening of the

Journal of Physics: Conference Series **309** (2011) 012005 doi:10.1088/1742-6596/309/1/012005

transition but it is not clear how to identify any Ba++ production. The environment about the ions is very different from the environment in which our Ba will be created.

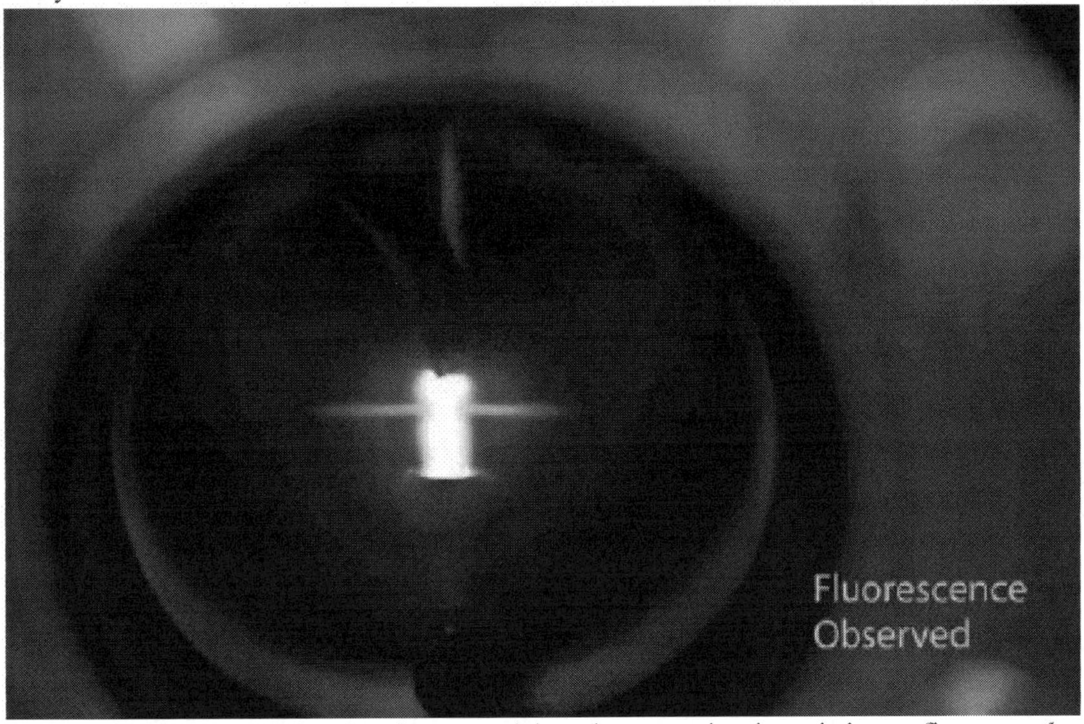

Figure 1. Production of barium ions in an arc. A laser beam passing through the gas fluoresces the Ba+ ions as they diffuse away from the arc.

A second source technique that was investigated is laser ablation. Barium metal and barium compounds were ablated using a focused, pulsed UV laser beam. In vacuum the products can be identified by measuring the time of flight of the ions to a detector. A typical spectrum is shown in Figure 2. In can be see that this is a convenient source for production of Ba+ but there is no evidence for production of Ba++.

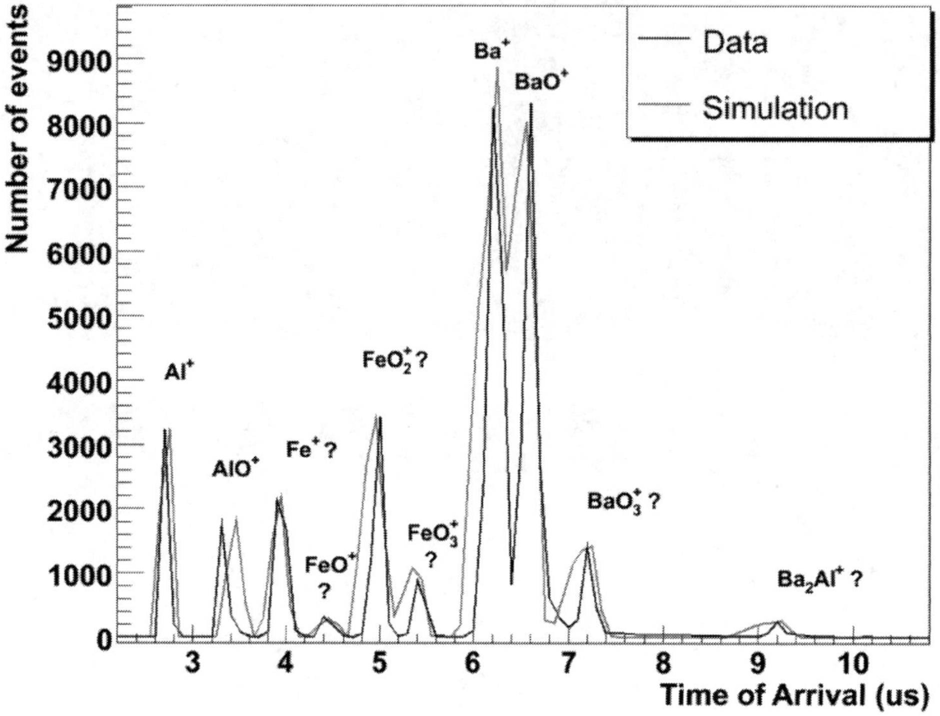

Figure 2. Ions produced by a laser ablation source using BaAl as the target and identification of ions by time of flight. The prediction is based on the use of the SIMION code with a source temperature taken to be 30,000K.

A third source was based on production of barium ions by alpha decay recoil. An alpha emitting material is deposited on a silicon detector and a thin barium salt is deposited over the source. When an alpha decay takes place, the alpha can be detected in the silicon detector and the recoil may knock barium out of the covering layer. Some of these atoms may emerge as ions. The ion spectrum can be measured by time of flight using the silicon detector as the start signal. A spectrum quite similar to Figure 2 above is found. No evidence is seen for the production of Ba++. This source has been described in more detail in [6].

The final source which we investigated, and found to work well, is electrospray. In an electrospray source of barium, a salt of barium is dissolved in water and mixed with methanol. The liquid is sprayed through a very fine nozzle which is which is maintained at a high negative potential. The flow of heated gas at atmospheric pressure causes the droplets to evaporate and as they reach the Rayleigh limit they produce a copious stream of ions. As the barium in the water tends to be in the form Ba++, it is expected that this will be the dominant ion produced.

A mass spectrometer (a Fisons Quattro II) with an electrospray source was used to test this mechanism. Ions from the source pass through a small orifice through a roughing pumped region into a high vacuum maintained by a turbo pump. The ions are trapped in a sextupole ion guide and travel to a quadrupole spectrometer. When the source conditions are optimized for Ba++ production the spectrum shown in Figure 3 is obtained. The dominant peak is Ba++. Figure 4 shows a portion of the spectrum expanded about m/z=69. We can confirm that this group is Ba++

by noting that all of the isotopes of Ba are seen in the proper intensity ratio and the group must be a z=2 group because the peaks are spaced by ½ unit in m/z.

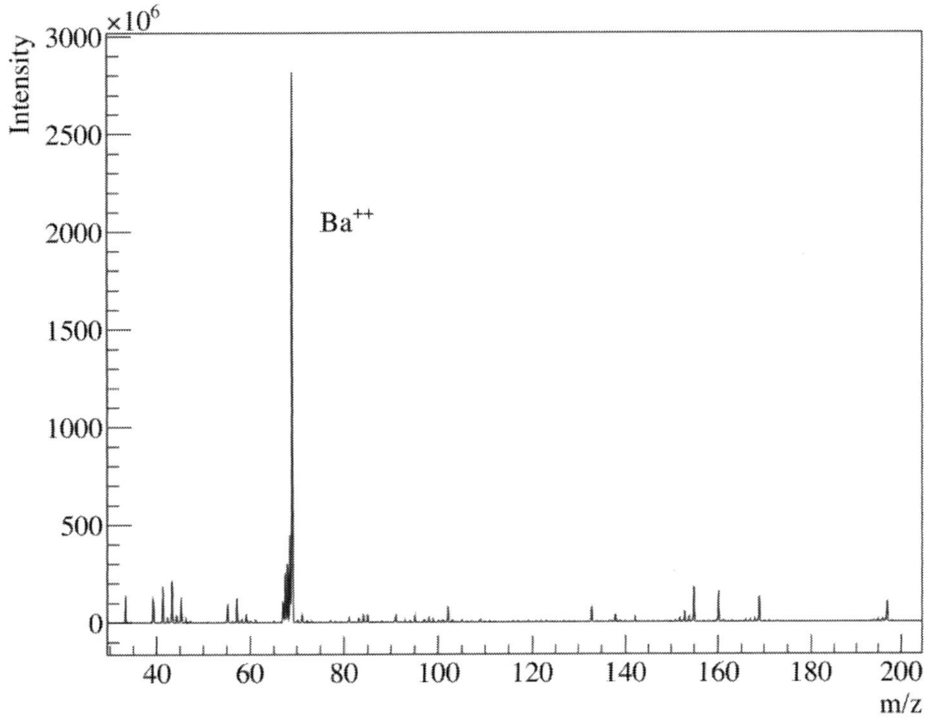

Figure 3. The spectrum obtained using a quadrupole mass spectrometer with an electrospray source. The source solution was 1 mM barium acetate in an equal mix of water and methanol.

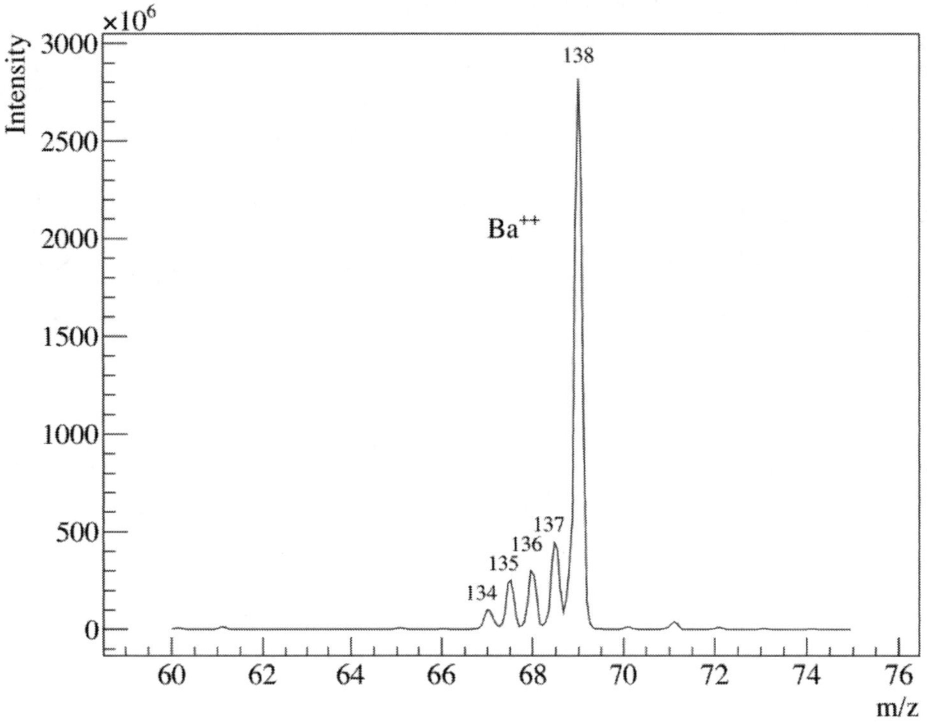

Figure 4. The spectrum of Figure 3 expanded about the m/z=69 region.

3. A Charge Changing Technique

The Quattro II contains two quadrupole spectrometers separated by a gas collision cell where the ions are contained using an RF sextupole. We have used this system to look at conversion of Ba++ to Ba+. Our first attempts used triethylethane (TEA) as a low ionization potential gas in the cell. For this measurement the first quadrupole was set to pass only m/z = 69. Ions were accelerated into the cell and the products were measured with the second quadrupole. Measurements were taken for a range of pressures and acceleration voltages. An example spectrum at high pressure and voltage is shown in Figure 5. We see that all of the Ba++ has been converted to Ba+ with the production of TEA+. The Ba peak is broad because the ion enters the spectrometer with a high velocity while the TEA picks up very little momentum in the transfer and is sharp. There is no evidence for the formation of any other species or molecules.

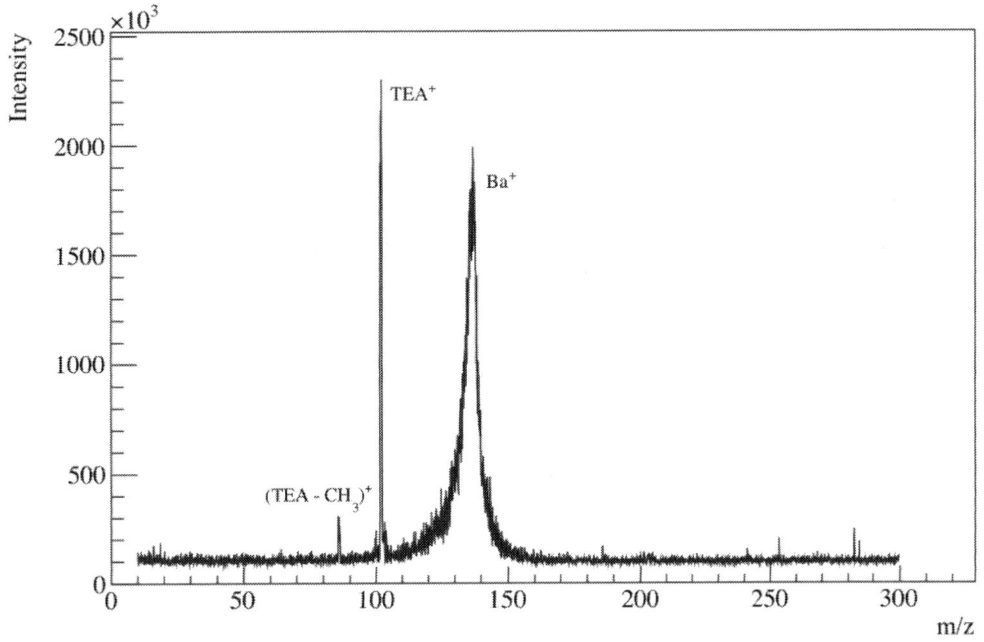

Figure 5. The spectrum of particles produced when Ba++ passes through TEA. The Ba++ ions are accelerated through 100 volts and the TEA pressure is 10^{-3} mbar. The length of the cell is about 20 cm.

4. Implications of the Measurements

It is clear from the above discussion that electrospray is a convenient mechanism for producing Ba++ is gas at atmospheric pressure. It is likely that the source will perform just as effectively at higher pressures. Even though the source region has water vapour and methanol vapour at the 0.3% level, there is no evidence for molecular formation. The use of TEA offers an effective method to achieve the charge state conversion. The overall design of the ion extraction from high pressure to vacuum is very similar to the scheme proposed for the final detector and this appears to work well although the efficiency is not yet determined,

5. Future Program

The next step in our investigation will be to operate an electrospray source in xenon gas. To do this a special source is being fabricated which will allow efficient re-circulation of the gas. A feature of the source will be a drift region where the mobility of the ions can be investigated by countering the motion due to the gas flow with motion due to an electric field. A sketch of the source is provided in Figure 6 below.

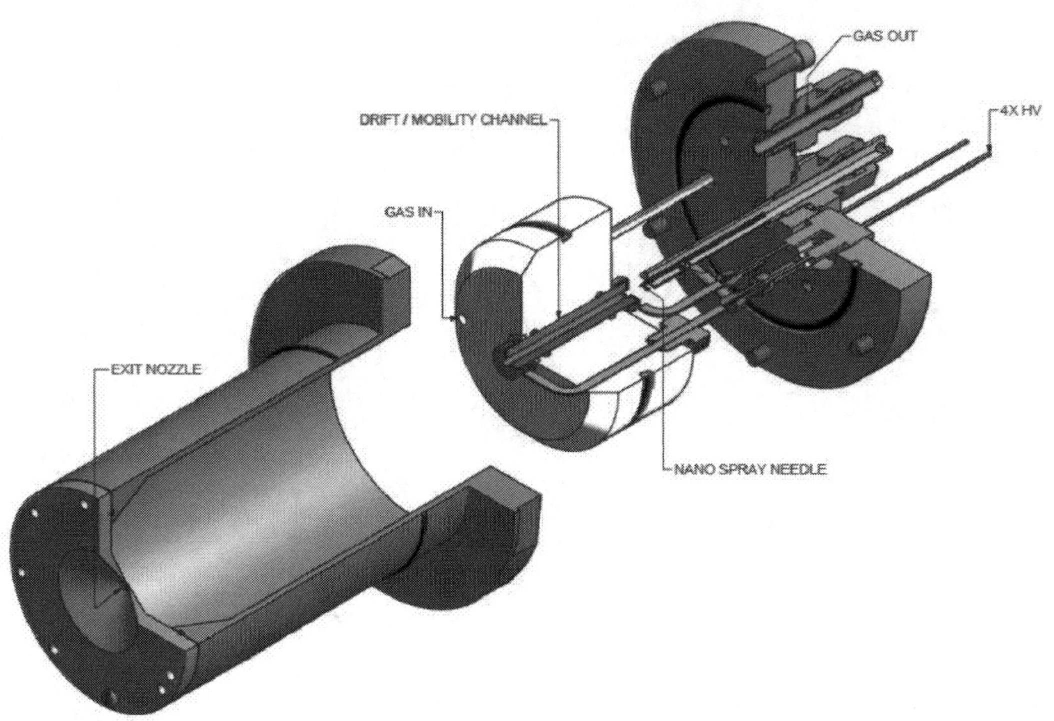

Figure 6. Exploded view of the nano-spray and mobility channel now under construction for further studies.

[1] M.K. Moe, *Phys. Rev.* **C44** (1991) R931.
[2] M. Danilov et al. *Phys. Lett* **B480** (2000) 12.
[3] R. Luescher et al. *Phys. Lett.* **B434** (1998) 407.
[4] W. Neuhauser et al. *Phys. Rev.* **A22** (1980) 1137.
[5] K. Graham et al. This conference
[6] M. Montero Díez et al. Review of Scientific Instruments, To be published

Journal of Physics: Conference Series **309** (2011) 012006 doi:10.1088/1742-6596/309/1/012006

Can the "intrinsic" energy resolution in xenon be surpassed?

David Nygren
Physics Division
Lawrence Berkeley National Laboratory
Berkeley, CA

E-mail: drnygren@lbl.gov

Abstract. Energy resolution is determined by fluctuations in the transformation of deposited energy to ionization and excitation. For noble gases, adding a molecular impurity that introduces a strong Penning effect reduces the Fano factor, improving energy resolution. I show that in xenon, adding trimethylamine (TMA) may provide a strong Penning effect and also permit electroluminescence (EL) at wavelengths characteristic of TMA. This combination may permit a large TPC to be operated with better than "intrinsic" energy resolution in the search for neutrino-less double beta decay in ^{136}Xe. This path is being explored within the context of the NEXT collaboration.

1. Introduction

A large EL TPC filled with high-pressure xenon gas appears to be an attractive approach for a neutrino-less double-beta decay (0-ν $\beta\beta$) search in ^{136}Xe [1-6]. This expected rarity of this decay requires best possible energy resolution and knowledge of event topology to suppress backgrounds [7]. A good tracking capability, such as provided by the TPC, has been shown to permit suppression of single electron backgrounds by a large factor [8,9].

With regard to energy resolution, the performance champions are based on bolometric or germanium diode systems. These detectors generally offer one or two parts per thousand resolution FWHM at MeV energies [10,11]. In contrast, large detectors based on liquid or gaseous xenon have typically shown substantially worse energy resolution [4,12,13]; the exception with competitive resolution uses signal detection based on electroluminescence (EL) [14]. I show here that an avenue may exist for a large tracking TPC to approach energy resolution comparable to the performance champions, perhaps within a factor of two or three. How can this be? To begin, the conventional "intrinsic" resolution needs to be defined.

2. The intrinsic energy resolution

For a fixed energy input E to some medium, the variance σ_I in the observed ionization signal is customarily defined as

$$\sigma_I{}^2 = FN_I \tag{1}$$

where F is the Fano factor [15] and N_I is the expectation for primary electrons for the energy E, N_I is the ionization signal to be detected, and is given by the ratio of E/w, where w is the average energy to make an electron-ion pair. Generally, in addition to E, w depends on density and other properties of the medium, and the strength of ambient electric field, if any. From this, the "intrinsic" energy resolution is

$$\delta E/E \equiv 2.35\ \sigma_I/N_I\ \ \text{FWHM} \tag{2}$$
$$= 2.35\ (Fw/E)^{1/2} \tag{3}$$

The Fano factor for the ionization signal for gaseous xenon ($\rho < 0.5$ g/cm^3) and MeV energies has been found experimentally to lie in the range $F_{gxe} = 0.15 \pm 0.02$ [16-19].[1] For energetic electrons, the parameter w falls in the range of 24 – 25 eV, depending on ambient electric field and density [20]. For ^{136}Xe, a Q-value of 2457 keV is released in a decay to ^{136}Ba [22,23]. From these numbers, the "intrinsic" energy resolution in gaseous xenon for the ionization signal is

$$\delta E/E = 2.7 \times 10^{-3}\ \text{FWHM} \tag{4}$$

This expression ignores all degrading effects of detection and measurement; the essential challenge is to realize something close to this in a large-scale system.

For E = 2457 keV, the ionization yield of electron/ion pairs is

$$N_I \sim 1 \times 10^5 \tag{5}$$

depending slightly on imposed electric field. At this energy, the fluctuation expected, from (1), is

$$\sigma_I \sim 122\ \text{electrons}\ \ \text{rms} \tag{6}$$

From an experimental perspective, this is a soberingly small number. N_I electrons must be detected while adding noise or fluctuations that are small on the scale of (6).

It is quite reasonable to expect that fluctuations in the energy deposition process are uncorrelated with additional fluctuations or noise added in the subsequent detection process. Therefore, (3) can be expanded to include a term G, encompassing all degrading effects of detection and measurement:

$$\delta E/E = 2.35\ ((F + G)E/w)^{1/2} \tag{6}$$

G, ideally, should be small compared to F, or at the very least, not larger. This challenge cannot be met with any avalanche gain device, since early gain fluctuations are amplified exponentially; G values around 0.7 are expected [24]. Nor can a unity gain approach suffice, since electronic noise and microphonic effects cannot be made sufficiently small in a large system.

3. Electroluminescence

The only ultra-low-noise gain process available appears to be electroluminescence (EL), in which an ionization signal is converted first to an optical signal, then reconverted to an electronic signal typically using photomultipliers (PMT) or other photon sensors [25-29]. While this approach may seem awkward, perhaps even irrational, excellent performance is realizable. An effective noise level much less than one electron rms can be achieved, offering single electron sensitivity, a remarkable circumstance in a large system. Space charge effects have no impact on gain or linearity, and microphonic effects are highly suppressed.

---

[1] As xenon density is increased above $\rho \sim 0.5$ g/cm^3, increasingly large fluctuations appear in the partitioning of energy between scintillation and ionization, seriously complicating the measurement of energy [14]. At densities near liquid, the Fano factor approaches 20 [21], resulting in a factor of ~11 worse resolution than low-density gas.

In EL, the primary ionization electrons drift into a high-field region where they rapidly gain energy. Collisions are entirely elastic up to the first excited state of xenon at 8.32 eV. If the EL field is high enough, very little energy is lost in collisions with xenon atoms, and the electron energy increases rapidly in the high field. When an electron energy exceeds 8.32 eV, an inelastic collision then transfers most of the electron's kinetic energy to a xenon atom. The electron begins again to acquire energy, producing another excitation, but under proper conditions never acquires energy sufficient to ionize xenon atoms.

The excited xenon gas atom rapidly forms an excimer. An infrared transition produces a relaxed excimer, for which the characteristic excitation energy is ~7.3 eV, (175 nm) somewhat less than the first excited state of xenon at 8.32 eV. The excimers emit VUV radiation in a band around 175 nm. This band is not absorbed by xenon and the light can be detected at large distances.

Each primary electron falls through the same gap potential V, producing M photons in a linear gain process. The variance of EL gain M can be described by a factor J_{CP}, for which $J_{CP} = 1$ would be characteristic of a Poisson distribution.[2] Due to the linearity of EL gain, the variance of M can reach values as low as $J_{CP} = 0.01$, remarkably sub-Poisson [30]. Well above the threshold field for producing EL (but below the threshold for additional ionization), the EL efficiency is high: about 60 – 80% of acquired energy is converted to light. In other words, for broadly optimal E/p conditions, only about ~10 - 12 V is expended per photon. For example, a gain M =1000 is possible with 10 – 12 kV across the gap supporting the EL field. In practice, due to limited geometrical coverage and PMT quantum efficiency, the detection efficiency of the EL light is less than 10%, and the number of detected photons n_{pe} is the quantum bottleneck dominating the statistical precision.

Nevertheless, in a real sense, EL offers a method to count individual primary electrons with high statistical precision. As photodetectors vary in their response to single photons, here denoted by σ_{pd}, the narrower that response, the better the electron counting precision. Taking these factors in quadrature, it is easy to show that [3]

$$G = J_{CP} / M + (1 + \sigma_{pd}^2) / n_{pe} \qquad (7)$$

A practical goal is that $G \leq F$. Perhaps, $G << F$ can be achieved someday. In this paper, I focus on the possibility that it may be possible to reduce F itself, substantially, while maintaining $G \leq F$ by a proportional increase in optical gain.

4. The Fano factor and the Penning effect

The Fano factor reflects the impact of a fixed energy input on the energy partition among major pathways of energy loss. The fluctuations in number populations of ionization, excitation, and heat are constrained by the fixed energy input. Providentially, the Fano factor for not-too-dense xenon, as foot-noted above, is much smaller than 1. The only apparent way to reduce F further is to eliminate or severely reduce the population of excitations through the Penning effect. Because one of two populations is severely reduced, the fluctuations in both populations is reduced, producing the desired result for ionization.

The Penning effect occurs in noble gases, to which an impurity atom/molecule with an ionization potential (IP) slightly smaller than the excitation energy of the long-lived noble gas atoms or excimers has been added. Collisional de-excitation of the majority noble gas by the impurity leads to

[2] The factor J_{CP} honors the pioneering contributions of C. A. N. Conde and A. J. P. Policarpo, University of Coimbra, Portugal to the development of EL as a technique.

ionization of the impurity. With xenon, no noble gas Penning effect candidates are possible, leaving only molecular species. For molecular impurities, the efficiency of the Penning process appears to depend not only on the IP, but on chemical or structural characteristics as well, as breakup reactions may also occur, instead of ionization. The choices appear to be few.

After the energy deposition process, an EL TPC with a strong Penning effect present will, by design, deplete the primary excitation population, reducing F and increasing N_I, as desired. Subsequently, after drift to the high-field EL region, the primary electrons will gain energy and collide with the molecular impurities as well as with xenon. It is essential that the energetic electrons never ionize the impurity molecules, since that would lead to a large G. This imposes a requirement that the Penning molecule also will provide efficient fluorescence, fluorescence that would be otherwise be provided by the xenon atoms. Paradoxically, the molecule ionizes freely when exposed to collisions with excited xenon atoms/excimers, but never when exposed to a population of electrons gaining energy in an increased electric field. Instead, the molecule must convert the kinetic energy of the electron population to light.

In short, I seek a molecule displaying a strong Penning effect with xenon, strong fluorescence, and low electronegativity at all relevant electron energies. Do such molecules exist? A starting point is the family of aliphatic amines, which are known to display strong fluorescence. The only candidates I have found that may meet all criteria are trimethylamine (TMA) and dimethylamine (DMA); but there may be others. The IPs of TMA and DMA are 7.85 ±0.05 eV and 8.32 ±0.03 eV, respectively [31, 32]. Both are claimed to display a strong Penning effect [31], inferred from a large reduction in operating voltage in xenon-TMA or xenon-DMA mixtures, relative to other similar xenon + molecular gas mixtures at one bar. The evidence for the Penning effect presence thus is indirect, but the effect is quite large. Interestingly, the response in xenon-triethylamine (TEA) mixtures did not suggest the presence of a significant Penning effect [31], even though the IP of TEA is ~7.5 eV.

TMA and TEA fluoresce strongly in 280 – 310 nm range [33]. Probably DMA does as well, but no studies appear to exist, but I have not found any studies of EL response in xenon-TMA or xenon-DMA mixtures. However, early work with a related mixture, argon – TEA showed a strong EL response, and a fairly wide range of optical gain before charge gain begins [29]. As the ~300 nm range corresponds to ~4 eV, less energy is needed per photon in a Penning – EL mixture; detection is easier as well.

5. EL TPC with Penning effect molecules?

A central element of this concept is that, as the electrons gain energy in the EL region, they transfer their excess energy efficiently to the TMA or DMA by excitation. The efficient transfer of energy is needed to ensure that no electrons reach energies corresponding to the IPs of TMA/DMA. The molecular density must be sufficient to scavenge energy from all electrons with energies above excitation, ~4 eV. So, in this scenario, TMA or DMA completely "replaces" xenon, even though the TMA or DMA is present only at the ~1% level, (perhaps much less). In other words, the role of xenon in EL is now non-existent, extracting negligible energy in collisions with energetic electrons as they gain energy up to the 4 eV range. Above 4 eV, collisional energy transfer efficiently extinguishes the electron population. Since the first excitation level of xenon is 8.32 eV, far above the EL range of TMA/DMA, the noble gas with TMA or DMA plays no role in EL [34]. There should be a range of electric field where strong TMA/DMA EL exists with no excess ionization. The range of electric field satisfying this requirement needs to be explored experimentally, varying the relative concentration of TMA or DMA in xenon, and total density ρ. Thus it should be possible to map out optimum conditions for EL in a small TPC. It could turn out that while DMA mixtures are also worth exploring, concern arises as to possibility of higher electronegativity in this gas since DMA is less symmetric.

What about the Fano factor F? Will the optimum conditions for EL production be sufficiently broad that an overlap exists for minimizing F as well? This question remains unanswered. Experimental work, hopefully coincident with EL exploration outlined above, will be required. A direct manifestation of the Penning effect should appear as increased charge yield for the same energy input, with improved energy resolution relative to the pure gas. The energy resolution of x-ray lines below 100 keV should provide a direct measure of the improvement as long as EL production is high enough to provide good statistical precision in detected photons.

The reduction of the Fano factor remains speculative until measurements are made and optimal conditions are determined. In some mixtures, a Fano factor as low as 0.05 has been calculated [35]. Possibly then, a reduction factor of 2 is not unreasonable. For this value and the assumption that $G = F = 0.075$, the "sub-intrinsic energy resolution" is found:

$$\delta E/E = 1.9 \times 10^{-3} \text{ FWHM} \tag{8}$$

This value is of course quite idealistic, as other effects, such as Bremmstrahlung and energy gained/lost due to electrons traversing the drift field, present obstacles to realizing this goal. Nevertheless, any advance in energy resolution adds to the sensitivity of the search in the case that background is expected in the energy region of interest. In the case of ^{136}Xe, the 2457 keV Q-value is quite close to a weak but dangerous ^{214}Bi γ-ray line just 10 keV below. It is at least conceivable that if resolution close to (8) is realized, then the discrimination against this background by energy resolution would be dramatically better.

Within the context of R&D for the NEXT-100 TPC, described at this Symposium, an experimental effort is being prepared to seek the optima for a Penning effect in xenon-TMA, and for EL in xenon-TMA. If the optima overlap well, a new pathway will emerge.

6. Discussion

Adding any molecule to xenon introduces profound changes for scintillation yields. With TMA present, the excitation population is extinguished by design. Hence the primary scintillation characteristic of pure xenon would be quenched. In this case, no fast signal will be available to provide the t_0 signal needed by the TPC to place an event in space. Even if all other aspects turn out to be optimally beneficial, the complete absence of the t_0 signal would make the goal of a Penning-enhanced EL TPC infeasible or unattractive. On the other hand, it is conceivable that not all of the excitation population is converted to ionization, and that some of the energy extracted by TMA is effectively wavelength-shifted to the more convenient 300 nm range. The understanding of remnant primary scintillation at either 175 or 300 nm is also an important part of the developing experimental effort within the context of the NEXT collaboration. That program will be described elsewhere.

Finally, the presence of complex molecules will reduce substantially diffusion of the primary ionization electrons during drift, even at the parts per thousand level. This is beneficial for event topology reconstruction. In the drift region, it is the lower-lying molecular levels that are important for cooling the electron population, not the excitation levels around 4 eV relevant to fluorescence. Some change in the drift velocity may also occur, but is not particularly important except for optimization of digitization rates for the tracking elements.

References

1. D. Nygren, NIM **A 581** (2007) p632-642
2. Diaz, Paris TPC Conference, (2008)
3. D. Nygren, NIM **A 603**, (2009) p337-348

4. R. Luescher *et al.,* "Search for ββ decay in ^{136}Xe: new results from the Gotthard experiment", Phys. Lett. **B 434** (1998) 407-414
5. E. Bellotti *et al.,* "A multi-element proportional chamber used in an experiment on ββ-decay of ^{136}Xe", Nucl. Inst. & Meth. **A** 315 (1992) 252-256
6. A. S. Barabash *et al.*, "Results of the experiment on the search for double beta decay of ^{136}Xe, ^{134}Xe, and ^{124}Xe", Phys. Lett. **B 223** no. 2 (8 June 1989) pp273-276
7. S.R. Elliot, J. Engel, "Double beta decay", J. Phys. G: Nucl. Part. Phys. 30 (2004) R183.
8. M.Z. Iqbal, B.M.G. O'Callaghan, H.T. Wong, Nucl. Instr. and Meth. A253 (1987) 278.
9. Letter of Intent, "NEXT, a HPGXe TPC for neutrinoless double beta decay searches", submitted to Canfranc Underground Laboratory by the NEXT Collaboration, 3 April 2009
10. CUORE project, Gran Sasso National Laboratory, Italy, at <http:// www.lngs.infn.it/>.
11. GERDA project, Gran Sasso National Laboratory, Italy at <http:// www.lngs.infn.it/>.
12. D. H. Beddingfeld *et al.*, "High-pressure xenon ion chambers for gamma-ray spectroscopy in nuclear safeguards", Nucl. Inst. & Meth. **A** 505 (2003) 474-477
13. EXO project, at /http://www-project.slac.stanford.edu/exo/
14. A. Bolozdynya *et al., Nucl. Inst. & Meth.* **A 385** (1999) 225-238
15. U. Fano, Phys. Rev. 72 (1947) 26.
16. D. F. Anderson, T. T. Hamilton, W. H. Ku, and R. Novick, NIM **163** (1979) p125,134
17. E. P, deLima *et al.,* NIM **192** (1981) p 575-581
18. T. H. V. Dias *et al.,* NIM **A 347**, (1991) p341-347
19. Sergio do Carmo *et al.,* IEEE Trans. Nucl. Sci. **55**, #5 (2008), p2637-2642
20. A. Bolotnikov & B. Ramsey, NIM **A 396** (1997) p360-370
21. E. Conti *et al., Phys. Rev.* **B 68** 054201 (2003)
22. P. M. McCowan, R. C. Barber, ``Q value for the double-beta decay of Xe-136," Phys. Rev. **C82**, 024603 (2010).
23. M. Redshaw, E. Wingfield, J. McDaniel, E. G. Myers, ``Mass and double-beta-decay Q value of Xe-136," Phys. Rev. Lett. **98**, 053003 (2007).
24. G.D. Alkhazov, Nucl. Instr. and Meth. **89** (1970) p155
25. C. A. N. Conde and A. J. P. L. Policarpo, *Nucl. Instr. & Meth.* **53** (1967) 7 -12
26. C. A. N. Conde, A. J. P. L. Policarpo, and M. A. F. Alves, *IEEE Trans. Nucl. Sci.* **15** (1968) 84-91
27. A. J. P. L. Policarpo, *Physica Scripta* **23** (1981) 539-549
28. G. Charpak *et al., Nucl. Inst. & Meth.* **126** (1975) 381-389
29. G. Charpak *et al., Nucl. Inst. & Meth.* **A 269** (1988) 142-148
30. Carlos A. B. Oliveira, J.F.C. Veloso, and A. L. Ferreira, "Electroluminescence simulations for NEXT", Internal document, University of Aveiro, Portugal
31. Brian Ramsey and P. C. Agrawal: NIM **A 278** (1989) p576-582
32. *NIST Chemistry WebBook*, NIST Standard Reference Database Number 69, http://webbook.nist.gov
33. C. G. Cureton *et al.*, Chem. Phys. **63** (1981) p31-49
34. V. Peskov *et al.*, NIM **A277** (1989) p547-556
35. G. F. Knoll, "Radiation Detection and Measurement" 3rd Ed. (2000) John Wiley and Sons; see pp 176-178 and references cited

Journal of Physics: Conference Series **309** (2011) 012007 doi:10.1088/1742-6596/309/1/012007

NEXT: The Electroluminescence Readout

Igor Liubarsky for the NEXT collaboration
Instituto de Física Corpuscular (IFIC), Edificio Institutos de Investigación, Valencia - Spain
E-mail: `Igor.Liubarsky@ific.uv.es`

Abstract. NEXT 100 is a neutrinoless double beta decay search experiment scheduled for installation at the Laboratorio Subterrneo de Canfranc (LSC) in 2013. The NEXT 100 detector is a Time Projection Chamber (TPC) utilizing the default technology solution of a electroluminescence signal amplification. NEXT 1 EL is an important milestone in the development of NEXT 100 which aims to demonstrate large size EL amplification and long track drift in 10 bar pressure Xe gas.

1. Introduction

Neutrinos, unlike the other Standard Model fermions, could be Majorana particles, that is, truly neutral particles indistinguishable from their antiparticles. The existence of Majorana neutrinos may have very deep physics implications. For example, they provide a natural explanation for the smallness of neutrino masses, the so-called seesaw mechanism [1]-[4]. Furthermore, Majorana neutrinos, through leptogenesis, could have induced the baryon asymmetry of the Universe. The only practical way to establish experimentally that neutrinos are their own antiparticles is the detection of neutrinoless double beta decay ($\beta\beta0\nu$). This is a hypothetical, very slow nuclear transition in which a nucleus with Z protons decays into a nucleus with Z + 2 protons and the same mass number A, emitting two electrons that carry essentially all the energy released ($Q_{\beta\beta}$). Such a process occurs if and only if neutrinos are massive, Majorana particles. Also, $\beta\beta0\nu$ violates lepton number conservation, and is therefore forbidden in the Standard Model. Any source of lepton number violation (LNV) can induce $\beta\beta0\nu$ and contribute to its amplitude. If we assume, nevertheless, that the dominant LNV at low energies is the Majorana mass term for the neutrino, the rate of $\beta\beta0\nu$ is proportional to the effective Majorana mass of the ν_e

$$m_{\beta\beta} = \left|\sum_i m_i U_{ei}^2\right| \tag{1}$$

where m_i are the neutrino mass eigenstates and U_{ei} are elements of the neutrino mixing matrix. Measuring the decay rate of ($\beta\beta0\nu$) process would provide direct information on neutrino masses. The most sensitive limit to date was set by the Heidelberg-Moscow (HM) collaboration [5]. They claimed to have observed neutrinoless double beta decay in ^{76}Ge and measured $T^{0\nu}_{\frac{1}{2}} \geq 1.9 \times 10^{25}$ years.

This result has been criticized [6]. However, together with the observation of neutrino oscillations [8] (which proves that neutrinos have a non-zero mass, which is an essential condition

Journal of Physics: Conference Series **309** (2011) 012007 doi:10.1088/1742-6596/309/1/012007

for the neutrinoless double beta decay to exist) have revived the interest in neutrinoless double beta decay searches, and several new experiments with sensitivity for $m_{\beta\beta}$ $100meV$ are presently under design and construction.

NEXT 100 is a 100 kg (enriched ^{136}Xe) Time Projection Chamber neutrinoless double beta decay detector scheduled for installation at the Laboratorio Subterrneo de Canfranc (LSC) in 2013. Electroluminescence readout default technology solution for NEXT 100 [7].

2. Methodology

Double beta decay experiments are, in general, calorimeters searching for a small peak at $Q_{\beta\beta}$ in their energy spectrum. This is a challenging task: backgrounds from natural radioactivity are abundant in this region of energies and can easily overwhelm the signal peak. Good energy resolution is therefore very important, because increases the signal-to-noise ratio, but it is not enough by itself. Indeed, experiments based on germanium diodes (such as HM), devices with excellent energy resolution, have been background-limited. Consequently, all new-generation experiments try to exploit other signatures, such as event reconstruction, to further discriminate between signal and background.

A high-pressure xenon gas (HPXe) TPC can provide both good energy resolution and event topological information for $\beta\beta0\nu$ searches. Double beta decay events have a distinctive topological signature in HPXe that can be used to reject backgrounds: a ionization track, of about 30 cm long at 10 bar, tortuous due to the multiple scattering, an with larger deposition (blobs) in both ends (see Figure 1). The Gotthard experiment [9], consisting in a small xenon TPC (5 kg) operated at 5 bars, proved the utility of this signature, achieving an impressively low background rate of about 0.01 counts/keV/kg/year. However, the detector suffered of a modest energy resolution, 6.6% FWHM at $Q_{\beta\beta}$, probably due to the the use of conventional avalanche amplification in a wire plane, and to the addition of methane (4%) to the xenon (in order to increase the drift velocity and to suppress diffusion), that quenched the primary signals. Measurements in other small HPXe systems [10, 11] have shown that optimal energy resolution, <0.5% FWHM at $Q_{\beta\beta}$, is possible using electroluminescence (EL) for the amplification of the signals: the charges from primary ionization are accelerated by a moderate electric field (3-4 kV/cm/bar), producing a proportional emission of UV light with sub-poissonian fluctuations. This performance seems independent of the gas pressure below some 50 bar.

Xenon is the only noble gas that has a $\beta\beta$-decaying isotope, ^{136}Xe. Its $Q_{\beta\beta}$-value is high enough (2458 keV) to be used in a $\beta\beta0\nu$ experiment. The natural abundance of the isotope is 9%, but it can be enriched by centrifugation at a reasonable cost. In addition, xenon does not have any other long-lived radioactive isotopes, and being a noble gas, it can be easily purified.

3. NEXT 1 EL

NEXT 1 EL is presently the largest operational Xe TPC. Operating at a designed pressure of 10 bar the TPC fits inside a chamber 600 mm long and 300 mm inner diameter. The fiducial volume consists of a hexagonal cross-section, defined by the PTFE reflector panels, 160 mm across the diagonal and 300 mm drift region. The electroluminescence region is made of two parallel grids separated by 5 mm. The maximum designed drift field is 1 kV/cm and the maximum electroluminescence field is 40 kV/cm. The optical readout is made up of a PMT plane, consisting of 19 Hamamatsu R7378A photomultiplier tubes, located 100 mm behind the cathode, and 248 SiPM located 2 mm behind the anode grid. The PMT plane gives the t_0 of the events and collects EL light to reconstruct the energy of the event. Figure 2 and Figure 3 show the internal view of the fiducial volume and the external view of the field cage. The field shaping rings of the field cage are made by cutting and machining aluminum pipe. The supports are manufactured from peek, a low outgassing plastic.

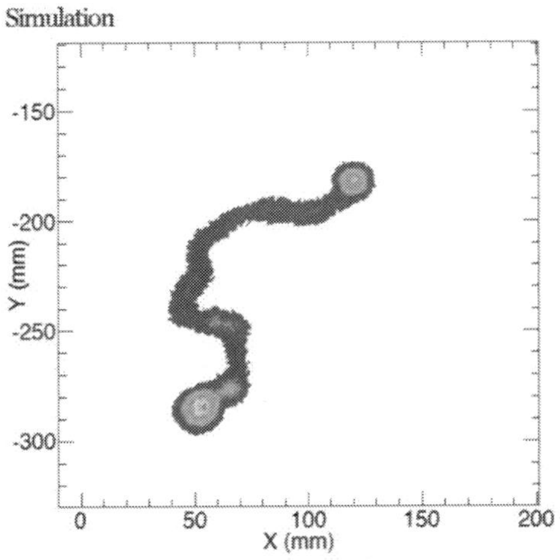

Figure 1. Example of a simulated double beta track. The two electrons terminate in a blob like structures due to rapid energy loss at the ends of their tragectories.

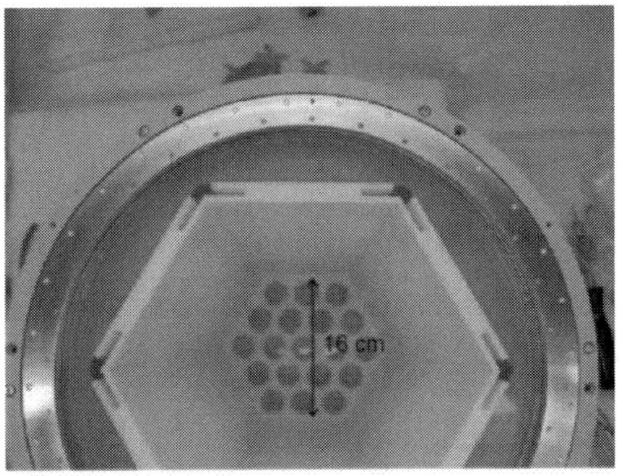

Figure 2. View inside the fiducial volume of NEXT 1 EL showing the PTFE reflector panels and the PMT holder seen in the far end of the hexagonal light tube

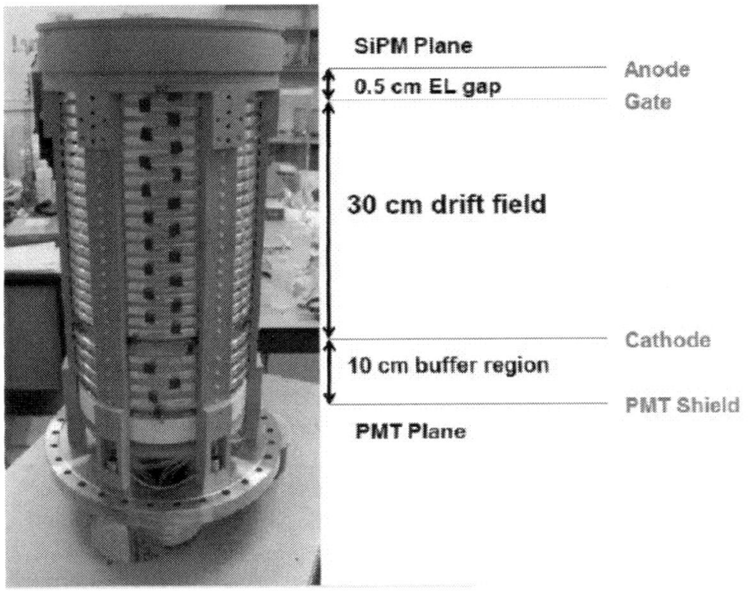

Figure 3. The external view of the field cage showing the field rings, divider resistor chain. The Cathode, EL region, PMT screen and drift region are indicated.

The SiPM plane is responsible for the tracking and provides topological information on the observed event such as the simulated track shown in Figure 1. The sensors are arranged on a square grid with a 20 mm pitch. Light emission in Xe takes place at 175 nm therefore PMTs with quartz windows are being used. However, SiPMs have no sensitivity to light at this wavelength. To gain sensitivity the SiPM have been coated with tetraphenyl-butadiene (TPB) by evaporation in vacuum. TPB is a molecular wavelength shifter capable of capturing the light emitted by Xe with high efficiency and re-emitting in blue. As a result the UV light, emitted by Xe, is shifted into the visible region where it now can be registered by the SiPM tracking plane.

The tracking plane consist of 18 daughter boards that plug into a single motherboard mounted onto one of the end-caps. In Figures 4-5 images of the motherboard and daughter boards are shown.

Figure 4. Motherboard for the tracking plane

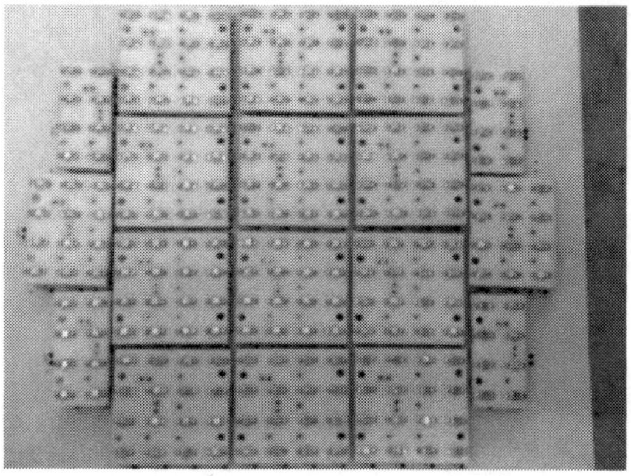

Figure 5. Daughter boards showing mounted SiPMs prior to coating with TPB. 12 boards with 16 SiPMs, 4 boards with 8 SiPMs and 2 boards with 12 SiPMs

The High Voltage is supplied to the Cathode and the Gate, in the electroluminescence region, through custom made HHV These have been tested to high vacuum and 100 kV without leaking or sparking.

The side of the chamber contains 8 CF40 size nipples. One set is located in the horizontal plane while the other 135°. these contain radioactive source ports used for calibration of the TPC. The ports are made by welding a 0.5 mm blank at the end of a 12 mm liquid feedthrough. The radioactive source is then located on the outside of the detector.

3.1. Gas System

The materials used for the vessel and the readout place outgas electro-negative impurities, which degrade the performance of the detector, into the Xe gas. The role of the gas system is to remove these. This is achieved by continuously re-circulating the Xe gas through a SAES Getters (MC500). All the gas piping, save for the inlet gas hoses and Getter fittings, are 1/2 inch diameter with VCR fittings. The re-circulation loop is powered by a KNF diaphragm pump with a nominal flow of 100 standard liters per minute. At a 10 bar operating pressure of NEXT 1 EL this translates to an approximate flow of 10 liters per minute. Figure 6 and Figure 7 show the schematic and photographs of the gas system respectively. The photograph shows the main elements used in the purification of the Xe gas. The Gas system was vacuum evacuated to a pressure of 10^{-5} mbar and baked to a temperature of 150°C. Additionally the system was flushed with Ar gas and again evacuated and baked.

3.2. Commissioning

At the time of TPC Symposium the commissioning of the NEXT 1 EL just began. Figure 8 shows first light in NEXT 1 EL. The detector was operated at a 5 bar absolute pressure and Cathode and Gate operated at -15 kV and -8 kV respectively. A source of ^{137}Cs positioned so as to emit gammas along the central axis of the detector. The trace give clear illustration of the

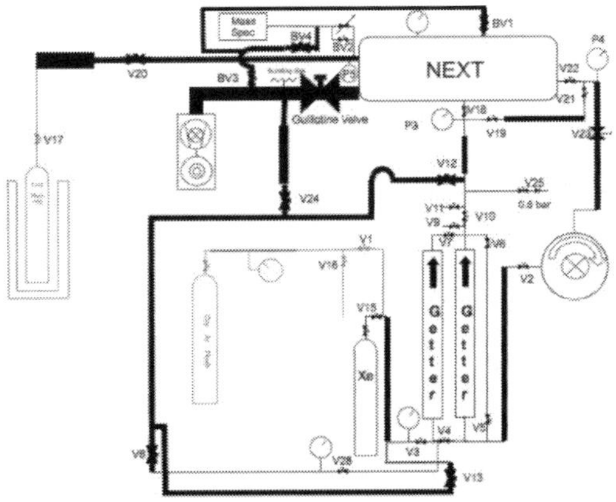

Figure 6. The schematic diagram of the NEX 1 EL gas system. The diagram shows the gas lines as well as vacuum lines used for evacuation of the system prior to use.

Figure 7. The photograph of the main elements of the gas system mounted on its frame. This was taken prior to the instillation of the vacuum lines used to evacuate the system.

primary scintillation pulse which is followed by the larger electroluminescence pulse produced by the ionized track drifting into the EL region.

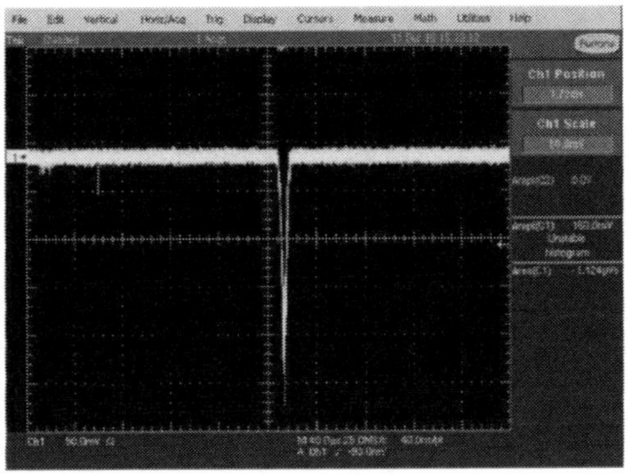

Figure 8. Showing first light of the NEXT 1 EL. The trace shows a clear primary scintillation pulse followed by the larger electroluminescence signal.

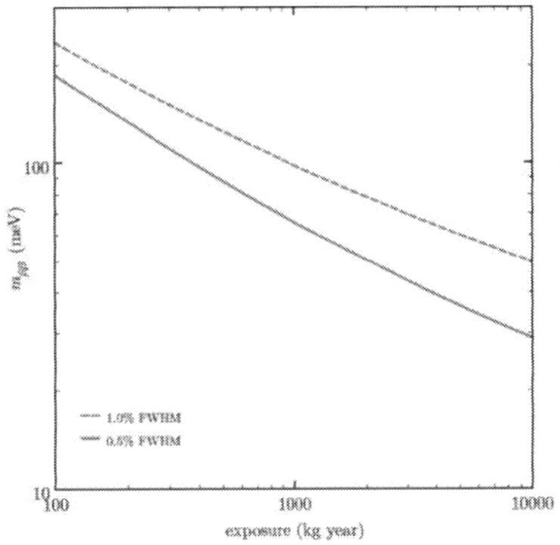

Figure 9. Expected sensitivity of NEXT 100 for two different energy resolution values.

4. The Future

NEXT 1 EL was an important milestone in the development of the large Electroluminescence TPC, which is presently the default solution for NEXT 100, although other technologies are under investigation[13, 14, 15] NEXT 100 is an approved and funded experiment scheduled to complete construction and installation at the Canfrac underground laboratory in 2013. The primary aim of the project is to instrument 100 kg of enriched ^{136} Xe in order to search for the neutrinoless double beta decay. The 90 % enriched gas has been obtained and is available at the

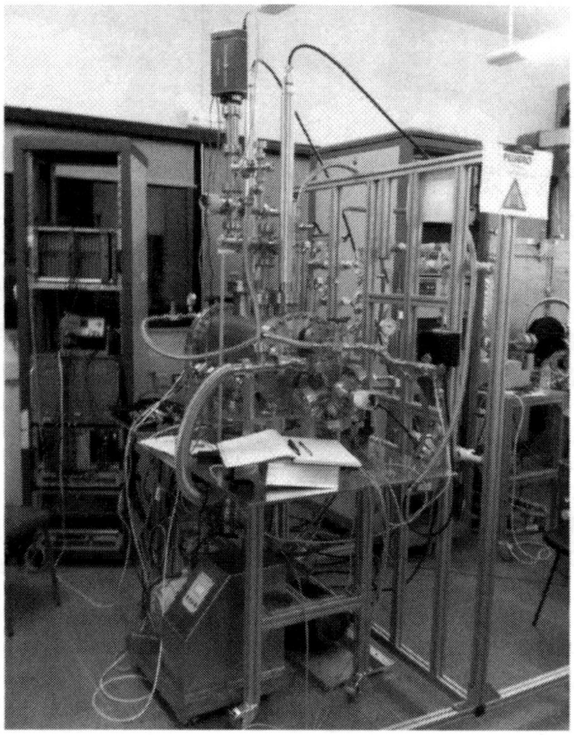

Figure 10. The assembled and operating NEXT 1 EL with the Gas system shown on the right.

Canfrac laboratory. The expected sensitivity of NEXT 100 is shown in Figure 9[12]. The plot was obtained assuming the shell model matrix element[16].

5. Acknowledgment

Special thank you to Justo Martín-Albo. This work was supported by the Spanish Ministerio de Ciencia e Innovación under grants Consolider-Ingenio 2010 CSD2008-0037 (CUP) and FPA2009-13697-C04.

References

[1] Minkowski P 1977 *Phys. Lett.* **B67** 421
[2] Gell-Mann M, Ramond P and Slansky R 1979 (*Preprint* Print-80-0576 (CERN))
[3] Yanagida T 1979 *Proc. Works. on the Baryon Number of the Universe and Unied Theories* (Tsukuba, Japan)
[4] Mohapatra R N and Senjanovic G 1980 *Phys. Rev. Lett.* **44** 912
[5] Baudis L *et al.* 1999 *Phys. Rev. Lett.* **83** 41-44 (*Preprint* arXiv:hep-ex/9902014)
[6] Aalseth C E *et al.* 2002 *Mod. Phys. Lett.* **A17** 1475-1478 (*Preprint* arXiv:hep-ex/0202018)
[7] Granena F *et al.* 2009 (*Preprint* hep-ex/0907.4054
[8] Gonzalez-Garcia M C and Maltoni M 2008 *Phys. Rept.* **460** 1-129 (*Preprint* arXiv:0704.1800 [hep-ph])
[9] Luscher R *et al.* 1998 *Phys. Lett.* **B434** 407-414
[10] Akimov D Y *et al.* 1996 *Nucl. Instrum. Meth.* **A379** 484-487
[11] Fernandes L M P *et al.* 2010 *JINST* **5** P09006 (*Preprint* arXiv:1009.2719 [astro-ph.IM]
[12] Martín-Albo J *et al. J. Phys.: Conf. Ser.* **259** (2010) 012040
[13] Nygren D R *et al.* 2010 *These proceedings*
[14] Dafni T *et al.* 2010 *These proceedings*
[15] Lux T *et al.* 2010 *These proceedings*
[16] Meendez J, Poves J A, Caurier E, and Nowacki F *Nucl. Phys.* **A818** (2009) 139151, [arXiv:0801.3760]

A Multi-APD readout for EL detectors

T. Lux[1], O. Ballester[1], J. Illa[1], G. Jover[1], C. Martin[1], J. Rico[1,2], F. Sanchez[1]

[1] Institut de Física d´Altes Energies (IFAE), Edifici Cn, Universitat Autònoma de Barcelona, 08193 Bellaterra (Barcelona), Spain
[2] Institució Catalana de Recerca i Estudis Avançats (ICREA), 08010 Barcelona, Spain

E-mail: Thorsten.Lux@ifae.es

Abstract. Detectors with an electroluminescence readout show an excellence performance in respect of energy resolution making them interesting for various applications as X-ray detection, double-beta and dark-matter experiments, Compton and γ cameras, etc. In the following the study of a readout based on avalanche photo diodes to detect directly the VUV photons is presented. Results of measurements with 5 APDs in xenon at pressures between 1 and 1.65 bar are shown indicating that such a readout can provide excellent energy and a moderate position resolution.

1. Introduction

In conventional gas detectors the electrons released due to ionization within the active volume are drifted to the readout plane where a gas amplification takes place to amplify the primary signal before it is processed by the electronics. The disadvantage of this approach is that the exponential amplification introduces fluctuations in the measurement limiting the achievable energy resolution. In an electroluminsence (EL) readout this charge amplification process is replaced by a light amplification. Primary electrons drift to a region which is terminated by two wire meshes. Between the two meshes a voltage is applied such that the electrons excite the gas, normally pure noble gases as xenon or argon, without ionizing it. In the de-excitation of the atoms, photons in the VUV region are emitted isotropically. Since this is not an exponential process as for charge amplification, the gain fluctuations introduced are smaller and therefore the achievable energy resolution is significantly better. The amount of photons produced depends on the applied voltage difference, the distance between the two meshes and also on the operation pressure. It turns out that the performance improves with increasing the pressure. In [1] a detailed description of the process for xenon can be found.
The double beta (DB) experiment NEXT (*Neutrino Experiment with a Xenon TPC*) is based on a xenon TPC operated at up to 10 bar. It is foreseen to use an EL readout to achieve the very good energy resolution of no more than 1 % FWHM required at the $Q_{\beta\beta}$ value (2.46 MeV). In the base design the chamber will be an asymmetric TPC in which the energy is measured by a PMT plane behind the cathode while the tracking information is obtained by a plane of MPPCs coated with TPB and covering about 1 % of the total area. A detailed description of the base design can be found in [2]. An alternative possibility using APDs to unify tracking and energy measurement in a single technology is discussed in this paper. The topic of the T_0 measurement in such a detector is not addressed here but will require additional R&D studies.

In the following, we present the performance study of an EL readout with 5 APDs as a first stage towards large readout areas. It turns out that such a readout based on APDs might not be only suitable for DB experiments like NEXT but also for other applications as e.g. Compton cameras. The use of APDs for the readout of EL chambers were already studied intensively by a research group at the University of Coimbra [3]. Their results (fig. 1) indicate that an excellent energy resolution of less than 5 % can be achieved at 22 keV for pressures between 4 to 6 bar. The authors believe that the rise for higher pressures is caused by experimental effects (micro sparks) in their setup. If this result could be scaled with $1/\sqrt{E}$ to the $Q_{\beta\beta}$ of xenon (2.46 MeV), an energy resolution of better than 0.5 % FWHM at 2.46 MeV would be possible. They also compared directly the performance of reading out the chamber with an APD to the one achieved with a PMT [4]. The result is shown in fig. 2. As one can see the energy resolutions with these two kind of sensors are very similar under the same experimental conditions. However, these measurements were performed with one single APD with a diameter of 16 mm (from Advanced Photonics Industries), while for the coverage of larger readout areas a multi-APD will be necessary.
We discuss here the result of a prototype to evaluate the performance of such readouts using an array of 5 APD sensors.

2. Experimental Setup

2.1. The chamber

The steel chamber is a cylindrical pressure vessel with a diameter of 18 cm and a height of 8 cm. It has two CF16 flanges for the connection to the gas system. The gas system consists of a loop between the two flanges which includes a hot getter (SAES ST707) for purification and acts at the same time as recirculation pump due to convection. The interior contains a tower made of PEEK. It consists of a copper-coated Kapton cathode, two EL meshes (wire diameter: 80 μm, wire pitch: 900 μm) and the holder for the 5 APDs. The drift distance between the cathode and the upper EL mesh is 15 mm, the gap between the meshes is 7 mm and the distance between lower mesh and APD surface about 5 mm. Horizontally the sensitive volume has the size of about 30x30 mm^2. The signal is carried out by 9-pin Lemo feedthroughs while the Macor HV connectors was built at IFAE. To allow X-rays to enter the chamber a Kapton window is located in the center of the end cap behind the cathode. Due to the distance of about 15 mm between the window and the cathode, low energetic X-ray sources cannot be used with this setup. The setup is shown in fig. 3, 4 and 5.

2.2. The APDs

For the readout of the chamber, 5 APDs (S8664-55-SPL) from Hamamatsu were used. These are standard APDs without the protection window which normally absorbs VUV photons. The APDs have a size of 5x5 mm^2 and were placed with a pitch of 10.3 mm between them. These APDs are directly sensitive to the EL light of xenon (172 nm) with a quantum efficiency of about 80 % according to the data sheet. Measurements performed together with the research group at the University of Coimbra gave a result of $70 \pm 16\%$. They are supposed to have also a high quantum efficiency to argon light (128 nm) which could amplify their field of application even more. Additional studies in respect of this are planned. The same bias voltage is applied to all 5 APDs, while the signal from each is read out independently. This powering scheme introduces a high correlation between the noise of the APDs. Fig. 6 shows this for two APDs. Proper treatment of the pedestal correlation might help improve further the energy resolution and it is under investigation.

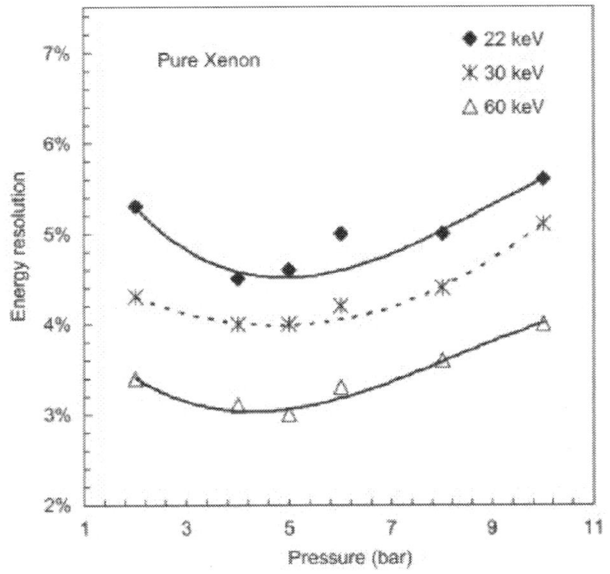

Figure 1. Energy resolution and gain as function of the pressure for three different energies.[3]

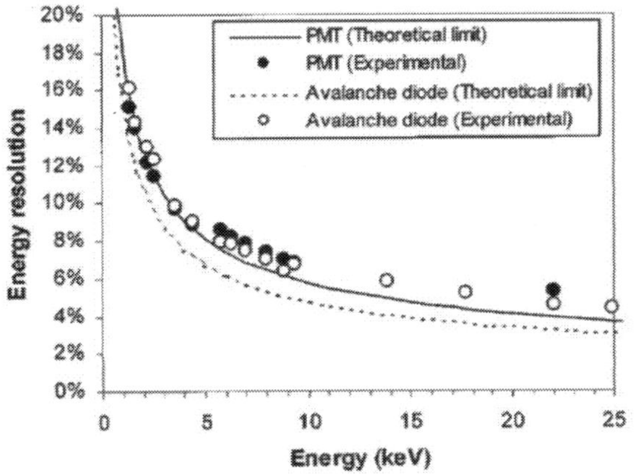

Figure 2. Direct comparison for the achieved energy resolution with one PMT and one APD.[4]

2.3. Electronics and DAQ

While for single sensor setups commercial electronics can be used, the development of a multi-sensor readout requieres the design of custom made electronics. The APD readout electronics developed at our institute is based on a transimpedance amplifier which converts the output current at the anode of the APD to voltage. This first stage of the readout has two main goals, convert the current signal to voltage and amplify it in a single step. This amplifier has the highest gain of about 10^5 which does not reduce the amplifier bandwidth below the signal bandwidth yielding to the full collection of the signal. Following the first amplifier stage there is a semi-Gaussian shaper which filters the signal eliminating both high and low frequency noises. Afterwards the signal is digitized by a comercial ADC. The DAQ system to read the digitized

Figure 3. Photo of the chamber with the gas system loop.

Figure 4. Side view of the inner tower.

Figure 5. Photo of the 5 APDs with the numbering scheme.

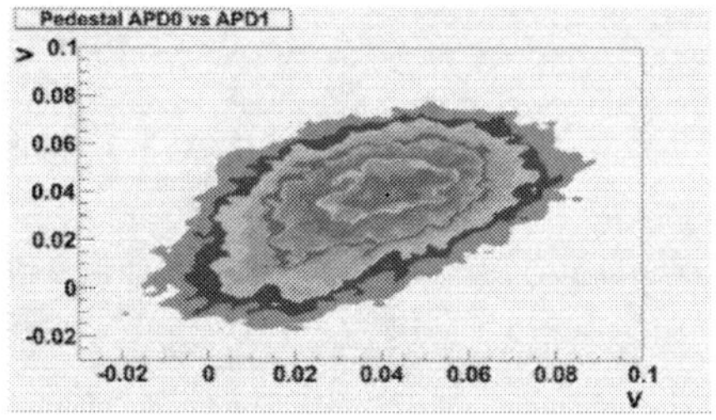

Figure 6. Pedestal of APD 1 versus pedestal of APD2 showing a high correlation (≈70 %) between them.

data from the ADC and to store it, is based on the DAQ framework MIDAS [5].

3. Data taking and analysis

For the signal creation, a ^{109}Cd source was used which emits mainly X-rays at 22.4 keV and 25 keV. The source is located on top of the central APD. The data taking is triggered by a threshold requirement on the energy deposition in the central APD while no requirements are applied to the outer APDs. After 1000 data events, 10 events with random trigger are taken to determine the pedestal and noise continuously during the data taking. The gain and energy resolution were measured for a wide range of parameters as drift field, EL field, APD voltages at various pressures between 1 and 1.65 bar.

3.1. Intercalibration

The operation of several APDs requires a careful intercalibration among them. The intercalibration was performed in various steps. In a first step, the two outer APDs on one axis were intercalibrated from the plot of the asymmetry, defined as $(E_i - E_{i+2})/(E_i + E_{i+2})$, between them versus the energy deposition in the central APD as shown in fig. 7. If no intercalibration is required the curve fitted to this distribution will have its maximum at 0, otherwise the observed shift from 0 can be used for the intercalibration. In the next step, the two axis were intercalibrated using the ^{109}Cd spectrum obtained with the outer APDs on each axis (see fig. 8). In the final step, the central APD is intercalibrated with the outer ones by minimizing the width of the 22 keV peak of the spectrum of all 5 APDs obtained from $E = E_0 + \alpha E_{out}$. The calibration factors depend on the gain of the APDs and therefore on the APD bias voltage but also external factors as the temperature[1]. However, for the moment, as long as not explicitly mentioned, the calibration factors for an APD voltage of 400 V are applied to all data.

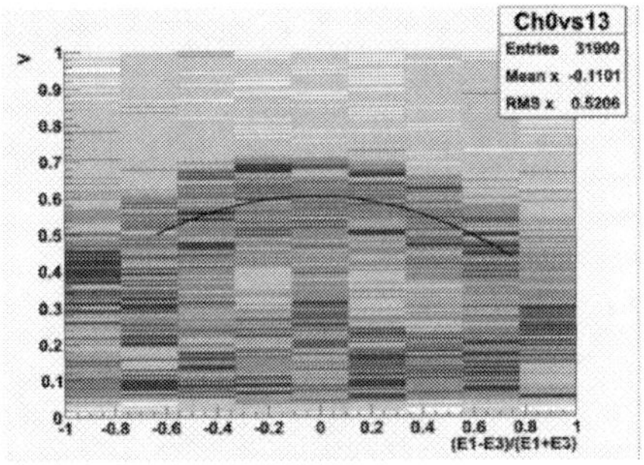

Figure 7. Plot of the energy in the central APD versus the asymmetry between APD 1 and 3.

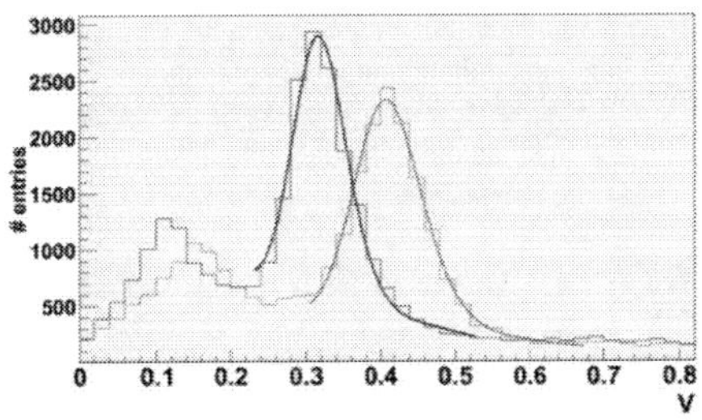

Figure 8. The spectrum of APD 1 and 3 (solid line) and of APD 2 and 4 (dotted line).

3.2. Position determination and first tracks

The position of the X-ray conversion in the xy-plane is determined by the center-of-gravity method. The spatial distribution is shown in fig. 9. Most of the conversions happen within the central region. The deviation from the expected rotation-symmetric distribution might be caused by a slight misalignment between the source and the center of the central APD or by imperfections in the used collimator.

Cosmic runs without source are used for first tracking attempts. The center-of-gravity of the charge distribution is calculated for each time bin. Fig. 10 shows one of the tracks reconstructed in this way. A point like charge deposition would produce only a straight line along the z-axis.

3.3. Energy resolution and threshold

The energy is measured with the following method:

- Determining the time bin position of the maximum of the central APD.
- Calculating the average charge deposition per voxel in a range of $\pm 3\mu$s around the maximum for each APD.
- Subtracting the pedestal for each APD.
- Summing over all APDs.

[1] not monitored until now

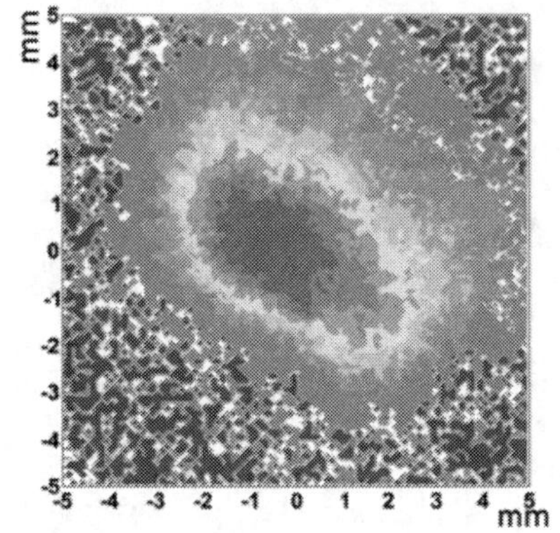

Figure 9. The xy distribution of the X-ray conversions. Most of the conversions happen in the central region (red area).

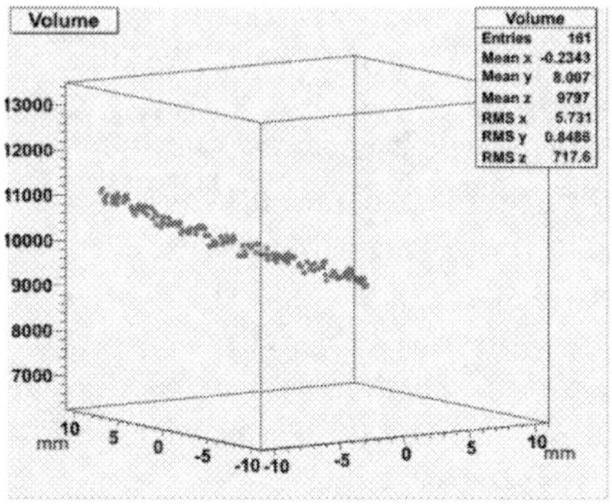

Figure 10. A track crossing the active volume under a large angle in respect of the z-axis.

- Correcting for the total energy depending on the up-down/left-right asymmetry.

Fig. 11 shows the total energy deposition as function of the asymmetry. One clearly sees a quadratic dependency caused by the smaller acceptance towards the edges.

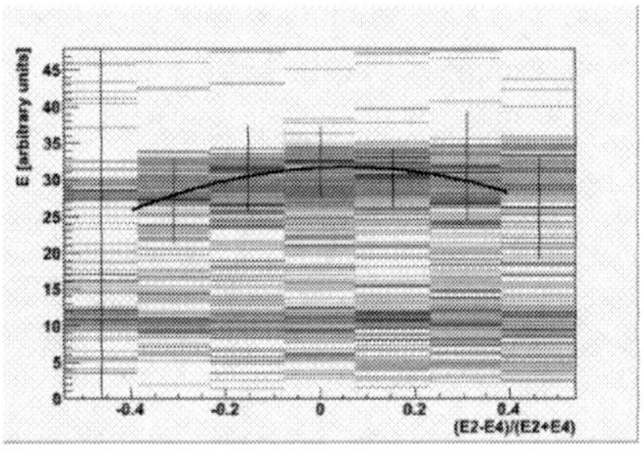

Figure 11. The asymmetry between APD 2 and 4 versus the summed energy over all APDs. The same was done for the asymmetry between APD 1 and 3.

First systematic scans of the energy resolution as function of various parameters were performed to find the optimal settings. Fig. 12 shows the scan of energy resolution as function of the drift field for 3 different pressures. The energy resolution saturates for drift fields above 200 V/cm/bar for all pressures. As expected, it was found that the result improves with increasing pressure. Fig. 13 shows the energy resolution as function of the applied EL field. The energy resolution improves significantly with higher EL fields with potential for further improvements of the results by running at fields higher than 4 kV/cm/bar. For the moment, the field is limited to this value to avoid HV instabilities within the chamber. The scan of the APD bias voltage is shown in fig. 14. The curve has a steep falling shape for low APD voltages and starts to saturate at a voltage of about 390 V. For this scan, the maximal APD bias voltage was conservatively chosen to 405 V. However, a further improvement of the energy resolution might be achieved by exploiting the full phase space of the voltages up to 425 V.

The best result for the energy resolution of (7.7 ± 0.1) % FWHM is shown in fig. 15. The data are restricted to an area of ± 5 mm around the center of the chamber. The calibration factors applied correspond to the applied voltage of 410 V to the APDs. The normalization is chosen

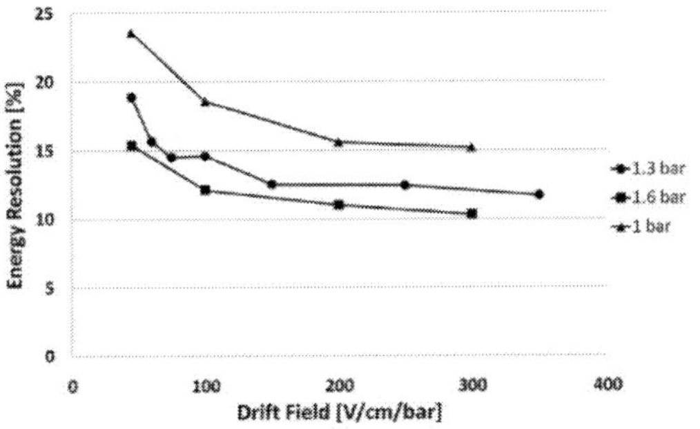

Figure 12. Scan of the drift field. The settings for the other parameters were fixed for this measurement series (EL field: 3.5 kV/cm/bar, APD bias voltage: 405 V).

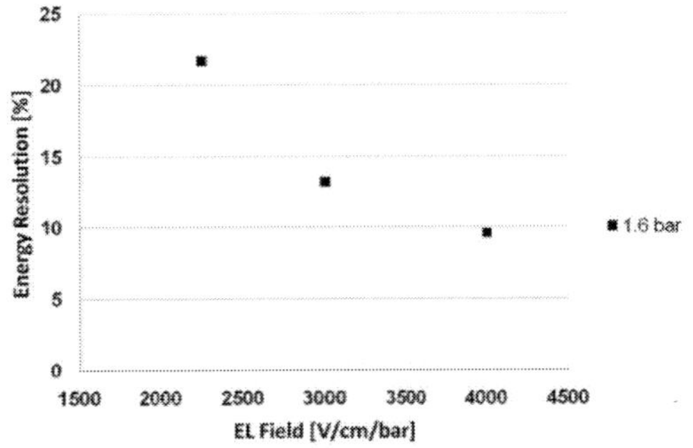

Figure 13. Scan of the EL field. The settings for the other parameters were fixed for this measurement series (drift field: 200 V/cm/bar, APD bias voltage: 405 V).

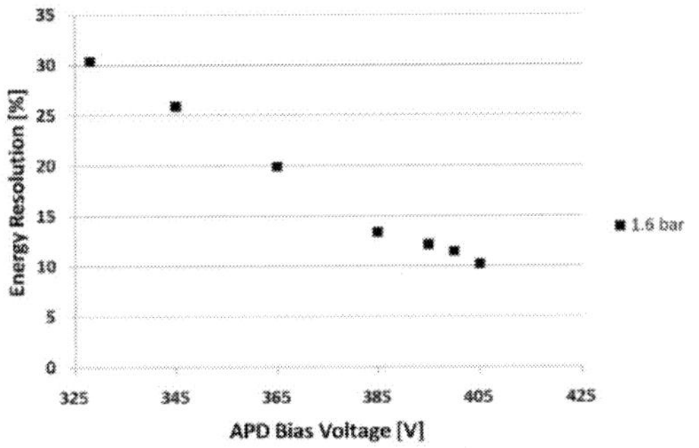

Figure 14. Scan of the bias voltage. The settings for the other parameters were fixed for this measurement series (EL field: 3.5 kV/cm/bar, drift field: 200 V/cm/bar).

such that the main peak corresponds to the literature value of 22.4 keV for the main line of ^{109}Cd. Also the peak of 25 keV is clearly visible. The additional peak which can be found at 8 keV is caused by X-ray fluorescence of the copper of the cathode. This peak has an energy resolution of 19.1 % which is worse than expected when scaled with $\sqrt{E}$, we assume due to a higher noise contribution. The red area shows the "'signal"' created by the pedestal when the same method as for the data is applied to a random time interval of 6 μs. One can conclude that a low threshold of about 2 keV can be achieved with such a read out at 1.65 bar. In respect of the achievable energy resolution, it is also interesting to investigate the two extreme cases: The energy resolution achievable when only the central APD is used for the energy measurement and the case when only the outer APDs are used. The latter case is of interest to study the dependency of the energy resolution on the pitch between the APDs when the conversion occurs within the gap between the APDs. The results are shown in fig. 16 and fig. 17 for the same data sample as before. Energy resolutions of 17.6 % FWHM for the central APD and 12.4 % FWHM for the outer APDs were achieved. The latter value suggests that even with a larger pitch an energy resolution of ≤ 1 % at 2.46 MeV could be reached.

4. Conclusion and outlook

A 5 APD readout system was successfully commissioned and operated for the first time to readout a small EL detector filled with xenon at pressure of up to 1.65 bar. Very good energy resolutions of 7.7 % FWHM at 1.65 bar were achieved for 22.4 keV X-rays. This corresponds to about 0.7 % FWHM at 2.46 MeV. Further improvement of this result seems to be likely by operation at higher pressure (see fig. 1), further optimization of the APD bias voltage and the

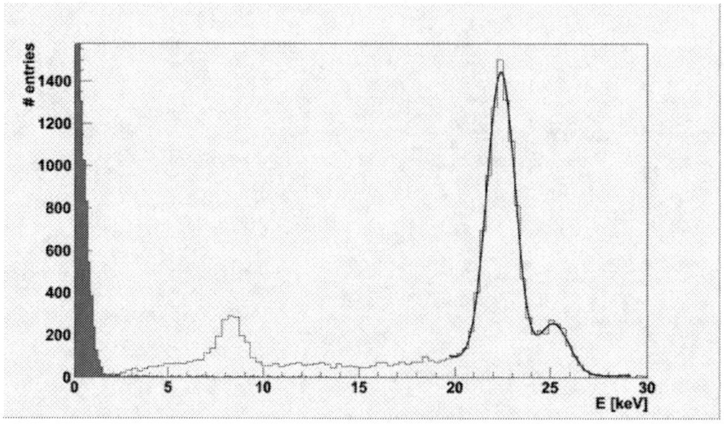

Figure 15. The best result achieved for the energy resolution was: $(7.7 \pm 0.1)\%$ FWHM for the 22.4 keV line. The settings for this run were: drift field: 300 V/cm/bar, EL field: 4 kV/cm/bar, APD bias voltage: 410 V, pressure 1.65 bar.

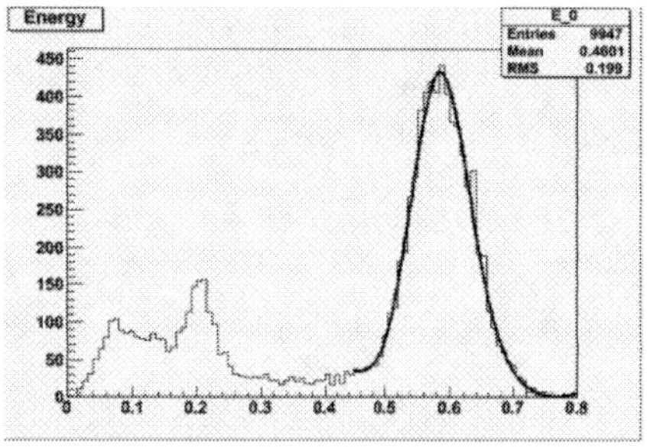

Figure 16. The spectrum of the central APD alone. No corrections were applied to the data. The energy resolution in this case is 17.5 %.

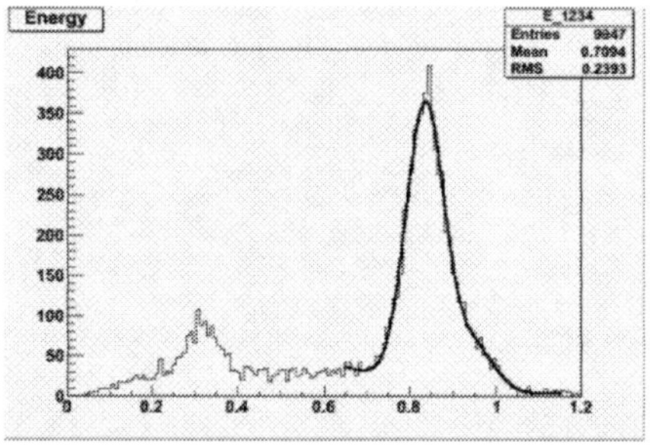

Figure 17. The spectrum of the outer APDs. Also here, no corrections were applied to the data and an energy resolution of 12.4 % was achieved.

EL field and temperature corrections. It was shown that such a readout could also provide a very low detection threshold of a few keV. In addition position determination and first tracking were performed.
For the future it is planned to extend the existing system of 5 APDs to 30 APDs in a chamber with an active volume of 20 cm diameter and 11 cm length. The system is currently under construction and should give more information about the possibilities of such a readout system for EL detectors.

Acknowledgments
The authors acknowledge the support received by the CONSOLIDER INGENIO Project CSD2008-0037 (CUP) and the FPA2009-13697-C04-03 project. We also want to thank Javier Gaweda for his technical support.

References

[1] Monteiro C M B *et al.* 2007 *JINST* **2** P05001 (*Preprint* physics/0702142)
[2] Granena, F *et al* 2009 (*Preprint* 0907.4054)
[3] Coelho L, Lopes J, Covita D, Conceio A and dos Santos J 2007 *Nuclear Instruments and Methods in Physics Research Section A: Accelerators, Spectrometers, Detectors and Associated Equipment* **575** 444 – 448 ISSN 0168-9002 URL http://www.sciencedirect.com/science/article/pii/S0168900207004639
[4] Lopes J A M, dos Santos J M F, Morgado R E and Conde C A N 2001 *IEEE TRANSACTIONS ON NUCLEAR SCIENCE* **48**
[5] Amaudruz P, Olchanski K and Ritt S 2010 http://midas.psi.ch/htmldoc/index.html

Journal of Physics: Conference Series **309** (2011) 012009 doi:10.1088/1742-6596/309/1/012009

Micromegas planes for the neutrinoless double beta decay search with NEXT

T. Dafni, on behalf of the NEXT Collaboration
Laboratorio de Física Nuclear y Astropartículas, Universidad de Zaragoza, 50009 Zaragoza, Spain

E-mail: Theopisti.Dafni@unizar.es

Abstract. NEXT is a project to look for the neutrinoless double beta decay of ^{136}Xe in the Canfranc Underground Laboratory (LSC) in the Spanish Pyrenees. The goal of the project is to construct NEXT-100, which will hold 100 kg of pressurized Xe gas, planned to run in late 2013. The project is currently progressing through its R&D phase: the first big (1 kg) prototypes will be taking data during 2011, exploring different options for the readout. One of them is reading the charge with the help of microbulk Micromegas planes. This option is discussed here.

1. Introduction

Although they have been known to exist since the 1950s, there is still important information on the neutrinos that eludes us, such as their mass and whether they are of Majorana or Dirac type. The field of neutrino physics is very active in the late years and several experiments that are trying to address this kind of questions are looking for the answers through the double beta decay process. The current generation of these experiments aims at detector target masses at the 100 kg scale. The next step would require their scaling up to few-tonnes in order to completely explore the inverse hierarchy models of neutrino mass. The main requirements are a very good energy resolution and ultra-low background levels.

NEXT (Neutrino Experiment with a Xenon TPC) is one of these experiments [1]. It is a project approved by the Canfranc Underground Laboratory (Laboratorio Subterraneo de Canfranc, LSC) in the Spanish Pyrenees. It has been funded by the Consolider Program of the Spanish Ministry of Science to carry out the construction of a 100 kg enriched high-pressure xenon gas (HPGXe) TPC.

The merit of the proposal lies on the excellent prospects that it has regarding the requirements of such a project: the excellent energy resolution which can be achieved, the background rejection power provided by the topological information of the electron tracks, the ability to define a t0 (thus a fiducial region away from any surface), and the scaling-up to large masses.. The final detector, NEXT-100, will hold 100 kg of Xe gas enriched in ^{136}Xe, its $\beta\beta$-decaying isotope.

The project baseline is focused on the EL technique using a PMT plane for the energy measurement and SiPMs for the topological information[2]. Another option studied is using APDs[3], while here we will discuss the option of the charge collection with Micromegas.

2. NEXT with Micromegas

Since their conception 15 years ago, the Micromegas detectors [4] have been used in several experiments in which they have presented outstanding performance in spatial, temporal and energy resolutions, as well as robustness and stability of operation.They are known to provide a very stable amplification factor due to the particular geometry of the avalanche, which makes it rather insensitive to the amplification gap imperfections and other external parameters like temperature and pressure [5].

The latest generation of the Micromegas planes, called microbulk Micromegas, is fabricated in such a way that provides superior mechanical homogeneity of the gap and mesh geometry [6]. Energy resolutions of 11% FWHM at 5.9 keV are routinely achieved with the new planes in mixtures of Argon-Isobutane. Because the structure is mainly made out of copper and kapton, the radioactivity levels are not expected to be high, something confirmed by the radioactivity measurements performed with microbulk planes, as commented later on. These characteristics, open the way for the application of the Micromegas in rare event searches, such as NEXT.

In the present paper we report on the results achieved with microbulk Micromegas in a small prototype called NEXT-0-MM using different gas mixtures. The commissioning of a 1 *kg* prototype, NEXT-1-MM is also described.

2.1. Tests in NEXT-0-MM

NEXT-0-MM is the first small prototype built in order to test microbulk planes in high pressure (Figure 1). It is a 2 litre chamber made of Stainless Steel, with a diameter of approx. 14 cm and a drift region of 6 cm. The vessel has been tested to hold pressures of up to 12 bar. With the help of resistors fitted at its exterior, the vessel can be heated up to 110 °C. A field cage consisting in 6 copper rings held together with three columns made of PEEK and interconnected by resistors. The structure is topped by a copper disk as a cathode. A hole is made in the cathode to accommodate the sources used in the tests (Figure 2).

Figure 1. A photo of NEXT-0-MM. The vessel is covered with an insulator layer (black).

Figure 2. The field cage of NEXT-0-MM: six copper rings interconnected via resistors, held in place by three PEEK bars and a copper disk as a cathode on top.

Two types of microbulk Micromegas have been used for the tests in NEXT-0-MM: circular readouts with an amplification gap of 50 μm and 35 mm diameter whose anode was not segmented and a bigger one, with an active area of $10\times10\,\mathrm{cm}^2$ (Figures 3 and 4). The latter has a pixelized anode with a total of 144 pixels and at the time was the largest Micromegas built with the microbulk technique. During the tests both signals of the anode and of the mesh are recorded.

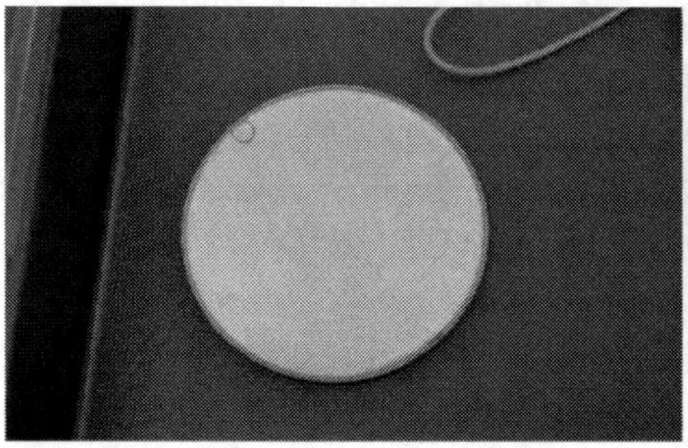

Figure 3. The first generation of microbulk Micromegas used. They had a diameter of 35 mm and an amplification gap of 50 μm.

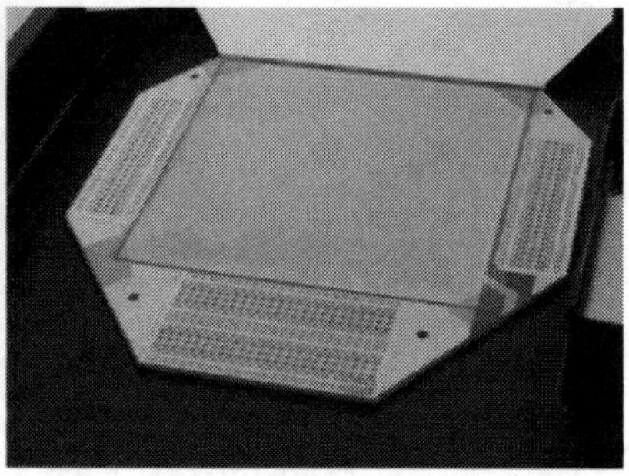

Figure 4. The first large microbulk built, with a segmented anode. It has an active area of $10 \times 10\,\mathrm{cm}^2$ divided into 144 pixels.

At the first phase of the measurements, the response of the detectors in different mixtures of Ar-Isobutane was studied, with the concentration of Isobutane in the mixture varying from 0.1% to 5%. The pressure inside the vessel was raised up to 10 bar and the tests were done with the alphas coming from an ^{241}Am source. The best energy resolution achieved in that configuration was 0.7% (FWHM) for the 5.5 MeV of the alphas coming from an source, at 4.75 bar in Ar-2% Isobutane gas mixture [7].

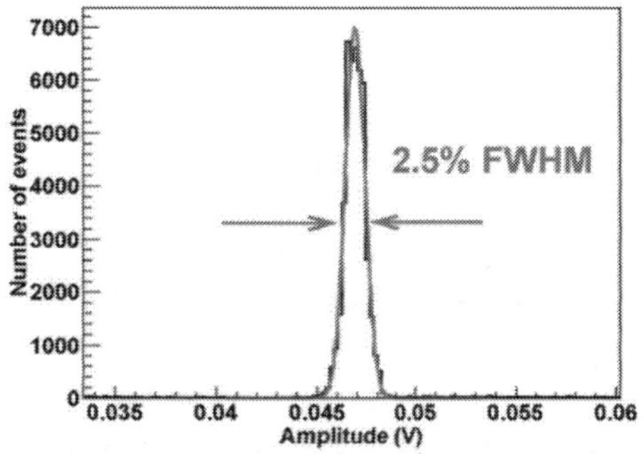

Figure 5. Energy spectrum of the ^{241}Am 5.5 MeV alpha peak measured at 4 bar of pure Xe, showing an energy resolution of 2.5% FWHM.

At a later stage, measurements were performed introducing pure Xenon in the chamber increasing the pressure up to 5 bar. In the range between 2 and 5 bar energy resolutions around 3% (FWHM) for the 5.5 MeV alphas of the ^{241}Am source were obtained [9], the best one being 2.5% for the 4 bar case (Figure 5). Using a selection of events based on risetime the aforementioned values improve to $\sim$2% (FWHM) (the best value at 1.8% FWHM for the 4 bar case). The system was working mainly in a closed mode, meaning that the gas was introduced and then the vessel sealed. However at times it was working in a recirculating mode, in which case the gas was passing through a purifying filter. In all the cases, the limitation in the increase of pressure was the gas quality; the energy resolution measurements started to be clearly affected by attachment.

The second campaign of measurements was focused on pure Ar and pure Xe gases. The quality of the gas was improved after long pumping periods but as well after prolonged bake-out periods in vacuum and circulating warm Nitrogen gas through the system. Figures 6 and 7 show the best results achieved in each case, for the 5.5 MeV α of the ^{241}Am, namely 2.03%

Journal of Physics: Conference Series **309** (2011) 012009 doi:10.1088/1742-6596/309/1/012009

(FWHM) at 8 bar of Ar and 1.94% (FWHM) at 4 bar of Xe. Under these conditions the effect of attachment in the gas started to be evident only in higher pressures, allowing thus data-taking up to 8 bar in pure Ar and pure Xenon. At 8 bar of Xe, the attachment effect was already apparent and therefore the resolution achieved of 4.8% (FWHM) at 5.5 MeV was not as good.

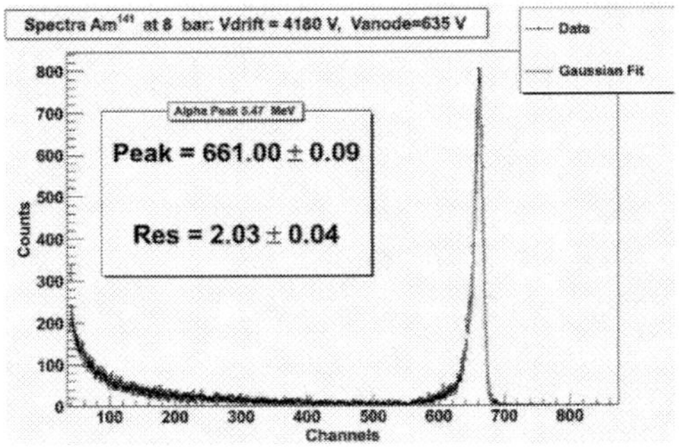

Figure 6. Energy spectrum of the ^{241}Am 5.5 MeV alpha peak measured at 8 bar of pure Argon, showing an energy resolution of 2.03% FWHM.

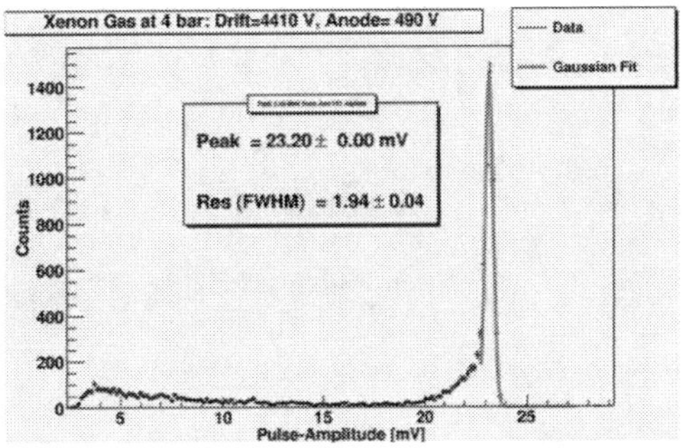

Figure 7. Energy spectrum of the ^{241}Am 5.5 MeV alpha peak measured at 4 bar of pure Xenon, showing an energy resolution of 1.94% FWHM.

By this time, when the pixelized microbulk detector was used, part of the pixels were read with a reduced version of the T2K electronics, allowing the 3D recontruction of the tracks gathered. Figure 8 shows an ^{241}Am alpha track in pure Ar at 1.23 bar, as recorded in the two-dimensional plane, and its reconstruction.

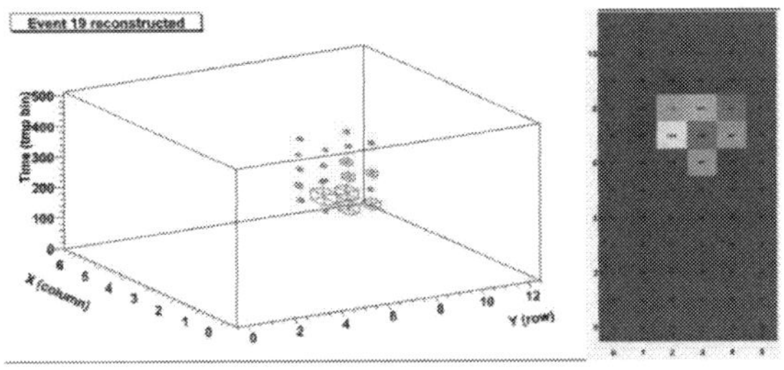

Figure 8. Projection of an alpha track obtained with the ^{241}Am source in NEXT-0-MM. On the right is the 2d mapping of the event, as read by a reduced version of the T2K electronics. On the left, a 3d reconstruction of the same event.

Measurements with γ in pure Xe Flipping the ^{241}Am source upside down, the α emitted from the deposition remain blocked. However, the source emits a γ line at 59.54 keV, which served to probe the lower energy region. Measurements were done at 1, 2 and 3.5 bars. These first measurements at high pressure, yield an energy resolution of 7.8% at 2 bar and 9.3% (FWHM) for the 60 keV γ of the ^{241}Am at 3.5 bar (Figure 9). If extrapolated to the $Q_{\beta\beta}$ value of Xe (2458 keV), they would correspond to 1.2% and 1.45% (FWHM) respectively. Interesting results inhigh pressures were obtained by other colleagues in a smaller setup [10]. We plan to extend the measurements with γ in the upcoming tests.

2.2. NEXT-1-MM

During the tests performed in NEXT-0-MM a bigger prototype was prepared, NEXT-1-MM, made of Stainless Steel. The vessel has an inner volume of approximately 75 litres, measuring 600 mm of height and almost 400 mm in diameter (Figure 10). The vessel was tested up to

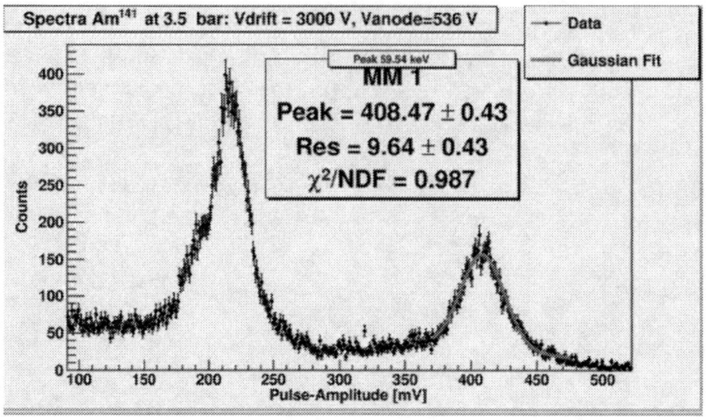

Figure 9. Spectrum of 59.54 keV γ in pure Xenon at 3.5 bar with an energy resolution of 9.6% (FWHM). The spectrum is acquired with an ^{241}Am source from which the alphas were blocked. The peak on the left is composed mainly of escape peaks from the 60 keV α, due to the fluorescence of Xe at 30 and 33 keV and partly of the 26 keV γ line of ^{241}Am.

12 bar and should hold 1 kg of Xenon at a pressure of 10 bar. Following the example of NEXT-0-MM, NEXT-1-MM is equipped with a field cage which defines a drift length of 35 cm. The structure, shown in Figure 11, is composed of 34 copper rings, interconnected with 10 MΩ resistors. The cathode is a copper disk, machined to accommodate the possible sources to be used in the tests. In order to reach the needed drift fields voltages of the order of 50 kV are needed. There are no commercial connectors for such high voltages which comply with the radiopurity requirements for the experiment. As a consequence, a dedicated study for the development of suitable feedthroughs was started.

Figure 10. A photo of the NEXT-1-MM prototype and its supporting structure.

Figure 11. The field cage of NEXT-1-MM: 34 copper rings, connected through resistors and held by PEEK bars.

The first readout plane fabricated for NEXT-1-MM was a bulk detector. It has a diameter of 30 cm and the anode is segmented into 1252 pixels. The pixels are individually read with the use of a full chain of the electronics used in the T2K experiment. Preliminary tests were performed introducing a ^{222}Rn diffuse source in Ar-2% Isobutane at atmospheric pressure. In these tests only the signal from the mesh was read, not the anode. After several hours a clear spectrum of the Rn alpha decays was gathered (Figure 14), proving that all of the detector's volume was active.

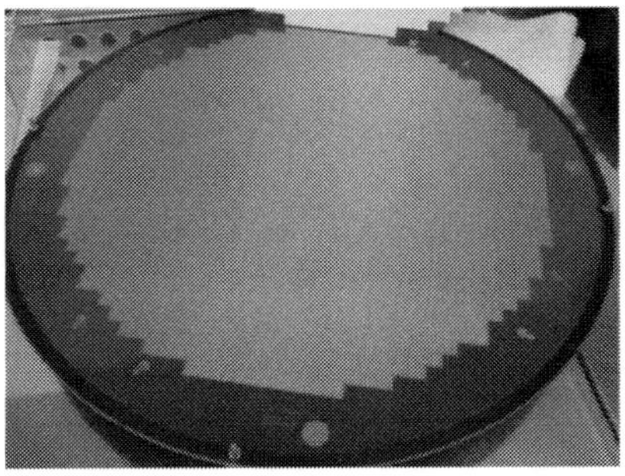

Figure 12. A photo of the bulk Micromegas detector installed in the NEXT-1-MM. It has a diameter of ~30 cm and its anode is segmented to ~1200 pixels.

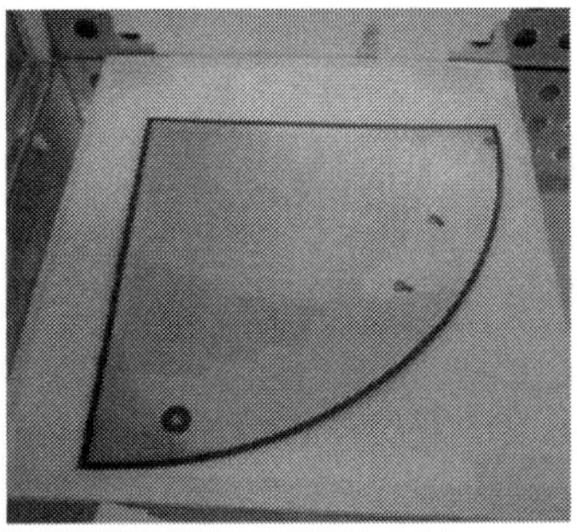

Figure 13. A photo of the new microbulk Micromegas recently constructed. It is one of the four sectors that will be installed in order to cover the sensitive surface of NEXT-1-MM.

The first tracks were seen when the T2K electronics were implemented in the chain. At the level of the readout plane, the signal of the pixels is grouped in 4 parts of 288 pads each. Flat cables are used to carry each group outside the vessel with the help of home-made feedthroughs. Outside the chamber, the 288 pads are divided into four groups of 72 with the help of an interface

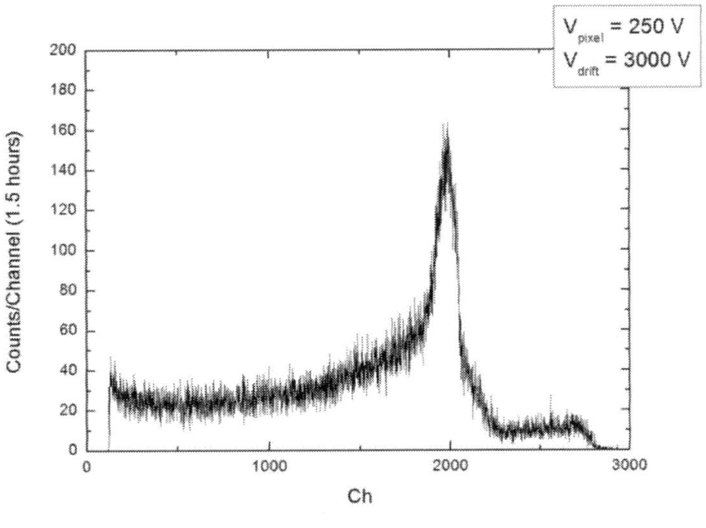

Figure 14. Spectrum of a gas diffused Rn source in Ar-2% Isobutane at 1 bar obtained with the bulk detector inside NEXT-1-MM. The main peak comes from α emission of the ^{222}Rn chain at 5.5 MeV. Two more alphas are expected from later on the chain at 6 MeV (^{218}Po) and 7.7 MeV (^{214}Po), which are collected in the tail of the spectrum.

card. This division fits the expected input of the T2K electronics, which is then easily connected to the chain. An example of an alpha track from the ^{222}Rn source can be seen in Figure 15.

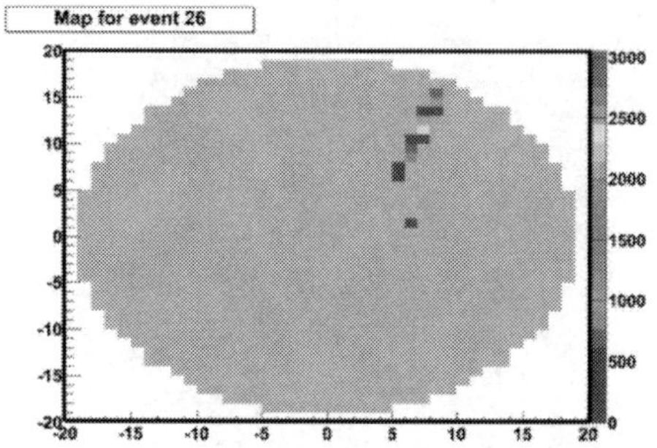

Figure 15. An alpha track of approximately 4 cm obtained with the ^{222}Rn source in NEXT-1-MM. The shaded area corresponds to the active surface of the bulk detector.

In late December the first microbulk detectors to be tested in NEXT-1-MM were received. The detectors have the form of a quarter of a circle (Figure 13), therefore four of them will be used to cover the sensitive area of the detector.

Next steps More measurements are ongoing with the NEXT-1-MM setup. The behaviour of the bulk detector at higher pressures is one of the first things to study. Currently, the system is prepared for measurements with pure Xe. The system is designed in a way that allows the incorporation of a plane where a small number of Photomultiplier tubes would be put. This will allow the t_0 measurement of the event. In the meantime, the newly arrived microbulk quarters will be characterized in a test bench prior to their installation in NEXT-1. Then the capabilities of large microbulk planes with respect to tracking and regarding energy resolution in Xe will be studied.

2.3. Radiopurity of Micromegas planes

As mentioned at the introduction, in experiments like NEXT, looking for the lowest background possible, a limiting factor could be the readout plane introduced. The first measurement of the radiopurity of Micromegas planes were performed in the low background facilities of the Canfranc Underground Laboratory (LSC) with a high purity germanium detector [9]. The obtained results prove that Micromegas readouts of the microbulk type are currently manufactured with radiopurity levels below 30 μBq/cm^2 for Th and U chains and $\sim$60μBq/cm^2 for ^{40}K, already comparable to the cleanest detector components of the most stringent low background experiments at present. Considering that the readouts measured were manufactured without any special care from the radiopurity point of view, after dedicated development these results will most probably be improved.

3. Conclusions

We have here presented the status of the Micromegas readout planes as an option for the equipment of a high pressure gas TPC for double beta decay searches such as NEXT. Current results show excellent energy resolutions achieved with the microbulk Micromegas like 1.94% (FWHM) at 4 bars of pure Xenon for the 5.5 MeV alphas of ^{241}Am, while measurements with the 60 keV photons of the same source have yield an energy resolution of 9.3% (FWHM) at 3.5 bar which would correspond to 1.45% (FWHM) if extrapolated to the $Q_{\beta\beta}$ value of Xe (2458 keV). At the same time, they have the advantage of rather low radioactivity levels, an important requirement in this field of searches. An intensive R&D program is underway for a complete study of the behaviour of these planes in even higher pressures. A big prototype able to hold approximately 1 kg of Xe, NEXT-1-MM, is already commissioned with a bulk Micromegas. Four new Micromegas of the microbulk type are being characterized in order to install them in NEXT-1-MM. The micromegas readouts offer a simple and compact way to equip an experiment like NEXT, since both the energy measurement and the tracking could be done in the same plane.

Acknowledgments

We want to thank our colleagues of the groups of the University of Zaragoza, CEA/Saclay and the RD-51 collaboration for helpful discussions and encouragement and especially D. Fortuño, J. Castel and A. Ortiz from Zaragoza and J. P. Mols from Saclay for their technical contributions to this work. We thank R. de Oliveira and his team at CERN for the manufacturing of the microbulk readouts. We acknowledge support from the Spanish Ministry of Science and Innovation (MICINN) under contract ref. FPA2008-03456, as well as under the CUP project ref. CSD2008-00037 and the CPAN project ref. CSD2007-00042 from the Consolider-Ingenio 2010 program of the MICINN. Part of these grants are funded by the European Regional Development Fund (ERDF/FEDER). We also acknowledge support from the European Commission under the European Research Council T-REX Starting Grant ref. ERC-2009-StG-240054 of the IDEAS program of the 7th EU Framework Program. Finally, we acknowledge support from the Regional Government of Aragón under contract PI001/08.

References

[1] Granena F et al. (The NEXT Collaboration), 2009 *Preprint hep-ex/0907.4054*
[2] Liubarsky I for the NEXT Collaboration,"NEXT: The Electroluminescence Readout", proceedings of this conference.
[3] Lux T et al.,"A Multi-APD readout for EL detectors", proceedings of this conference.
[4] Giomataris Y, Rebougeard P, Robert JP and Charpak G, 1996 *Nucl. Instr. and Meth.* A **376** 29
[5] Giomataris Y, 1998 *Nucl. Instr. and Meth.* A **419** 239
[6] Andriamonje S et al., 2010 *J. Instrum.* JINST 5 P02001
[7] Dafni T, Ferrer-Ribas E, Giomataris I, Gorodetzky P, Iguaz F, Irastorza IG, Salin P and Tomás A, 2009 *Nucl. Instr. and Meth.* A **608** 259–66
[8] Tomás A et al., 2009 2009 *J. Instrum.* JINST 4 P11016
[9] Cebrián S et al., 2010 *J. Cosmol. Astropart. Phys.* JCAP10(2010)010
[10] Balan C, Freitas EDC, Papaevangelou T, Giomataris I, Natal da Luz H, Monteiro CMB and dos Santos JMF, 2011 *J. Instrum.* JINST 6 P02006
[11] Cebrián S et al., 2011 *Astropart. Phys.* **34** 354–9 (Preprint arXiv:1005.2022v1)

A background free double beta decay experiment

I. Giomataris [1]
IRFU, Centre d'études de Saclay, 91191 Gif sur Yvette CEDEX, France

E-mail: Ioannis.giomataris@cea.fr

Abstract. We present a new detection scheme for rejecting backgrounds in neutrino-less double beta decay experiments. It relies on the detection of Cherenkov light emitted by electrons in the MeV region. The momentum threshold is tuned to reach a good discrimination between background and good events. We consider many detector concepts and a range of target materials. The most promising is the high-pressure ^{136}Xe emitter where the required energy threshold is easily adjusted. Combination of this concept and a high pressure Time Projection Chamber could provide an optimal solution. A simple and low cost effective solution is the use of the Spherical Proportional Counter that provides, using a single read-out channel, two delayed signals from ionization and Cherenkov light. In solid-state double beta decay emitters, because of its higher density, the considered process is out of energy range. An escape will be the fabrication of double decay emitters having lower density by using for instance the aerogel technique. It is surprising that a technology used for particle identification in high-energy physics becomes a powerful tool for rejecting backgrounds in such low-energy experiments.

1. Introduction

The discovery of neutrino-less double beta decay ββ(0ν) is one of the top priorities in contemporary neutrino physics. Observation would require Lepton Number Violation, would show that neutrinos are Majorana particles [1] and would provide precious information on the origin of neutrino masses. This process requires the neutrino is its own anti-particle and provides the most sensitive laboratory technique for measuring the neutrino mass.

Double beta decay occurs when a nucleus is energetically forbidden to decay through single beta decay. Earlier experiments [1] are providing sensitivities down to 0.2–0.8 eV scale for the effective neutrino mass using detector masses at the few tens of kg scale while present projects aim to achieve sensitivities down to 50–100 meV by increasing detector mass at the 100 kg scale. To explore the inverse hierarchy models of neutrino masses [1] will require much higher mass at the few tons scale to reach sensitivities down to 10–20 meV.

For a detector of a few tons a background level down to 10^{-5} c/keV/kg/year could be needed. In order to separate the tail of the ββ(2ν) distribution from the ββ(0ν) peak, which constitute an irreducible background for the latter, the energy resolution of an experiment would be kept as low as possible of the order of 1%.

The experimental challenge will include:

- A clear peak at the ββ(0ν) energy endpoint with zero background in the region surrounding the

[1] To whom any correspondence should be addressed.

peak i.e. no other lines in the neighborhood

- A proof of the two electron nature of the ββ(0ν) event,
- A demonstration that two electron kinematics match those of ββ(0ν)

Current experiments using solid-state emitters, such as germanium or adequate bolometer detectors [2,3], dispose a particular strength to provide a record energy resolution. An example is the Heidelberg-Moscow experiment [2] which searched for the ββ(0ν) decay using high-purity Ge semiconductor detectors enriched to 87% in ^{76}Ge. This experiment set the most stringent limit on the lifetime of the ββ(0ν) process.

To accommodate the other three items the use of a lighter target such as gas is necessary to identify the vertex of the interaction and track the two decaying electrons.

The NEMO detector [4], still in operation at the LSM in Modane, is combining solid target and tracking in gas. Its principle of operation is based on the use of thin foils of the double beta emitter surrounded by a tracking chamber and a calorimeter. The strength of the design is a good track reconstruction which provides a topological signature useful to discriminate signal and background. The weakness is the poor energy resolution that is ultimately limited from the fact that NEMO is not a fully active detector: fluctuations are dominated by the energy loss in the target.

A competitive double beta decay candidate is the ^{136}Xe gas whose natural abundance is rather high (9 %). Xenon gas can be easily enriched by centrifugation methods to high concentrations of ^{136}Xe: for example, an enrichment of 80% is being used by the EXO-200 experiment [5].

A well-designed Time Projection Chamber (TPC) filled with ^{136}Xe is a good candidate. The Gothard TPC [6] in the 90's is a pioneering experiment, with some drawbacks due the old read-out technology, like the modest energy resolution achievable. It however demonstrated the potential of a gaseous TPC to powerfully utilize the rich topological signature of ββ(0ν) events in the gas; recognize the signature as two electrons, coming from a common vertex, to further reduce background. The most dangerous background is a gamma interacting by photoelectric or Compton effect which produces an electron of an energy in the vicinity of the end point (2480 keV).

Recent progress with the Micromegas detector [7-9], with a 2% FWHM achieved in the MeV energy range for Xenon pressures up to 5 bar [10], together with their recently measured radio purity [11], prove that such readout is competitive for ββ(0ν) searches. This type of readout is currently used in the CAST experiment [12] providing a record low-background level.

A second read-out scheme followed by the NEXT collaboration is the electroluminescence (EL) process that presents a potentiality to improve the energy resolution [13]. Indeed both theoretical arguments [14] and a variety of experimental measurements suggest that this process could reach an energy resolution as small as predicted by the Fano factor. However the experimental result was reached in a small laboratory set-up at lower energy and therefore must be verified at the MeV energy scale using a large prototype. A drawback is that the EL process is achieved in pure Xenon which induces a degradation of the tracking accuracy; pure Xenon exhibits large diffusion coefficient and ionization electrons drifting in the long gaseous TPC are highly diffusing before reaching the read-out plane.

In this paper we propose a new read-out concept, which can be used by future double beta decay experiments, with some modifications in their experimental set-up. The purpose is to fully reject gamma backgrounds that could generate a fake ββ(0ν) peak.

2. The new concept

Xenon Time Projection Chamber (TPC) is a fully active detector, in which high pressure noble gas acts simultaneously as target and detector. We will consider a pressure range between 10 and 20 bars as proposed by the various projects. At such pressures the xenon refractive index at the far UV (λ=150 nm) is n=1.03 and n=1.06. These values could be extracted from figure 1 and the fact that n-1 is proportional to the pressure.

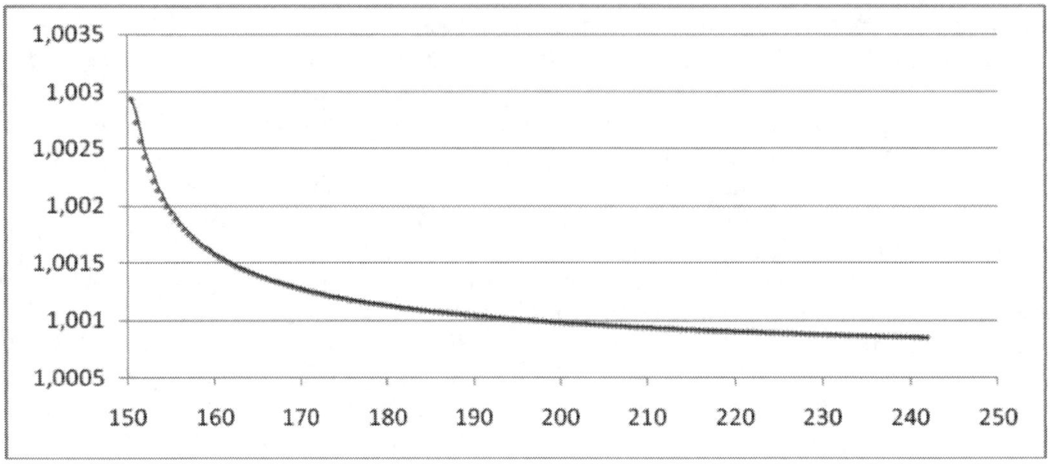

Figure 1. The Xenon refractive index as function of the light wavelength at atmospheric pressure

A ββ(0ν) event will produce two electrons having each a kinetic energy of about T=1240 keV. At a Xenon pressure of 10 bar each electron will travel a path of about 15 cm with a trajectory completely dominated by multiple scattering. The challenge is then to precisely measure a peak at the endpoint i.e. twice the energy of each electron (2460 keV). Backgrounds in the vicinity of this energy and especially gamma rays must be identified and rejected.

We found that this energy range is astonishing in the neighbourhood of the momentum threshold required to enable Cherenkov radiation in high pressure Xenon. The main idea of our proposal is to exploit this effect in order to discriminate background electrons having energy of about 2460 keV from two electrons at half energy produced in ββ(0ν) process.

Cherenkov radiation is electromagnetic radiation emitted when a charged particle (such as an electron) passes through an insulator at a constant speed greater than the speed of light in that medium.

The momentum threshold in a gaseous radiator is given by

$$p = \frac{m}{\sqrt{2(n-1)}} \tag{1}$$

where m is the particle mass and n the refractive index of the radiator.

Using relation (1) and taking the refractive index from Figure 1 at the far UV (λ=150nm) and at pressures of 10 and 20 bar of xenon the momentum threshold for an electron is at a kinetic energy of T=2.1 MeV and T=1.5 MeV respectively. By adjusting carefully the pressure in this range we would obtain the desired result: electrons from ββ(0ν) are below momentum threshold for Cherenkov emission while background electrons will emit photons by Cherenkov radiation. Detecting these photons we will be able to recognize between the two classes of electron events.

The intensity and spectrum of the radiation is given by the Frank-Tamm relation

$$\frac{dN}{dE} = \left(\frac{\alpha}{hc}\right) L(1 - 1/(n\beta)^2) \tag{2}$$

Where dN is the number of photons with energy between E and E+dE , α is the fine structure constant and L the particle path in the medium. By integrating relation (2) and taking into account the quantum efficiency QE of the photodetector we can calculate the number of detected photoelectrons N_d.

$$N_d = \int \frac{dN}{dE}(QE)dE \tag{3}$$

In the approximation of a constant n the integral over the energy bandwidth is simplified

$$N_d = N_0 L(1 - 1/(n\beta)^2) \quad (4)$$

Where N_d is the number of detected photoelectrons and N_0 is the Cherenkov quality factor of the detector (mean number of photoelectrons per cm):

$$No = \left(\frac{\alpha}{hc}\right)\int(QE)dE \quad (5)$$

From (4) we could have a first estimation of the expected signal produced by a relativistic electron (β=1) trajectory in a 10 bar Xenon gas, assuming a quality factor N_0 of 60 (a value easily reached by typical Cherenkov systems) and L=30 cm. Under these conditions the number of detected photons is 40, a large signal that should allow reaching full detection efficiency. A more precise estimation will be made in a next section but we would like to point out that such large yield will allow to efficiently reject background 'single' electron events having energies near the endpoint of $\beta\beta(0\nu)$.

In the next sections we will give some examples where the new concept could be added in existing or future experiments.

3. High pressure Xenon TPCs

EXO collaboration is studying two options in view of the implementation of a ton scale detector; an extension of the liquid TPC with improved energy resolution and barium tagging and a high-pressure gas TPC with similar features.

In the case of a liquid Xenon target the refractive index is two high (n = 1.6) and therefore our concept could not be used. The momentum threshold for Cherenkov emission is too low and therefore the pair of electrons generated from $\beta\beta(0\nu)$ process and background electrons, both, they will emit Cherenkov light; there is not discrimination between the two processes. The second option, a high pressure gaseous Xenon is based on a Time Projection Chamber (TPC) with separated-function capabilities for calorimetry and tracking.

NEXT collaboration is also considering a similar high pressure TPC. Charges produced by ionization process are drifting then collected in the end-cap readout system as it is used in conventional TPCs. The charge readout detector is serving for measuring the total electron energy and precisely reconstructing tracks.

In both experiments the two end caps of the TPC are covered by UV photodetectors.

In order to improve the energy resolution the NEXT project is studying the possibility of reading out light crossing the cylindrical vessel.

To incorporate our concept in these experiments we need a full 4π light collection coverage. Such system is under study by the NEXT collaboration in order to fully collect UV scintillation light produced by the electroluminescence process in pure Xenon. Under these conditions it will be difficult to separate the Cherenkov light from the very-intense scintillation light. In the case of using photodetectors sensitive only to visible and near UV light the Cherenkov light could be possibly separated from the scintillation. A second solution is the use of an adequate quencher, such as CH_4, mixed with the Xenon gas. Such mixture will improve the gain of the detector and will block scintillation light emission; the light collection system will then devoted to the detection of Cherenkov light.

4. The case of the spherical detector

The Spherical Proportional Counter (SPC) is a new type of radiation detector based on a spherical geometry [15]. It consists of a large spherical gas volume with a central electrode forming a radial electric field. Charges deposited in the conversion volume drift to the central sensor where they are amplified and collected. The sensor is a small spherical ball located at the center acting as a proportional amplification structure. It allows high gas gains to be reached and operates in a wide range of gas pressures.

The main advantage of this structure is the use of a single electronic channel to read-out a large volume. This single information still allows the determination of the radial coordinate of the conversion point through the measurement of the time dispersion of the detected charge pulse. Such information is serving to localize the depth of deposits, which is a key point for applying fiducial cuts.

A good energy resolution of the order of 2.5 % (FWHM) is achieved in the MeV energy range as it is shown in figure 2.

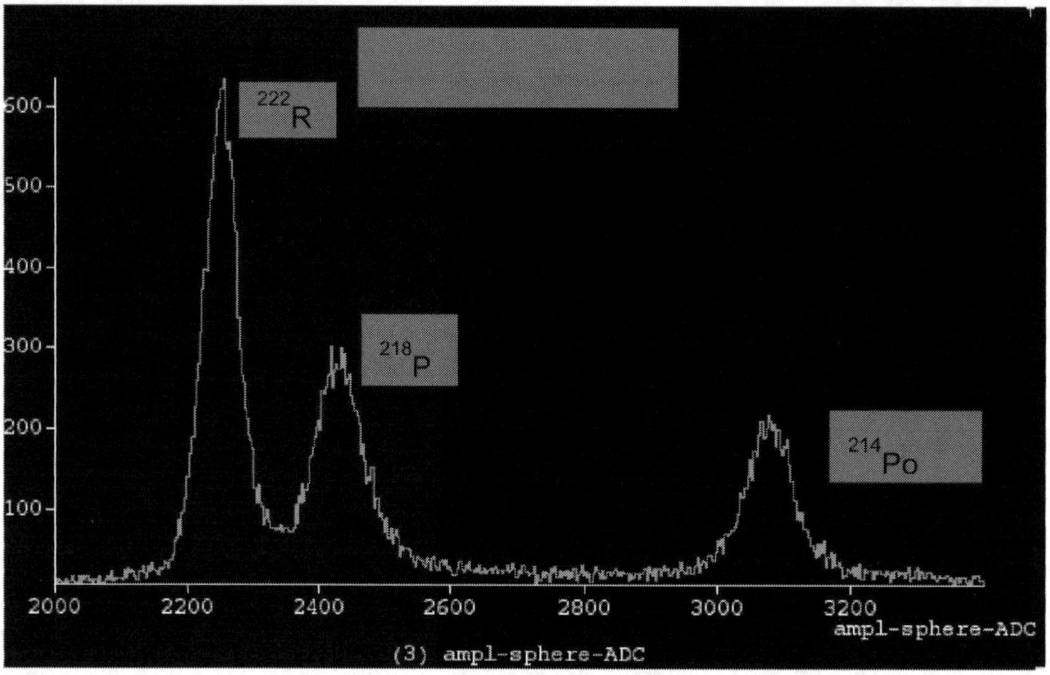

Figure 2. Peaks observed from a ^{222}Rn radioactive source. From left to right we observe the ^{222}Rn peak at 5.6 MeV, the ^{218}Po and ^{214}Po at 6.1 MeV and 7.8 MeV respectively in Argon - 2% CH4 gas mixture.

In view of this excellent performance, to be confirmed with high-pressure Xenon gas, it is natural to consider this detector for a future ββ(0ν) investigation. A second motivation is the capability of transforming this device to a full light collection photodetector.

The idea is to deposit on the internal surface a thin layer of CsI photocathode. This could be easily achieved by the vacuum deposition technology, a standard process which uses an electric resistance heater to melt the material and raise its vapour pressure to a useful range. This is done in a high vacuum, both to allow the vapour to reach the substrate without reacting with or scattering against other gas-phase atoms in the chamber, and reduce the incorporation of impurities from the residual gas in the vacuum chamber. Our spherical metallic vessel is well suited for this process with an excellent vacuum level we are routinely reach ($<10^{-7}$ mbar). The heater carrying the CsI sample will be installed at the center of the vessel and the photocathode will be evaporated and uniformly deposited in the internal surface of the vessel.

Such deposit provides a quantum efficiency (QE) [16-17] in the far UV bandwidth (limited by the Xenon transmission at 150nm) as it is shown in figure 3.

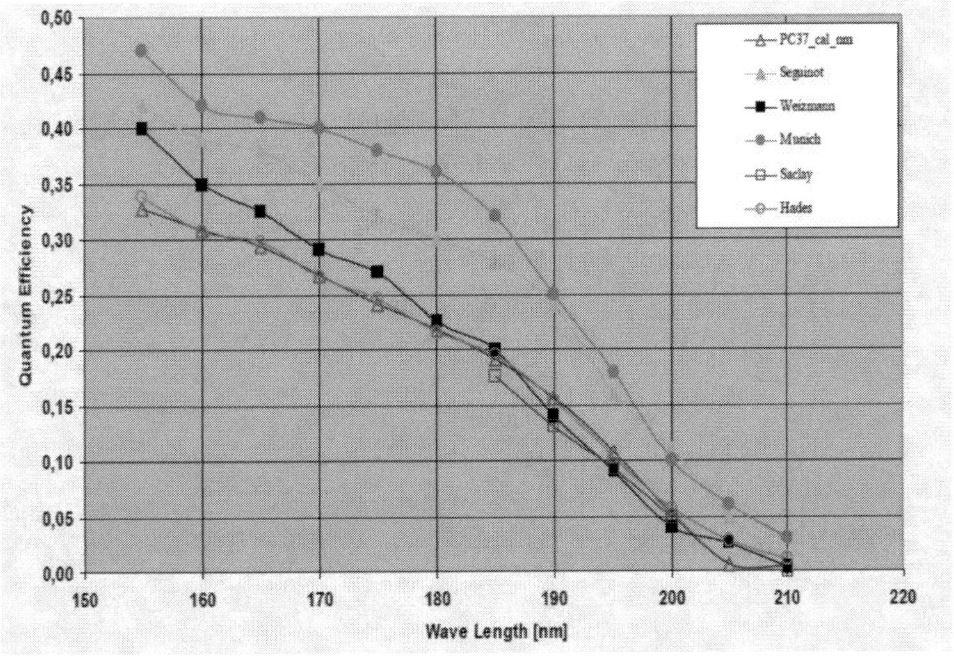

Figure 3. Quantum efficiency of CsI versus photon wavelength.

A schematic view of the set-up is presented in figure 4. The volume is filled with Xenon at high pressure.

Two electrons from the ββ(0ν) with an energy of 1.23 MeV each, because they are below Cherenkov threshold, they will deposit their energy by only the ionization process. Charges are then will drift and will be collected by the central ball. The charge signal (S1) will serve for the energy measurement.

An electron of about 2.46 MeV (background at the ββ(0ν) endpoint) will deposit its energy inside the gas volume in a typical range of 30 cm. Energy loss is mainly due to ionization but at the same time UV light will be emitted by the Cherenkov process that will extract photoelectrons from the internal vessel at the CsI layer. Therefore we expect two delayed signals to be developed in the preamplifier:

- the first is a large signal (S2) due to the ionization and it is expected after a time equal to the electron drift from the primary vertex to the central sensor.
- the second due the UV Cherenkov light will induce a weaker signal (S3) at fixed delay equal to the time required for electrons to drift a radius of the sphere. The drift time has a typical value of 1 ms and depends on the size of the sphere, the gas mixture and the gas pressure.

In order to precisely calculate the yield from the Cherenkov radiation we use the Xenon refractive index from figure 1 and the quantum efficiency of CsI from figure 3. The efficiency for single electron detection is assumed to be 80% as it has been measured in [15].

Under these considerations we proceed to the calculation of photoelectron yield by integrating equation (3) over the available energy band width and using an electron length equal to the half of its range. Results are presented in figure 5: the number of detected photoelectrons as a function of the electron kinetic energy for various pressures of Xenon gas. For pressures ranging from 15 bar to 28 bar the photoelectron yield the energy varies between 22 and 38 at 2.46 MeV electron energy.

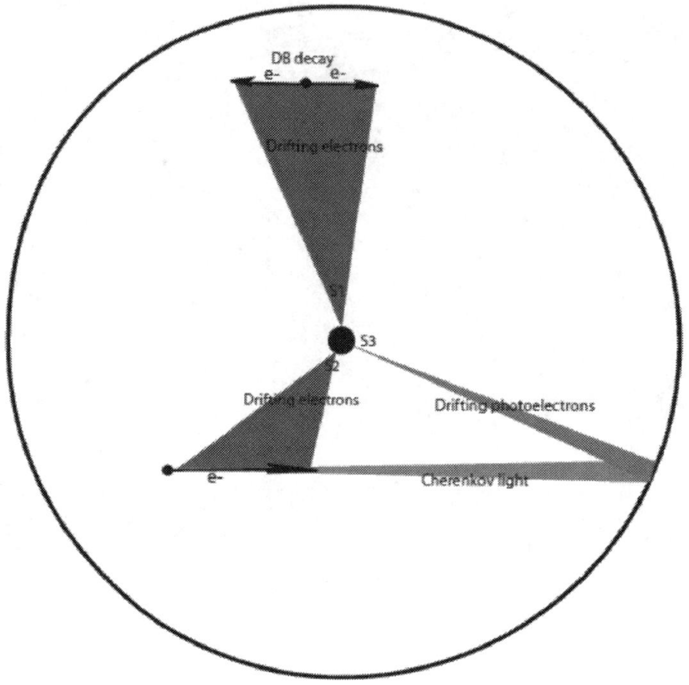

Figure 4. Two electrons from ββ(0ν) with an energy of 1.23 MeV are ionizing the Xenon gas. Secondary electrons (in red) are drifting to the central ball where they are amplified giving rise a signal (S1). A background electron of 2.46 MeV with larger range in the gas will generate a signal (S2) by similar process and since it is above Cherenkov threshold a second signal (S3) will be generated with a delay by the drifting photoelectrons created in the internal vessel by the Cherenkov radiation (in green) interacting with CsI photocathode.

So the delayed signal produced only by single electrons near the endpoint will be identified and rejected with a high efficiency, whereas the pair of electrons from ββ(0ν) are under the Cherenkov threshold.

To get a more precise photoelectron yield we need to measure the photoelectron extraction efficiency from the CsI under realistic conditions of ^{136}Xenon gas at high pressure. A better estimation requires a Monte Carlo program to simulate multiple scattering and energy loss of the beta particle in the gas. Such program will be investigated in the near future.

An obvious advantage of the proposed set-up is the simultaneous reading out of charge and UV light using a single electronics channel.

A large pressurized detector volume would easily provide the required mass for the ββ(0ν) experiment. For instance the present prototype, having a diameter of 130 cm, filled with Xenon at 20 bar will contain a target mass of 137 kg. Building a larger 3m in diameter detector will provide a total mass of 1691 kg.

In summary combining the new concept with the spherical detector provide a simple and cost effective solution and the scaling up to several tons is not a big challenge. In this case we could employ a second identical system filled with raw Xenon gas running in parallel under similar conditions. Subtracting the acquired spectra from the two systems will be a way to further reject backgrounds and improve the sensitivity to observe double beta decay.

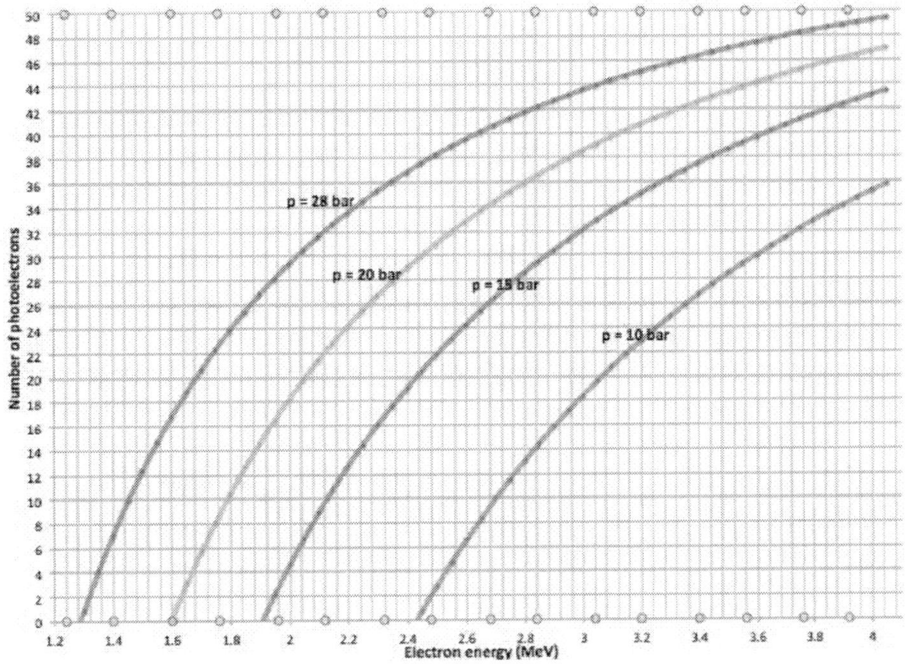

Figure 5. The number of photoelectrons as a function of the electron energy for various pressures of Xenon gas.

5. Solid state or liquid detectors

In the case of solid or liquid state detectors the main requirement to have a transparent medium with a refractive index of the order of 1.05 is not fulfilled. Such value stands much below the range provided by solid or liquid media. Cherenkov radiation will be emitted for all electrons in the MeV energy range and therefore the discrimination between beta decay electrons and background electrons is not possible. However this signal could serve to reject heavier particles such as alphas, protons, heavy recoils.

We could argue that, because of lack of adequate refractive index, solid targets are not good candidates unless we could manage to reduce their density using for instance aerogel technology. Aerogel is a manufactured material with the lowest bulk density of any known porous solid. It is derived from a gel in which the liquid component of the gel has been replaced with a gas. The result is an extremely low-density solid with several remarkable properties, transparent in visible light and sometimes in the near UV light where the transmission is determined by the light scattering length.

Studies on production of aerogel with refractive index in the range of interest, for use as the Cherenkov light radiator in RICH detectors, have being carried out and such products have been successfully developed by the industry. Silica aerogel having a colloidal form of SiO_2 was usually employed as bulk material. The deal will be to get them with other elements that are double beta emitters. More difficult is to preserve the good energy resolution, which could not be provided by the weak Cherenkov process. In the case of a bolometer, for instance, the accurate energy evaluation should be performed through the heat measurement. In the case of a semiconductor, after the aerogel fabrication, it will be certainly difficult to preserve the good semiconductor properties.

In NEMO like experiments because the target material is not active it could be replaced by an aerogel without affecting the detection scheme.

A second possibility is to keep the solid or liquid material as it is and proceed to a slightly different read-out approach as it is illustrated in figure 6.

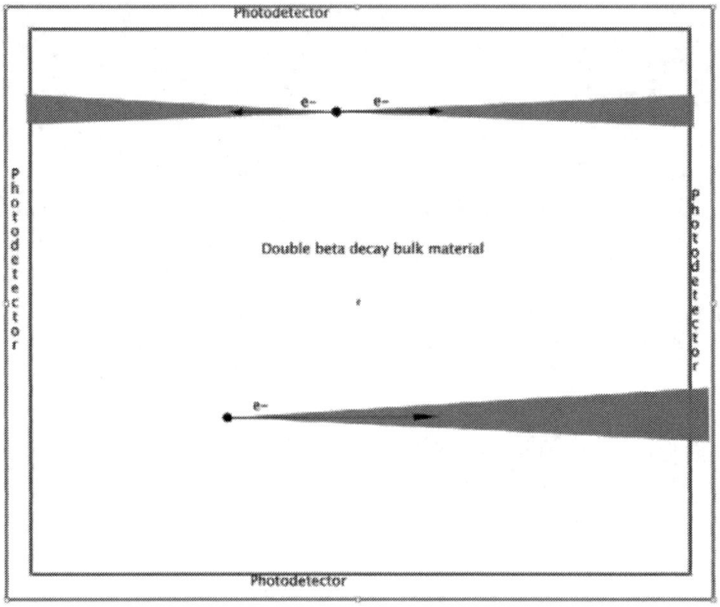

Figure 6. The internal box is the double beta emitter surrounded by photodetectors. The process at the top of the box illustrates the Cherenkov light emission by two back-to-back electrons from ββ(0ν) while at the bottom a single background electron generates a single light spot on the photodetector.

The double beta decay emitter is presented as a box surrounded by a photodetector system. In principle the device could be any of existing solid of liquid systems. The only requirement is to be transparent in order to transmit Cherenkov light. Now the two electrons from the ββ(0ν) process will produce Cherenkov light as shown in the figure (top of the box) because of the high density, high refractive index of the medium. Therefore we expect two back-to-back detected spots in the photodetector. This is another way of discriminating two electrons signal from a single background electron producing a single spot at the photodetector. Notice the the two well separated spots is just an illustration of the process. Because of multiple scattering of the electron in the dense medium the presented spot will be highly diluted. This configuration needs a careful study with an appropriate simulation program.

6. Conclusions and outlook

A novel experimental approach, for eliminating gamma backgrounds in future ββ(0ν) experiments, is proposed. The new approach rely on the detection of Cherenkov light emitted by electrons with energy close to the end point of the ββ(0ν) process to sort between various processes.

This is an additional signature to discriminate between signal and backgrounds, which is independent and will be added to other identification cuts such as event topology or daughter ion tagging.

A powerful rejection of gamma backgrounds is expected.

This scheme is well adapted to high-pressure ^{136}Xe target which is UV transparent Cherenkov radiator and at the same time the optimal momentum threshold could be easily tuned by choosing the right gas pressure.

Because of the high refractive index the concept is useless in the case of liquid or solid detectors but it could serve to reject heavy particles. An issue is to fabricate low-density double beta decay emitters using for instance aerogel technology.

References

[1] Avignone III FT, Elliott SR and Engel J, Double Beta Decay, Majorana Neutrinos and Neutrino Mass, 2008 *Rev. Mod. Phys.* **80** 481

[2] Klapdor-Kleingrothaus HV et al., *Eur. Phys. J. A* 12 (2001)147.

[3] Arnaboldi C et al., *2005 Phys. Rev. Lett.* **95** 142501.

[4] Arnold R et al.,2005 *Nucl. Instr. and Meth. A* **536**:79-122,2005.

[5] Akimov D. et al., 2005 Nucl. Phys. Proc. Suppl. 138 (2005), 224.

[6] Luscher R et al., 1998 *Phys. Lett. B* **434** (1998) 407

[7] Giomataris Y, Rebourgeard R, Robert JP and Charpak G, 1996 *Nucl. Instr. and Meth. A* **376** 29

[8] Giomataris Y, 1998 *Nucl. Instr. and Meth. A* **419** 239

[9] Giomataris I et al., 2006 *Nucl. Instr. and Meth. A* **560** 405

[10] Cebrian S et al., 2010 *J. Cosmol. Astropart. Phys.* JCAP10(2010)010

[11] Cebrian S et al., 2011 *Astropart. Phys.* **34** 354-9 (Preprint arXiv:1005.2022v1)

[12] Galan J et al.,2010 *J. Instrum.* JINST 5 P01009

[13] Granena F et al. (The NEXT collaboration) 2009 *Preprint hep-ex/0907.4054*

[14] Nygren D, 2009 *Nucl. Instr. and Meth. A* **603** 337

[15] Giomataris I et al., 2008 *J. Instrum.* JINST 3 P090007

[16] Sequinot J et al., 1990 *Nucl. Instr. and Meth. A* **297** 133-147

[17] Bream A et al., 1994 *Nucl. Instr. and Meth. A* **343** 163-172

[18] Bougamont E et al., 2010 *arXiv*:1010.4388v1[hep-ph]

TPC2010 IOP Publishing
Journal of Physics: Conference Series **309** (2011) 012011 doi:10.1088/1742-6596/309/1/012011

XENON100 – Results and Prospects

Marc Schumann
for the XENON100 collaboration
Physik Institut, University of Zurich, Switzerland
E-mail: marc.schumann@physik.uzh.ch

Abstract. The XENON100 Dark Matter experiment, installed in the Laboratory Nazionali del Gran Sasso (LNGS, Italy), is searching for WIMP type Dark Matter particles scattering off a 62 kg liquid xenon target in a dual phase (liquid/gas) time projection chamber. Careful material selection, a novel detector design, and an upgrade of the passive shield, together with capitalizing the self-shielding power of liquid xenon, are crucial in order to achieve a background of less than 0.01 events kg^{-1} keV^{-1} day^{-1} in the fiducial volume. This background, which has been verified experimentally, is lower than in any other dark matter experiment.
The analysis of 11.2 live days of background data taken during a commissioning run in fall 2009 leads to the first science result of XENON100: No events are observed in a pre-defined fiducial volume of 40 kg mass, excluding spin-independent WIMP-nucleon scattering cross sections above 3.4×10^{-44} cm^2 (at 100 GeV/c^2). Below 80 GeV/c^2, this is the most sensitive exclusion limit so far, disfavouring the interpretation of DAMA and CoGeNT being due to spin-independent, elastic interactions of light mass WIMPs.

1. Introduction

Various indirect astronomical observations at all cosmological length scales suggest that the vast majority of the energy content of the Universe is dark, i.e. invisible over the whole electromagnetic spectrum. Precise measurements of the anisotropies in the cosmic microwave background show that only about 4% of the Universe is made from baryonic matter, 21% is Dark Matter, and 74% is Dark Energy [1] . While the nature of the Dark Energy remains speculative, there are strong indications that Dark Matter is made up from heavy, non-relativistic (cold) particles that build large scale cosmological structures. Since no Standard Model particle has the right properties to be the Dark Matter particle, it must be from *new physics* and many experiments aim to find it. A well motivated candidate is the WIMP (Weakly Interacting Massive Particle), a stable particle arising naturally in many theories beyond the Standard Model, such as supersymmetry or theories with extra dimensions [2].

There are three general approaches to search for Dark Matter: The first is to create the particles in high energy collisions. The second is to search for indirect astronomical evidences such as gamma-lines from Dark Matter annihilation. In the following we will focus on the third method, the direct detection of WIMPs scattering off target nuclei in the laboratory.

Because they are neutral, WIMPs are expected to interact with the target nucleus (nuclear recoil interactions) whereas the main background comes from electromagnetic interactions of gammas and electrons with the atomic electrons (electronic recoil interactions). From the assumed WIMP mass and the velocity distribution one expects to measure a steeply falling, featureless nuclear recoil spectrum at energies of a few keV only. The predicted rates are tiny, much less than one interaction per kg and day of exposure. Therefore, the experiments have to decrease their background as much as possible in order to be sensitive to WIMPs.

2. Mini-Review on Direct Dark Matter Detection

In this section we want to give a very short overview on the current status of the field of direct Dark Matter searches. This can be done by separating the parameter space for spin-independent WIMP-nucleon scattering into two regions of high and low WIMP masses.

At higher WIMP masses (> 20 GeV/c^2), the field is dominated by CDMS-II and XENON100. The CDMS experiment, located in the Soudan mine (USA), measures the ionization and the phonon signal (heat) deposited by particle interactions in ultra-pure germanium crystals at very low temperatures ($<$50 mK). The ionization/heat ratio is used to discriminate very effectively between signal and background. Imperfect charge collection at the detector surface might mimic valid signal events, however, this problem is solved with an additional cut on the shape of the charge signal, which allows to run the experiment free of background. In the latest analysis of 612 kg$\times$days blind background data, CDMS observed 2 events in the pre-defined WIMP search region [3]. However, with an expected background of 0.9 ± 0.2 events, the likelihood to observe two or more events is 23% and the statistical significance not large enough to consider these events as WIMP signals. The combined limit from all CDMS-II data taken at Soudan excludes spin-independent cross sections of 3.8×10^{-44} cm^2 for 70 GeV/c^2 WIMPs at 90% CL [3].

The EDELWEISS-II experiment also uses germanium bolometers, however, their design of so-called InterDigit electrodes allows fiducialization within the crystals and rejection of surface events. The collaboration has recently published new preliminary results [4] using 322 kg$\times$days effective exposure, leading to a sensitivity of 5×10^{-44} cm^2 for 80 GeV/c^2 WIMPs (90% CL), which is appoaching the CDMS and XENON sensitivities.

XENON100 gives the best limits between $\sim$15 and 80 GeV/c^2. It is described in greater detail in the following section of this paper.

At very low WIMP masses (in the order of 10 GeV/c^2), the experimental situation is more complicated, see Fig. 1: Since many years, the DAMA/LIBRA Collaboration claims experimental evidence for an annual modulation signal due to Dark Matter particles, caused by the Earth's motion around the Sun. This modulation, observed from 2–6 keV in ultra-pure NaI(Tl) crystals, has reached a statistical significance of almost 9σ [5], however, without any possibility to discriminate signal from background on an event-by-event basis.

CoGeNT is a prototype germanium detector of 330 g mass operated underground at Soudan. It uses p-type point contacts which allow to reach a very low threshold of 0.4 keV (gamma energy), before the detector is limited by electronic noise. The spectrum taken with an exposure

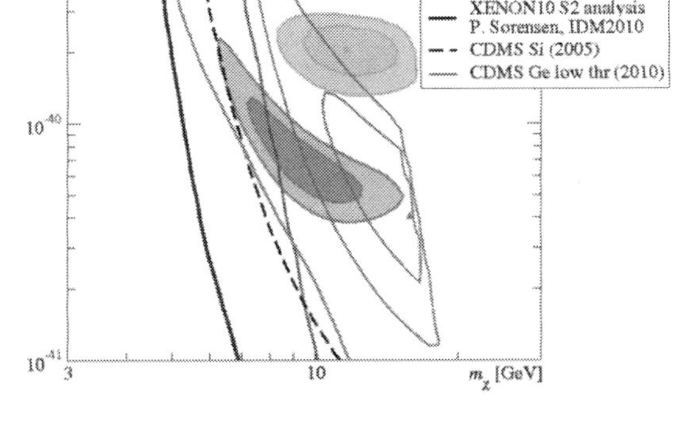

Figure 1. Current experimental situation at low WIMP masses for spin-independent WIMP-nucleon cross sections. The DAMA, CoGeNT, and CRESST-II regions are given at 90% and 3σ CL. The exclusion limits from XENON100, XENON10, and CDMS are 90% CL. Figure reproduced with kind permission from [6].

of 15.5 kg×days rises exponentially at very low energies, a shape which could be explained by a light mass WIMP of ~10 GeV/c^2. The null-hypothesis, however, has the same χ^2 [7].

The CRESST-II experiment is located at LNGS und uses $CaWO_4$ crystals as target material. The collaboration has recently presented preliminary results [8], observing an excess of nuclear recoils in the oxygen-band which cannot yet be explained by the expected backgrounds. However, the situation and its interpretation remains quite speculative until there is a publication of these results.

Fig. 1 shows the DAMA and CoGeNT results when interpreted as being due to elastic spin-independent WIMP nucleon scattering. They are severely constrained by the non-observation of signals by other experiments: XENON100 (details in Sect. 3) basically excludes DAMA and severly constrains CoGeNT, and a recent re-analysis of CDMS-II data acquired in 2008 with a lowered threshold of 2 keV recoil energy excludes even more parameter space [9].

A preliminary re-analysis of the XENON10 data using the ionization only has been presented recently [10]. While this requires giving up precise information about the position of an interaction as well as signal/background discrimination, it allows to considerably lower the detector threshold for nuclear recoils to about 1 keV. While this analysis sacrifices lots of target mass and does not allow to place competitive results at higher WIMP masses, it provides very strict constraints at low WIMP masses and the DAMA and CoGeNT regions.

3. XENON100

In the remaining sections of this paper, we will focus on the XENON100 experiment and give an outlook into the future.

The XENON collaboration uses liquid xenon (LXe) as target material. Xenon is a heavy target ($A \sim 131$) which enhances the sensitivity for spin-independent WIMP-nucleon scattering. Being an efficient scintillator ($\lambda = 178$ nm), xenon also serves as detector material. Its high density ($\rho \sim 3$ g/cm^3) allows to build compact detectors with excellent self-shielding capabilities. Furthermore, xenon has no long-lived radioactive isotopes, and small admixtures of the radioactive ^{85}Kr can be removed to the ppt level [11].

Liquid xenon detectors provide 3-dimensional interaction vertex reconstruction and signal to background discrimination when operated as a dual-phase (liquid/gas) time projection chamber (TPC), see Fig. 2: An interaction in the LXe generates prompt scintillation light (the S1 signal) and ionizes the target. The ionization electrons are drifted towards the liquid gas interface by a strong electric field. Here, the electrons are extracted into the gas phase and accelerated towards the anode while generating proportional scintillation light (S2). Both signals, S1 and S2, are detected by two arrays of photosensors, one immersed in the liquid for optimal light collection, and one located in the gas phase above the target. The position of the interaction can be reconstructed using the S2 signal distribution on the top array (x, y) and the time difference between prompt S1 and delayed S2 signal (z). Due to their different track densities in the medium, the ratio S2/S1 can be used to discriminate between signal (nuclear recoil interactions) and background (electronic recoils).

XENON100 is the current detector at the 100 kg scale within the phased program of the XENON collaboration. It follows the XENON10 phase which has proven that liquid xenon detectors are able to deliver very competitive results [12]. XENON100 has a total mass of 161 kg of LXe, out of which 62 kg are in the cylindrical target volume which is viewed by two arrays of Hamamatsu R8520 photomultipliers from above (98 PMTs) and below (80 PMTs). The remaining 99 kg of LXe are surrounding the target in 4π. This volume is instrumented with another 64 PMTs and acts as an active veto.

The detector, made from low-radioactivity stainless steel, is enclosed by a passive shield made from 5 cm of ultra-pure OFHC copper, followed by 20 cm of polyethylene, 5 cm of low radioactivity lead, and another 15 cm of standard lead. The whole shield sits on a 20 cm thick

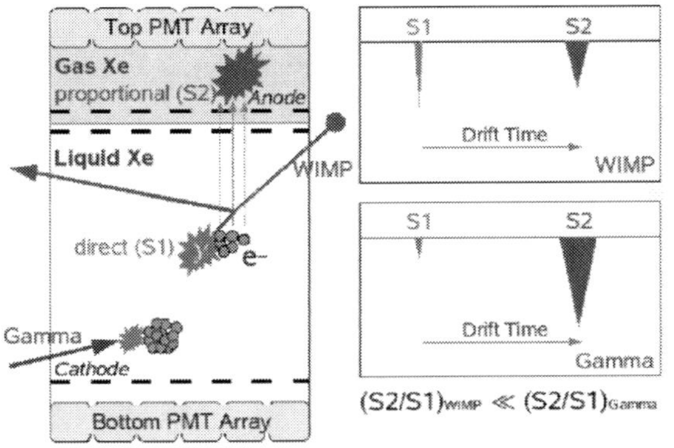

Figure 2. Working principle of a double-phase liquid xenon time projection chamber (TPC): An interaction generates prompt scintillation light (S1) and ionizes the target. The ionization electrons are drifted upwards and extracted into the gas phase where they generate proportional scintillation (S2). The S2 light pattern on the top photosensor array, together with the drift time of the electron cloud, is used to reconstruct the event vertex in (x, y, z). The ratio S2/S1 is different for electronic recoil interactions (background) and nuclear recoil interactions (signal) and is used for discrimination.

slab of polyethylene and is additionally shielded against neutrons with ~20 cm of water on four sides. All parts with a potentially higher intrinsic radioactivity, such as the pulse tube refrigerator (PTR) to cool the xenon down to $-92°$C and the feedthroughs, are placed outside the shield.

3.1. Data Analysis and Result

XENON100 is installed underground at Laboratori Nazionali del Gran Sasso (LNGS, Italy) since spring 2008. After extensive calibrations and studies to characterize the detector response, first science data has been taken in fall 2009. This article summarizes the results of this run, which have been published in [13].

Detector Calibration The PMT response is calibrated by stimulating single photoelectron (PE) emission using blue light. The PMT gains were measured regularly and were found to be stable in time within $\pm 2\%$ (σ/μ). The standard gamma line used to obtain the nuclear recoil energy scale from quenching factor measurements (see below) is from ^{57}Co at 122 keV. Since 122 keV gamma rays cannot penetrate deep enough into the detector, the average light yield at this energy was derived to be 2.20(9) PE/keV from calibrations with several gamma lines of higher and lower energy: 40 keV and 80 keV from inelastic neutron scattering processes, 164 keV and 236 keV from neutron activated xenon, and with the higher-energy lines from ^{137}Cs and ^{60}Co.

Because of the geometry of the TPC, the light (S1) and the charge signal (S2) have to be corrected according to their position in the detector. The correction functions have been measured with different sources at various energies, with a mutual agreement of better than 3%. The largest correction for the S2 signal is due to the finite electron lifetime in the LXe: while drifting towards the liquid-gas interface, a certain fraction of ionization electrons is lost to electronegative impurities in the LXe. The charge loss is described by an exponential function with the electron lifetime τ as the main parameter. The data presented here were taken with a maximum lifetime of $\tau \sim 200$ μs.

The drift time of the electrons is used to determine the z-position of the interaction. It is measured with a resolution better than 2 mm, limited by the possibility to separate two independent S2 peaks. For the xy-position, three independent position reconstruction algorithms (neural network, support vector machine, minimal χ^2) yield consistent results for $r < 140$ mm. The precision is better than 3 mm as verified experimentally using a collimated gamma source.

In order to use the $\log_{10}$(S2/S1) parameter to discriminate between electronic and nuclear recoils, the low energy response of the detector to these interactions has to be calibrated.

This has been done using the Compton continuum of ^{60}Co and elastic neutron interactions from an ^{241}AmBe source, respectively. The two populations are well separated and lead to a discrimination of $> 99\%$ at 50% nuclear recoil acceptance.

Background All materials considered for the XENON100 detector and the shield were screened in high purity germanium spectrometers in order to determine their intrinsic radioactivity. Only materials with a reasonable radioactivity were accepted for the detector construction. The detector cavity is constantly purged with boil-off nitrogen in order to avoid radon penetrating the shield. All measured radioactivity values are used as input parameters for a detailed Monte Carlo model of the detector, and the result of the simulation agrees remarkably well with the measured background spectrum over the full energy range [14].

In a fiducial volume of 30 kg mass, XENON100's background rate in the low energy range is 5.3×10^{-3} events keVee^{-1}kg^{-1}day^{-1} (electron recoil equivalent energy, indicating that the energy scale is based on a ^{57}Co calibration) when the active LXe veto is employed. This is a factor 100 lower than XENON10, thus achieving one of the design goals of XENON100. In fact, XENON100 is currently the experiment with the lowest background level of all running dark matter detectors, more than a factor 100 below any other experiment at low energies.

Below 200 keVee, in the energy range of interest, the XENON100 background is dominated by ^{60}Co mainly from the PMTs and ^{85}Kr. The latter has been removed from the LXe by cryogenic distillation [11], and the remaining concentration in this run has been measured to be around 150 ppt via a delayed coincidence technique. This result is consistent with the value from the comparison of the Monte Carlo spectrum and the data [14].

The nuclear Recoil Energy Scale Based on the cut acceptance and the expected recoil spectrum from WIMPs, the energy region of interest for the first Dark Matter analysis has been chosen from 4–20 PE (S1 signal). This region has to be converted to a nuclear recoil equivalent energy scale (given in keVnr) which takes into account the signal losses for nuclear recoil interactions due to quenching.

The relative scintillation efficiency $\mathcal{L}_{\text{eff}}$ of a nuclear recoil interaction of energy E_{nr} is the ratio of the light yield at this energy, $\text{LY}(E_{\text{nr}})$, and the light yield of the 122 keV gamma line from ^{57}Co,

$$\mathcal{L}_{\text{eff}}(E_{\text{nr}}) = \frac{\text{LY}(E_{\text{nr}})}{\text{LY}(E_{\text{ee}} = 122\text{ keV})}.$$

In the presence of an electric field, the right hand side of the equation is modified by a factor S_e/S_n to take into account field quenching. The formula allows the conversion of a light signal (measured relative to 122 keV electron recoil equivalent energy E_{ee}) to E_{nr}. The typical measurement to determine $\mathcal{L}_{\text{eff}}$ is illustrated in Fig. 3: The true elastic energy deposition of a neutron from a monoenergetic neutron beam in a LXe chamber is measured via its scattering angle. At the same time, the scintillation light yield is measured in the LXe which gives access to $\mathcal{L}_{\text{eff}}$. For lowest energies, however, the scattering angle θ is close to zero, the energy deposition in the LXe is close to the detection threshold, and the measurement becomes very difficult. Therefore, the results of various groups [15, 16] differ more than expected from the stated error bars, see Fig. 4. For a recent discussion of the systematics of these measurements, see [17].

In order to derive a nuclear recoil energy scale from this data, we have chosen to use all $\mathcal{L}_{\text{eff}}$ data from scattering measurements and to employ a statistical approach: The thick line in Fig. 4 is the best fit description of the data which was extrapolated to energies below 5 keVnr using a constant function. This choice was motivated by the flat trends seen in some of the data points. The thinner lines give the upper and lower 90% CL contours, the lower contour was very conservatively extrapolated to zero near 1 keVnr using a logarithmic function. In order to

Journal of Physics: Conference Series **309** (2011) 012011 doi:10.1088/1742-6596/309/1/012011

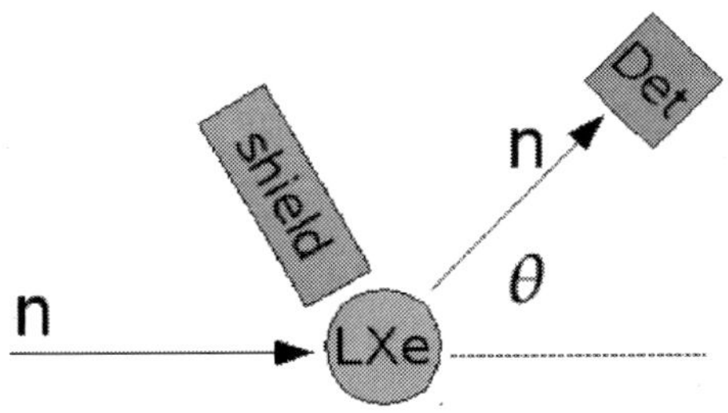

Figure 3. Principle of the measurements of the relative scintillation efficiency $\mathcal{L}_{\text{eff}}$: The elastic energy deposition of a monoenergetic neutron beam is measured in a LXe chamber (as S1 light) and via the scattering angle. The observed amount of light relative to the gamma standard ^{57}Co is the relative scintillation efficiency.

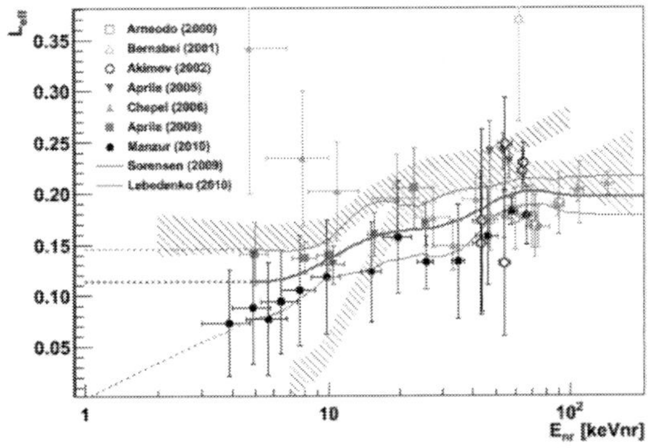

Figure 4. Compilation of all existing $\mathcal{L}_{\text{eff}}$ data [15]. The shaded regions are indirect determinations from comparisons of data with simulations [16]. The thick line is the best fit $\mathcal{L}_{\text{eff}}$ with a constant extrapolation to lower energies. The thinner lower line is the lower 90% CL contour with a logarithmic extrapolation to zero.

reflect the uncertainty currently present in $\mathcal{L}_{\text{eff}}$, we will show the results for both functions, the best fit and the lower 90% CL contour.

In order to clarify this situation, a new measurement of $\mathcal{L}_{\text{eff}}$ with improved systematics is currently ongoing at Columbia University. Another independent measurement is in preparation at Zurich University.

Result The data leading to the first XENON100 results were taken in stable conditions in October and November 2009. Altogether 11.2 life days of data have been used for the analysis. The data has not been blinded, however, the analysis has been performed in a quasi-blind way and all cuts and selections were developed on calibration data only.

The WIMP search region between 4 and 20 PE corresponds to 8.7–32.6 keVnr (best fit $\mathcal{L}_{\text{eff}}$). The upper bound in $\log_{10}(S2/S1)$ space was the median of the nuclear recoil band from the neutron calibration, the lower bound a software threshold of $S2 > 300$ PE. A simple cylindrical fiducial volume with 40 kg of xenon was chosen for this analysis, and < 0.2 background events were expected in this volume for the given exposure.

Only events with a single S2 peak (single-scatter events) were selected for the WIMP analysis, since the interaction probability of WIMPs is far too small to scatter in the detector twice. Fig. 5 shows the distribution of all single-scatter events in the energy region of interest; the events below the nuclear recoil median are marked in red. No nuclear recoil events are observed in the fiducial volume, indicated in yellow. As expected, the vast majority of the events is located close to the borders of the TPC. Incomplete charge (S2) collection can happen in areas very close to the PTFE walls and the field shaping electrodes of the TPC and is the typical reason for electron recoil events being shifted below the nuclear recoil median. The distribution of the remaining 22 events in the fiducial volume seems to be quite uniform, in agreement with the expectation of a background with a sizeable admixture of ^{85}Kr.

The positions of the events remaining in the fiducial volume in $\log_{10}(S2/S1)$ space are shown in Fig. 6. All events are well separated from the WIMP search region. This figure is the most remarkable result of this analysis: It demonstrates that LXe detectors can be indeed used for

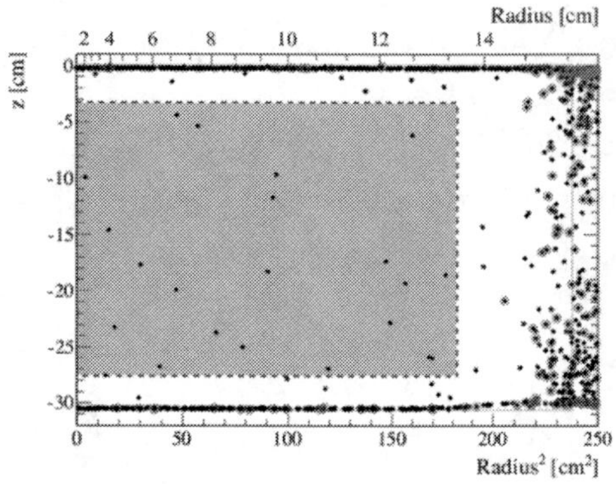

Figure 5. Distribution of the background events in the 11.2 days dataset in the energy region of interest. Most events are close to the edges of the TPC. Only 22 events end up in the fiducial volume of 40 kg mass which is indicated in yellow.

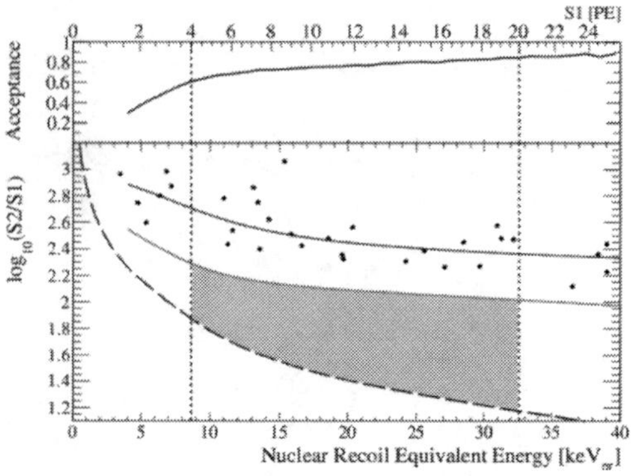

Figure 6. In the $\log_{10}(S2/S1)$ space, used for discrimination between signal and background, all remaining events have $\log_{10}(S2/S1)$ values well above the nuclear recoil median (red). No event falls in the predefined WIMP search region (yellow).

background-free WIMP searches. The upper part of Fig. 6 gives the energy dependent acceptance function not taking into account the 50% acceptance from $\log_{10}(S2/S1)$-based electronic recoil discrimination.

Poisson fluctuations determine the resolution at the S1 threshold and have to be taken into account to derive a limit from this data in order to treat the resolution properly. Since the expected recoil spectrum from spin-independent WIMP-nucleon scattering is a featureless falling exponential function, the resolution helps to increase the sensitivity at low WIMP masses, as more events from below the threshold fluctuate into signals above the threshold than vice versa. Fig. 7 shows the limits from this analysis, calculated for the two choices of $\mathcal{L}_{\rm eff}$ as discussed above, assuming an isothermal WIMP halo. The acceptance-corrected exposure, weighted by a spectrum of a 100 GeV/c^2 WIMP, is 172 kg$\times$days. For WIMP masses below 80 GeV/c^2, this result places the lowest limit so far. At light WIMP masses, DAMA [5, 18] and CoGeNT [7] are constrained even assuming the conservative $\mathcal{L}_{\rm eff}$.

3.2. Outlook

The previous sections summarized the first Dark Matter results of XENON100, derived from 11.2 life days of data taken during a commissioning run in fall 2009 [13]. In the meantime, about 10$\times$ more science data has been acquired and a blind analysis is ongoing. The ultimate sensitivity of XENON100 for spin-independent WIMP-nucleon scattering, based on the background predictions from Monte Carlo simulations, is shown in Fig. 8: For 200 life days exposure and a fiducial volume of 30 kg mass, we expect to reach a sensitivity of $\sigma = 2\times10^{-45}$ cm^2 (at 100 GeV/c^2).

In order to explore the WIMP parameter space down to lower cross sections, or to confirm a possible WIMP detection, we are already in the design phase for XENON1T. This future detector will be a LXe double-phase TPC with a fiducial mass of 1000 kg. Its expected sensitivity to spin-independent WIMP-nucleon cross sections is a few 10^{-47} cm^2 (see Fig. 8). A total amount of about 2.5 tons of xenon will be required for this detector. A large fiducial volume cut, together with careful material selection and LXe purification, will further decrease the background by a factor 100. Detector, LXe handling and purification systems, and shield are currently being designed. Two possible candidates for the light readout are tested: The QUPID [19], a novel

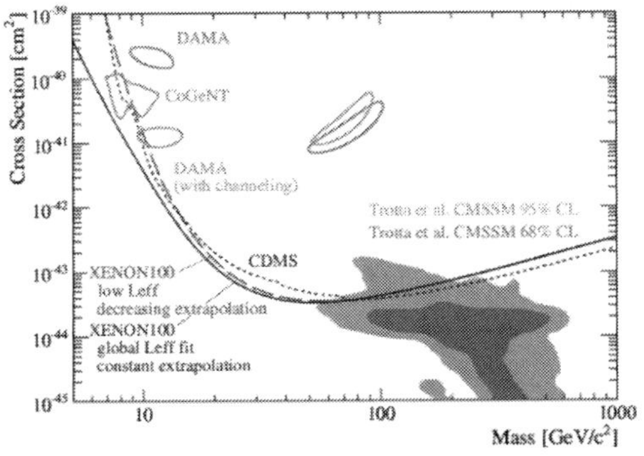

Figure 7. The limit on the spin independent WIMP-nucleon cross section derived from the first XENON100 data. The exclusion plot shows the 90% CL exclusion contours for the two $\mathcal{L}_{\rm eff}$ cases shown in Fig. 4.

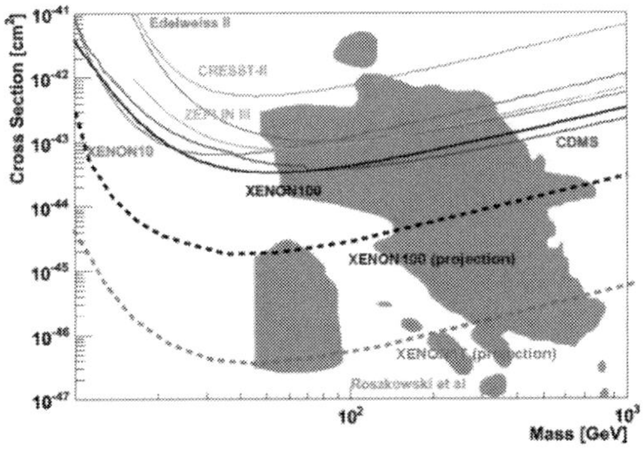

Figure 8. Sensitivity reach of XENON100 and XENON1T for spin independent WIMP-nucleon couplings. XENON1T will be able to explore most of the WIMP parameter space predicted by CMSSM models.

hybrid detector with very low intrinsic background, and the Hamamatsu R11410, a 3" PMT optimized for operation in LXe.

The new DARWIN project [20] aims for the time beyond XENON1T and for cross sections below 10^{-47} cm^2. This design study of a multi-ton Dark Matter detection facility with possibly two targets (xenon/argon) brings together the European groups of the XENON, WARP, and ArDM in one consortium in order to share knowledge and expertise. It is part of the ASPERA roadmap for Astroparticle Physics in Europe.

References

[1] Nakamura K et al. (Particle Data Group) 2010 *J. Phys. G* **37** 075021 and references therein.
[2] Bertone G (ed.) 2010 *Particle Dark Matter* (Cambride: Cambridge University Press)
[3] The CDMS II Collaboration 2010 *Science* **327** 1619
[4] Armengaud R (EDELWEISS-II) 2010 *Preprint* arXiv:1011.2319
[5] Bernabei R et al. (DAMA) 2008 *Eur. Phys. J. C* **56** 333
[6] Schwetz T 2010 *Preprint* arXiv:1011.5432
[7] Aalseth C E et al. (CoGeNT) 2010 *Phys. Rev. Lett.* **106** 131301
[8] Seidel W 2010 *talk presented at IDM2010*
[9] Ahmed Z et al. (CMDS II) 2011 *Phys. Rev. Lett.* **106** 131302
[10] Sorensen P 2010 *talk presented at IDM2010*; Sorensen P et al. (XENON10) 2010 *Preprint* arXiv:1011.6493
[11] Abe K et al. (XMASS) 2009 *Astropart. Phys.* **31** 290
[12] Angle J et al. (XENON10) 2008 *Phys. Rev. Lett.* **100** 021303;
Angle J et al. (XENON10) 2008 *Phys. Rev. Lett.* **101** 091301
[13] Aprile E et al. (XENON100) 2010 *Phys. Rev. Lett.* **105** 131302
[14] Aprile E et al. (XENON100) 2011 *Phys. Rev. D* **83**
[15] Arneodo F et al. 2000 *Nucl. Inst. and Meth. A* **449** 147; Bernabei R et al. 2001 *EPJ direct* **3** 11;
Akimov D et al. 2002 *Phys. Lett. B* **524** 245; Aprile E et al. 2005 *Phys. Rev. D.* **72** 072006;
Chepel V et al. 2006 *Astropart. Phys.* **26** 58; Aprile E et al. 2009 *Phys. Rev. C* **79** 045807;
Manzur A et al. 2010 *Phys. Rev. C* **81** 025808
[16] Sorensen P et al. (XENON10) 2009 *Nucl. Inst. and Meth. A* **601** 339;
Lebedenko V N et al. (ZEPLIN-III) 2009 *Phys. Rev. D* **80** 052010
[17] Manalaysay A 2010 *Preprint* arXiv:1007.3746
[18] Savage C et al. 2009 *J. Cosmol. Astropart. Phys.* **4** 10
[19] Arisaka K et al. 2009 *Astropart. Phys.* **31** 63
[20] Baudis L (DARWIN) 2010 *PoS(IDM2010)***122** *Preprint* arXiv:1012.4764

Journal of Physics: Conference Series **309** (2011) 012012 doi:10.1088/1742-6596/309/1/012012

GLACIER and related R&D

Alessandro Curioni
ETH Zürich
E-mail: alessandro.curioni@cern.ch

Abstract.
Liquid argon detectors, with mass up to 100 kton, are being actively studied in the context of proton decay searches, neutrino astrophysics and for the next generation of long baseline neutrino oscillation experiments to study the neutrino mass hierarchy and CP violation in the leptonic sector. The proposed Giant Liquid Argon Charge Imaging ExpeRiment (GLACIER) offers a well defined conceptual design for such a detector. In this paper we present the GLACIER design and some of the R&D activities pursued within the GLACIER.

1. Introduction

The development of very large liquid argon time projection chambers (LAr TPCs), with a fiducial mass up to 100 kton, is being actively pursued in the framework of tackling the following topics:

a. neutrino oscillations, in particular to study CP violation in the leptonic sector with a sensitivity far superior to running/approved experiments as T2K [1] and NOνA [2];
b. proton decay searches, where LAr TPCs can provide unprecedented sensitivity especially in the decay channels with a kaon in the final state, with good complementarity to searches with water Cherenkov detectors;
c. astrophysical neutrinos, i.e. neutrinos from stellar collapse, solar neutrinos etc.

This is essentially a continuation of the the physics program covered by a large water Cherenkov as Superkamiokande [3, 4, 5, 6, 7, 8, 9, 10], which has operated successfully for more than a decade, therefore setting strict requirements on the sensitivity of any future experiment. Liquid argon is a persistent candidate because the LAr TPC technology offers significant improvements on key parameters with respect to water Cherenkov detectors, namely: energy resolution, excellent control over background, good signal efficiency over the relevant energy range, and a detailed pictorial reconstruction on an event-by-event basis, much as in a bubble chamber. Another key parameter of LAr TPC is the possibility to scale up to the needed size, of the order of 100 kton in the fiducial volume; for a review, see [11].

So far, the ICARUS [12, 13, 14, 15] experience has produced the largest operating LAr TPC, the ICARUS T600, which comprises two separate modules for a total mass of about 600 ton and is now taking data in the LNGS. In order to deliver the needed sensitivity, a next generation LAr TPC has to be scaled up by a factor larger than 100 from the current state of the art. Various proposals exist to bridge the gap in detector mass; here we present the GLACIER concept and related R&D activities.

Journal of Physics: Conference Series **309** (2011) 012012 doi:10.1088/1742-6596/309/1/012012

2. GLACIER

The basic concept of GLACIER was originally put forward in [16], where many details can be found. The key points of the proposed design are:

- single module non-evacuable cryogenic tank based on industrial liquefied natural gas (LNG) technology, of cylindrical shape with excellent surface/volume ratio;
- simple, scalable detector design, possibly up to 100 kton;
- single very long vertical drift (up to 20 m) with full active mass;
- for the readout structure, a very large area, up to 3500 m^2, instrumented with Large Electron Multipliers (LEM) operating in double phase argon (liquid-vapor);
- possibly immersed visible light readout for Cherenkov imaging;
- possibly immersed (high T_c) superconducting solenoid to obtain a magnetized detector;
- reasonable excavation requirements (<250,000 m^3).

Such a design requires a suite of R&D activities to asses the feasibility of the concept. These activities are the topic of this work, and four of them are presented in more detail:

1. LEM R&D work at CERN;
2. the ArDM Dark Matter experiment;
3. experimental study of Ar purity in a non-evacuable 6m^3 vessel;
4. the T32@J-PARC experiment.

2.1. LAGUNA

The critically important item of a proper underground location has been extensively addressed in the framework of LAGUNA (Large Apparatus for Grand Unification and Neutrino Astrophysics) [17, 18, 19, 20]. The LAGUNA design study is dedicated to the feasibility of very large underground infrastructures able to host the next generation neutrino physics, astroparticle physics and proton decay experiments. It is conducted in the spirit of an European-wide coordination and is supported by the EC FP7 program [1]. The LAGUNA design study considers three different detector technologies (liquid scintillator, liquid argon and water Cherenkov), and seven potential underground sites, in order to identify the scientifically and technically most appropriate and cost-effective strategy for future large scale underground detectors. Seven technical reports have been produced and made available.

3. LEM R&D

In a LAr TPC, the charge produced by ionizing particles in liquid argon is drifted by an applied electric field towards the readout electrodes. In the case of a LAr LEM-TPC the readout electrodes are in the vapor phase, above the liquid surface. Therefore the electrons are extracted at the liquid-vapor interface by means of an appropriate electric field. In the vapor phase, Townsend avalanche takes place in the high electric field regions confined in the LEM holes, similar to the situation of the Gas Electron Multiplier (GEM) [21]. The LEM is a macroscopic hole electron multiplier built with standard PCB techniques. The amplified charge induces a detectable signal on a set of segmented electrodes, proportional in amplitude to the ionization charge, therefore giving both calorimetric and spatial information.

A 3L LAr LEM-TPC has been built and operated at CERN to test different LEM designs under realistic conditions. In this setup, the LAr drift volume has a cross-section of 10×10 cm^2 and the distance between cathode and LAr surface can be extended up to 30 cm. The LAr scintillation light is detected by a Hamamatsu R6237-01 photomultiplier tube, coated with

[1] http://www.laguna-science.eu/

the wavelength shifter tetraphenylbutadiene (TPB) and placed below the optically transparent cathode grid, electrically shielded by a grounded protection grid. The uniform drift field is defined by equally-spaced rectangular field shaping rings, one every 5 mm, connected to a chain of 940 MΩ resistors. On top of the drift volume there are two parallel extraction grids with a wire pitch of 5 mm and a gap of 10 mm between them, the lower one immersed in LAr and the upper one in Ar vapor, which allow to apply an electric field across the LAr surface, and to set it independently of the drift field. The level of the LAr is controlled with millimeter precision by a capacity measurement of the two grids. The LEM readout is mounted 10 mm above the upper grid. After testing several LEM configurations [22, 23, 24] , a very promising design has been successfully selected, with a single 1 mm thick LEM amplifying stage and a two dimensional projective readout anode with 3 mm pitch readout [25]. A stable gain larger than 25 has been achieved with this detector configuration corresponding to a signal-to-noise ratio exceeding 200 for minimum ionizing tracks. An example of an event recorded in this configuration is shown in Fig. 1. It should be stressed that with the 2D projective anode the two orthogonal views are completely equivalent, while in a LAr TPC with a wire chamber readout one typically has to distinguish between the collection view and the induction view, with different behavior in terms of signal shape and signal-to-noise ratio.

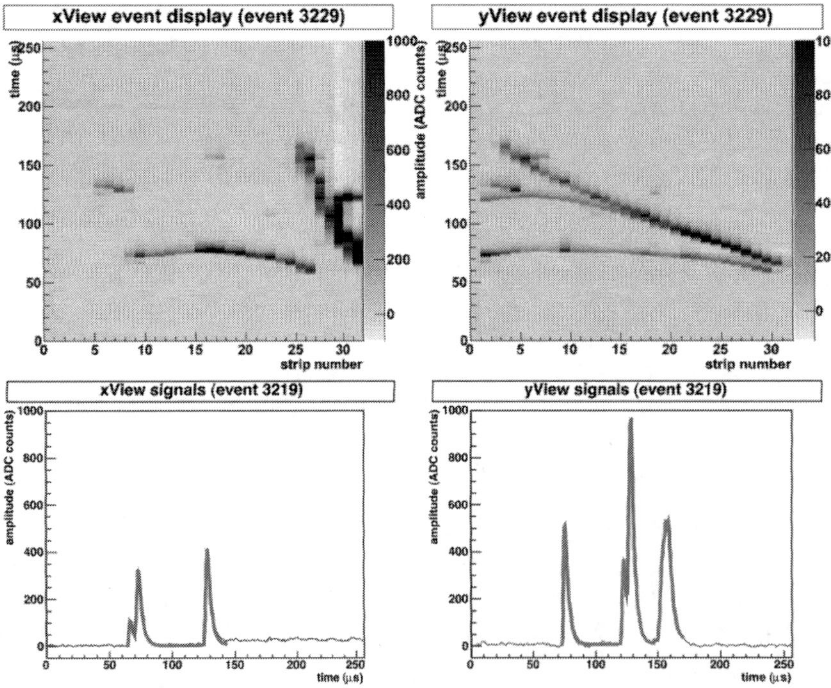

Figure 1. Multi-track event. Top: Event display: channel number vs. drift time, X-view left, and Y-view right. The gray scale is proportional to the signal amplitude. Bottom: Typical waveforms for the X-view left and the Y-view right. The thick curve shows the result of the signal fit using the response of the preamplifier. From [25].

For a GLACIER type detector, square-meter size LEM and projective anode are needed, in order to cover the entire readout plane with a large but still acceptable number of modules. A 40×80 cm^2 "readout sandwich", i.e. LEM plus matching 2D projective anode (see Fig. 2), has been produced, with the same geometry of the 10×10 cm^2 LEM tested in the 3L LAr LEM-TPC setup and described in [25], i.e. PCB thickness 1 mm, hole diameter 500 μm, hole pitch 800 μm, dielectric rim size $\sim 50\mu$m; it is presently being tested at CERN. It will be mounted on the 250L J-Parc@T32 LAr TPC (Sec. 6). Square-meter size readout planes are also being built for the ArDM Dark Matter detector (Sec. 4).

Figure 2. Left: Close up of the 40×80 cm^2 2D projective anode. Right: 40×80 cm^2 LEM, and close up of the holes. The well centered rims around the holes are visible.

4. ArDM

ArDM is an experiment designed for the direct search of WIMP Dark Matter [26]. It has been built and commissioned at CERN in the past few years and is now approved for installation in LSC (Laboratorio Subterráneo de Canfranc) in 2011. The ArDM detector is a 1-ton [2] two-phase LAr detector, with independent charge and light readout optimized for the detection of low energy events, in the 10-100 keV range. WIMP interactions on an argon nucleus are expected to produce a nuclear recoil, much similar to the ones from elastic scattering of MeV neutrons, with an exponentially falling recoil spectrum. With an energy threshold of 30 keV (for nuclear recoils) and WIMP-nucleus cross section of 10^{-44} cm^2, not yet excluded by current limits, about one event per day is expected in ArDM. Together with having a sizeable detector sensitive to low energy interactions, a key parameter for direct Dark Matter searches is also a high discrimination power against background in the relevant energy range. The LEM TPC readout, with mm position resolution on each spatial coordinate, helps background rejection through accurate fiducial volume cuts and separate detection of multiple interactions within the same event. In terms of rate, the background in a carefully designed 1-ton LAr detector will be dominated by the presence of ^{39}Ar, which is a beta emitter (Q-value 565 keV and a half-life of 269 years) present in natural argon with a concentration of 8×10^{-16}, resulting in an activity of about 1 Bq/kg [27]. LAr offers an extremely good discrimination between beta and nuclear recoils, based on a combined analysis of the charge/light ratio and pulse shape for the scintillation light. In ArDM, the present light readout of the prompt scintillation light is done using 14 cryogenic photomultipliers immersed in LAr. The detailed description of this system can be found in [28].

While ArDM is a stand-alone experiment pursuing a well defined physics program, some of the developed technical solutions are of immediate interest for GLACIER, in particular for

[2] In the field of direct Dark Matter searches, a 1-ton instrumented mass is a very large one.

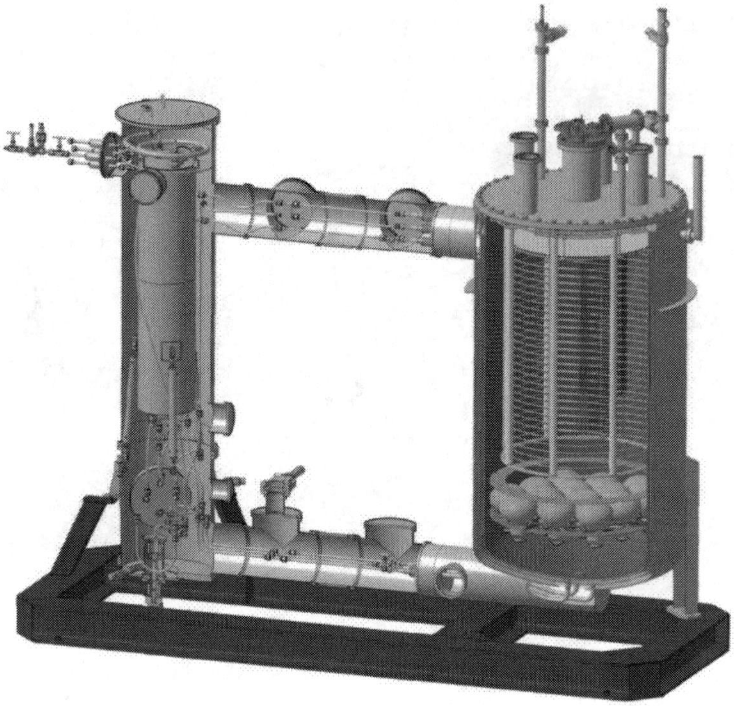

Figure 3. 3D of the fully assembled ArDM detector. On the left, the cooling system, on the right, the detector, with the photomultiplier tubes at the bottom of the active volume.

the cathode high voltage and the LAr LEM-TPC charge readout. The LEM charge readout is very similar to the one envisioned for GLACIER, and is studied in a common R&D program. The main difference, compared to the GLACIER case, is the large gain required in order to detect the few electrons which can be extracted following a nuclear recoil in LAr. In fact, the ionization from a nuclear recoil in liquid argon is highly quenched, and for the same energy loss a nuclear recoil produces about 4% of the free ionization charge when compared to a beta electron. Considering that, with optimized low noise electronics, the energy threshold for the detection of an ionization signal (beta electron) without charge multiplication (gain=1) is ~100 keV, it is clear that a gain of several hundreds is needed in order to detect a nuclear recoil of 30 keV. This large gain is in principle obtainable with a double stage LEM. The cathode high voltage of ArDM is provided using a 210 stages Greinacher (Cockroft-Walton) circuit, built to sustain up to 400 kV [29]. It has been stably operated in liquid argon at 70 kV (0.6 kV/cm drift field in ArDM) in fully operational configuration.

5. 6 m^3

Liquid argon purity in a non-evacuable vessel is one of the critical design parameters for a GLACIER type detector. In the ICARUS approach, the TPC is housed in a vacuum vessel, and obtaining a good vacuum allows to remove air from the vessel prior to filling with LAr, help outgassing from the TPC materials and the vessel walls, and verify the integrity (tightness) of the vessel itself. Commercially available LAr, which is an acceptable starting point for farther purification in the liquid phase, is handled, transported and stored without using high vacuum

Journal of Physics: Conference Series **309** (2011) 012012 doi:10.1088/1742-6596/309/1/012012

vessels, and has an O_2 content of about 1 ppm. Therefore it looks very reasonable to expect that at least the removal of the air can be easily achieved without vacuum, and air in a closed vessel can be efficiently displaced by a flow of Ar gas. This technique for reducing the air concentration down to the ppm level using a 6 m^3 vessel has been tested experimentally, and results are presented in detail in [30]. This work is the first step of a targeted experimental activity to verify the feasibility and to study the practical details of purging a large, fully instrumented, non-evacuable vessel from air down to ppm level air concentration through flushing with Ar gas, and to fill with ultra-pure liquid Ar (ppt level concentration of residual impurities).

6. T32@J-PARC

A 250L LAr TPC is being developed in a collaboration between KEK, Iwate University, Waseda University and ETH, for the T32@J-PARC experiment. The detector itself is an innovative step in the direction of a dual-phase, GLACIER-like LAr TPC. The goal of the experimental project is to perform measurements with well defined charged particle beams, in order to benchmark the performance of the detector and develop analysis and software tools for LAr TPC detectors, starting from large samples of positrons, pions, kaons and protons in LAr. A preliminary version of the detector has already been exposed to the K1.1BR beam in October 2010 and about 170,000 events have been collected. For the initial test, a cryogenic vessel from the MEG experiment[3] has been used, which has now been replicated and replaced with a new one. This preliminary version is equipped with a 1D readout plane, with 76 strips with 1 cm pitch, running orthogonal to the beam, working as single phase LAr TPC (see Fig. 4). As discussed, the full LEM readout is presently being commissioned at CERN.

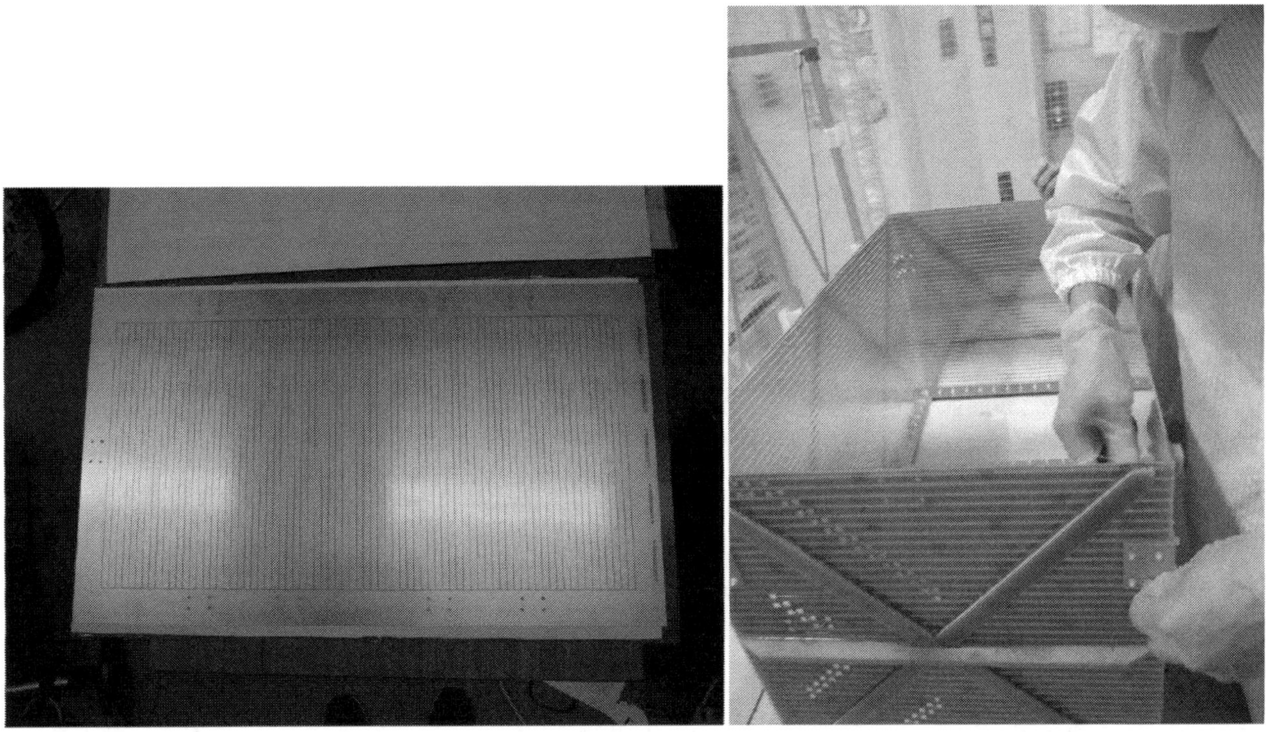

Figure 4. Left: Readout plane of the T32@J-PARC LAr TPC, with 76 1 cm wide strips. Right: Field cage.

[3] http://meg.web.psi.ch/

7. Conclusions and outlook

Several LAr detectors have been developed at CERN and KEK to address the R&D needs of GLACIER. In particular the 3L setup has been a workhorse for developing the LEM charge readout. We are now working on devices at the ton-scale, like ArDM and T32@J-PARC. In the near future, we plan to fully instrument a 6 m^3 device (10 ton scale) as a double phase imaging detector. On a longer timescale, we are considering a 1 kton detector with a well defined physics program, as the final step in a series of prototypes leading to the 100 kton scale.

References

[1] Y. Itow *et al.* [The T2K Collaboration], arXiv:hep-ex/0106019.
[2] D. S. Ayres *et al.* [NOvA Collaboration], arXiv:hep-ex/0503053.
[3] R. Wendell *et al.* [Kamiokande Collaboration], Phys. Rev. D **81** (2010) 092004 [arXiv:1002.3471 [hep-ex]].
[4] E. Thrane *et al.* [Super-Kamiokande Collaboration], Astrophys. J. **704** (2009) 503 [arXiv:0907.1594 [astro-ph.HE]].
[5] J. Hosaka *et al.* [Super-Kamiokande Collaboration], Phys. Rev. D **74** (2006) 032002 [arXiv:hep-ex/0604011].
[6] J. Hosaka *et al.* [Super-Kamkiokande Collaboration], Phys. Rev. D **73** (2006) 112001 [arXiv:hep-ex/0508053].
[7] K. Abe *et al.* [Super-Kamiokande Collaboration], arXiv:1010.0118 [hep-ex].
[8] K. Kobayashi *et al.* [Super-Kamiokande Collaboration], Phys. Rev. D **72** (2005) 052007 [arXiv:hep-ex/0502026].
[9] H. Nishino, J. Phys. Conf. Ser. **136** (2008) 042018.
[10] Y. Ashie *et al.* [Super-Kamiokande Collaboration], Phys. Rev. D **71** (2005) 112005 [arXiv:hep-ex/0501064].
[11] A. Curioni, Mod. Phys. Lett. A **24** (2009) 81.
[12] S. Amerio *et al.* [ICARUS Collaboration], Nucl. Instrum. Meth. A **527** (2004) 329.
[13] F. Arneodo *et al.* [ICARUS-Milano Collaboration], Phys. Rev. D **74** (2006) 112001 [arXiv:physics/0609205].
[14] P. Cennini *et al.*, Nucl. Instrum. Meth. A **345** (1994) 230.
[15] P. Benetti *et al.*, Nucl. Instrum. Meth. A **332** (1993) 395.
[16] A. Rubbia, "Experiments for CP-violation: A giant liquid argon scintillation, Cerenkov and charge imaging experiment?," arXiv:hep-ph/0402110.
[17] D. Autiero *et al.*, "Large underground, liquid based detectors for astro-particle physics in Europe: scientific case and prospects," JCAP **0711** (2007) 011 [arXiv:0705.0116 [hep-ph]].
[18] A. Rubbia [LAGUNA Collaboration], Acta Phys. Polon. B **41** (2010) 1727.
[19]
[19] J. Kisiel *et al.* [LAGUNA Collaboration], PoS E **PS-HEP2009** (2009) 283.
[20] D. Angus *et al.* [LAGUNA Collaboration], arXiv:1001.0077 [physics.ins-det].
[21] F. Sauli, Nucl. Instrum. Meth. A **386** (1997) 531.
[22] A. Badertscher *et al.*, arXiv:1010.2482 [physics.ins-det].
[23] A. Badertscher *et al.*, Nucl. Instrum. Meth. A **617** (2010) 188 [arXiv:0907.2944 [physics.ins-det]].
[24] A. Badertscher *et al.*, arXiv:0811.3384 [physics.ins-det].
[25] A. Badertscher *et al.*, arXiv:1012.0483 [physics.ins-det].
[26] A. Rubbia, J. Phys. Conf. Ser. **39** (2006) 129 [arXiv:hep-ph/0510320].
[27] P. Benetti *et al.* [WARP Collaboration], Nucl. Instrum. Meth. A **574** (2007) 83 [arXiv:astro-ph/0603131].
[28] C. Amsler *et al.* [The ArDM Collaboration], JINST **5** (2010) P11003 [arXiv:1009.3641 [physics.ins-det]].
[29] S. Horikawa *et al.*, arXiv:1009.4908 [physics.ins-det].
[30] A. Curioni *et al.*, arXiv:1009.4073 [physics.ins-det].

Journal of Physics: Conference Series **309** (2011) 012013 doi:10.1088/1742-6596/309/1/012013

Puzzles of Dark Matter in the Light of Dark Atoms

Maxim Yu. Khlopov[1,2,3], Andrey G. Mayorov[1] and Evgeny Yu. Soldatov[1]

[1] National Research Nuclear University "Moscow Engineering Physics Institute", 115409 Moscow, Russia
[2] Centre for Cosmoparticle Physics "Cosmion" 115409 Moscow, Russia
[3] APC laboratory 10, rue Alice Domon et Léonie Duquet
75205 Paris Cedex 13, France

E-mail: `khlopov@apc.univ-paris7.fr`, `mayorov.a.g@gmail.com`, `Evgeny.Soldatov@cern.ch`

Abstract. The nonbaryonic dark matter of the Universe can consist of new stable charged leptons and quarks, if they are hidden in elusive "dark atoms" of composite dark matter. Such possibility is severely constrained, since free positively charged stable species remaining after recombination can form anomalous isotopes with electrons. This problem can be solved, if there exist stable particles with charge -2 and there are no stable particles with charges +1 and -1. These conditions can be satisfied in several recently developed alternative scenarios, linking excess of -2 particles (leptons or heavy quark clusters with suppressed hadronic interactions) to baryon asymmetry. The excessive -2 charged particles are bound with primordial helium in O-helium "atoms", maintaining specific nuclear-interacting form of the Warmer than Cold Dark Matter. O-helium looks like Bohr-like atom with heavy -2 surrounded by extended helium shell. When nuclei approach O-helium, the nuclear attraction of the helium shell polarizes O-helium. It causes dipole Coulomb repulsion, forming a shallow well outside nucleus, in which levels of O-helium-nucleus bound states can exist. The range of parameters of nuclear interaction is found, at which the binding energy of sodium with O-helium is equal to 3 keV. Annual modulations of radiative capture of O-helium to this level in the matter of DAMA/NaI and DAMA/LIBRA detectors can explain positive results of these experiments. The puzzles of direct dark matter searches appear in this case as a reflection of nontrivial nuclear physics of O-helium.

1. Introduction

According to the modern cosmology, the dark matter, corresponding to 25% of the total cosmological density, is nonbaryonic and consists of new stable particles. Such particles (see e.g. [1, 2, 3] for review and reference), created in very early Universe, should be stable, provide the measured dark matter density and be decoupled from plasma and radiation at least before the beginning of matter dominated stage. The easiest way to satisfy these conditions is to involve neutral elementary weakly interacting massive particles (WIMPs). SUSY Models provide a list of possible WIMP candidates: neutralino, axino, gravitino etc.

However it may not be the only particle physics solution for the dark matter problem.

One of such alternative solutions is based on the existence of heavy stable charged particles bound in neutral "dark atoms". This idea of composite dark matter was first proposed by S.L. Glashow in [4]. According to [4] stable tera-U-quarks with electric charge $+2/3$ forms stable (UUU) $+2$ charged "clusters", which in combination with two -1 charged stable tera-electrons E produce neutral $[(UUU)EE]$ tera-helium "atoms" that behave like WIMPs. The main problem for this solution is the over-abundance of positively charged species bound with ordinary electrons, which behave as anomalous isotopes of hydrogen or helium. This problem turned to be unresolvable, because in the early Universe as soon as primordial helium is formed it would capture all the free E^- and form positively charged $(HeE)^+$ ion, preventing any further suppression of positively charged species [5]. Therefore, in order to avoid anomalous isotopes overproduction, stable particles with charge -1 (and corresponding antiparticles) should be absent, so that stable negatively charged particles should have charge of -2 only [6]. Stable particles with the charge of -2 turned to be the only solution that saved the idea of dark atoms of dark matter.

One should mention here that stable double charged particles can hardly find place in SUSY models, but there exist several alternative elementary particle frames, in which heavy stable -2 charged species, O^{--}, are predicted:

(a) AC-leptons, predicted in the extension of standard model, based on the approach of almost-commutative geometry [6, 7, 8, 9].

(b) Technileptons and anti-technibaryons in the framework of walking technicolor models (WTC) [10, 11].

(c) stable "heavy quark clusters" $\bar{U}\bar{U}\bar{U}$ formed by anti-U quark of fourth [6, 12, 13, 14] or 5th [15] generation.

All these models also predict corresponding +2 charge particles. If these positively charged particles remain free in the early Universe, they can recombine with ordinary electrons in anomalous helium, which is strongly constrained in the terrestrial matter. Therefore cosmological scenario should provide mechanism of suppression of anomalous helium. There are two possibilities, which require two different mechanisms of such suppression:

(i) The abundance of anomalous helium in the Galaxy may be significant, but in the terrestrial matter there exists a recombination mechanism suppressing this abundance below experimental upper limits.

(ii) Free positively charged particles are already suppressed in the early Universe and the abundance of anomalous helium in the Galaxy is negligible.

These two possibilities correspond to to two different cosmological scenarios of dark atoms. The first is realized in the scenario with AC leptons, which form WIMP-like neutral AC atoms.

The second, which is considered here following [10, 16, 17, 18, 19, 20, 21] assumes charge asymmetric case with the excess of O^{--} that form atom-like states with primordial helium. After it is formed in the Big Bang Nucleosynthesis, ^4He screens the O^{--} charged particles in composite $(^4He^{++}O^{--})$ *O-helium* "atoms" [13].

In all the forms of O-helium [6, 8, 10, 14, 19, 22], O^{--} behaves either as lepton or as specific "heavy quark cluster" with strongly suppressed hadronic interaction. Therefore O-helium interaction with matter is determined by nuclear interaction of its helium shell. These neutral primordial nuclear interacting objects contribute to the modern dark matter density and play the role of a nontrivial form of strongly interacting dark matter [23, 24].

If new stable species belong to non-trivial representations of electroweak SU(2) group, sphaleron transitions at high temperatures provide the relationship between baryon asymmetry and excess of -2 charge stable species. It makes possible to relate the density of asymmetric O-helium dark matter with the baryon density.

Here after a brief review of main features of OHe structure, interaction with matter and evolution in the Universe we concentrate on the nuclear radiative capture of O-helium in underground detectors that might explain puzzles of dark matter searches.

Journal of Physics: Conference Series **309** (2011) 012013 doi:10.1088/1742-6596/309/1/012013

2. O-helium formation, interaction and evolution of in the Universe

2.1. O-helium structure

Following [6, 10, 13, 14, 17, 18, 19, 21, 22] consider charge asymmetric case, when excess of O^{--} provides effective suppression of positively charged species.

In the period $100\,\mathrm{s} \le t \le 300\,\mathrm{s}$ at $100\,\mathrm{keV} \ge T \ge T_o = I_o/27 \approx 60\,\mathrm{keV}$, ^4He has already been formed in the SBBN and virtually all free O^{--} are trapped by ^4He in O-helium "atoms" ($^4He^{++}O^{--}$). Here the O-helium ionization potential is[1]

$$I_o = Z_o^2 Z_{He}^2 \alpha^2 m_{He}/2 \approx 1.6\,\mathrm{MeV}, \tag{1}$$

where α is the fine structure constant,$Z_{He} = 2$ and $Z_o = 2$ stands for the absolute value of electric charge of O^{--}. The size of these "atoms" is [8, 13]

$$R_o \sim 1/(Z_o Z_{He} \alpha m_{He}) \approx 2 \cdot 10^{-13}\,\mathrm{cm} \tag{2}$$

Here and further, if not specified otherwise, we use the system of units $\hbar = c = k = 1$.

The analysis [19] favors Bohr-atom-like structure of O-helium, assumed in [6, 10, 13, 14, 17, 18, 22]. However, the size of He, rotating around O^{--} in this Bohr atom, turns out to be of the order and even a bit larger than the radius r_o of its Bohr orbit, and the corresponding correction to the binding energy due to non-point-like charge distribution in He is significant.

Bohr atom like structure of OHe seems to provide a possibility to use the results of atomic physics for description of OHe interaction with matter. However, the situation is much more complicated. OHe atom is similar to the hydrogen, in which electron is hundreds times heavier, than proton, so that it is proton shell that surrounds "electron nucleus". Nuclei that interact with such "hydrogen" would interact first with strongly interacting "proton" shell and such interaction can hardly be treated in the framework of perturbation theory. Moreover in the description of OHe interaction the account for the finite size of He, which is even larger than the radius of Bohr orbit, is important. One should consider, therefore, the analysis, presented in [6, 10, 13, 14, 17, 18, 19, 22, 26], as only a first step approaching true nuclear physics of OHe.

2.2. O-helium interaction with nuclei

The approach of [17, 18, 19, 21] assumes the following picture:

(i) At the distances larger, than its size, OHe is neutral, being only the source of a Coulomb field of O^{--} screened by He shell. Owing to the negative sign of $Z_o = -2$, this potential provides attraction of nucleus to OHe.

(ii) Then helium shell of OHe starts to feel Yukawa exponential tail of attraction of nucleus to He due to scalar-isoscalar nuclear potential.

(iii) Nuclear attraction results in the polarization of OHe and the mutual attraction of nucleus and OHe is changed by Coulomb repulsion of He shell.

(iv) When helium is completely merged with the nucleus the interaction is reduced to the oscillatory potential of O^{--} with homogeneously charged merged nucleus with the charge $Z + 2$.

It should be noted that scalar-isoscalar nature of He nucleus excludes its nuclear interaction due to π or ρ meson exchange, so that the main role in its nuclear interaction outside the nucleus plays σ meson exchange, on which nuclear physics data are not very definite. The nuclear potential depends on the mass μ of the σ-meson, coupling to nucleon g^2 and on the relative distance between He and nucleus. It would imply axial symmetric quantum mechanical description. In the approximation of spherical symmetry nuclear attraction beyond the nucleus was taken into account in [19] in a two different ways:

[1] The account for charge distribution in He nucleus leads to smaller value $I_o \approx 1.3\,\mathrm{MeV}$ [25].

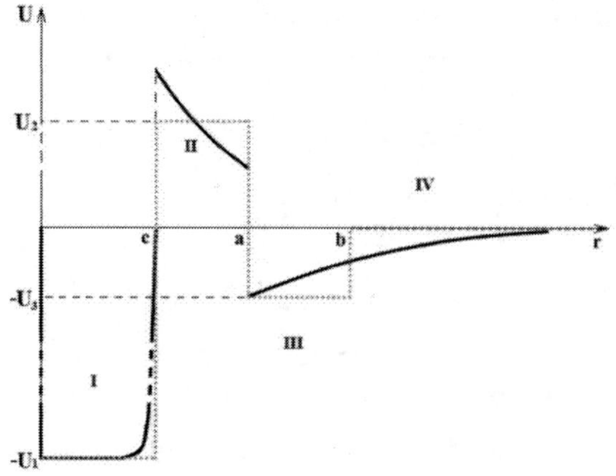

Figure 1. The approximation of rectangular well for potential of OHe-nucleus system.

(m) The nuclear Yukawa potential was averaged over the orbit of He in OHe,

(b) The potential was taken at the position of He most close to the nucleus.

These two cases (m) and (b) correspond to the larger and smaller distance effects of nuclear force, respectively, so that the true picture should be between these two extremes.

To simplify the solution of Schrodinger equation the rectangular potential was considered in [19] that consists of

(i) a potential well with the depth U_1 at $r < c = R$, where R is the radius of nucleus;

(ii) a rectangular dipole Coulomb potential barrier U_2 at $R \leq r < a = R + R_o + r_{he}$, where r_{he} is radius of helium nucleus;

(iii) an outer potential well U_3, formed by the Yukawa nuclear interaction and residual Coulomb interaction.

It lead to the approximate potential, presented on Fig. 1. Solutions of Schrodinger equation for each of the four regions, indicated on Fig. 1, are given in textbooks (see e.g.[27]) and their sewing determines the condition, under which OHe-nucleus bound states do exist.

Nuclear reactions with OHe can play interesting role in Big Bang and stellar nucleosynthesis, but these studies are only under way. The first results of these studies confirm the earlier guess about the dominance of OHe elastic collisions with nuclei and sufficiently strong suppression of the inelastic processes. That is why we follow qualitative picture of [6, 10, 13, 14, 17, 18, 19, 22, 26] in the description of OHe cosmological evolution.

2.3. O-helium dark matter

Due to elastic nuclear interactions of its helium shell with nuclei in the cosmic plasma, the O-helium gas is in thermal equilibrium with plasma and radiation on the Radiation Dominance (RD) stage, while the energy and momentum transfer from plasma is effective. The radiation pressure acting on the plasma is then transferred to density fluctuations of the O-helium gas and transforms them in acoustic waves at scales up to the size of the horizon.

At temperature $T < T_{od} \approx 200 S_3^{2/3}$ eV the energy and momentum transfer from baryons to O-helium is not effective [10, 13] because

$$n_B \langle \sigma v \rangle (m_p/m_o) t < 1,$$

where m_o is the mass of the OHe atom and $S_3 = m_o/(1\,\mathrm{TeV})$. Here

$$\sigma \approx \sigma_o \sim \pi R_o^2 \approx 10^{-25}\,\mathrm{cm}^2, \quad (3)$$

and $v = \sqrt{2T/m_p}$ is the baryon thermal velocity. Then O-helium gas decouples from plasma. It starts to dominate in the Universe after $t \sim 10^{12}\,\mathrm{s}$ at $T \leq T_{RM} \approx 1\,\mathrm{eV}$ and O-helium "atoms" play the main dynamical role in the development of gravitational instability, triggering the large scale structure formation. The composite nature of O-helium determines the specifics of the corresponding dark matter scenario.

At $T > T_{RM}$ the total mass of the OHe gas with density $\rho_d = (T_{RM}/T)\rho_{tot}$ is equal to

$$M = \frac{4\pi}{3}\rho_d t^3 = \frac{4\pi}{3}\frac{T_{RM}}{T} m_{Pl} (\frac{m_{Pl}}{T})^2$$

within the cosmological horizon $l_h = t$. In the period of decoupling $T = T_{od}$, this mass depends strongly on the O-helium mass S_3 and is given by [10]

$$M_{od} = \frac{T_{RM}}{T_{od}} m_{Pl} (\frac{m_{Pl}}{T_{od}})^2 \approx 2 \cdot 10^{44} S_3^{-2}\,\mathrm{g} = 10^{11} S_3^{-2} M_{\odot}, \quad (4)$$

where $M_{\odot}$ is the solar mass. O-helium is formed only at T_o and its total mass within the cosmological horizon in the period of its creation is $M_o = M_{od}(T_{od}/T_o)^3 = 10^{37}\,\mathrm{g}$.

On the RD stage before decoupling, the Jeans length λ_J of the OHe gas was restricted from below by the propagation of sound waves in plasma with a relativistic equation of state $p = \epsilon/3$, being of the order of the cosmological horizon and equal to $\lambda_J = l_h/\sqrt{3} = t/\sqrt{3}$. After decoupling at $T = T_{od}$, it falls down to $\lambda_J \sim v_o t$, where $v_o = \sqrt{2T_{od}/m_o}$. Though after decoupling the Jeans mass in the OHe gas correspondingly falls down

$$M_J \sim v_o^3 M_{od} \sim 3 \cdot 10^{-14} M_{od},$$

one should expect a strong suppression of fluctuations on scales $M < M_o$, as well as adiabatic damping of sound waves in the RD plasma for scales $M_o < M < M_{od}$. It can provide some suppression of small scale structure in the considered model for all reasonable masses of O-helium. The significance of this suppression and its effect on the structure formation needs a special study in detailed numerical simulations. In any case, it can not be as strong as the free streaming suppression in ordinary Warm Dark Matter (WDM) scenarios, but one can expect that qualitatively we deal with Warmer Than Cold Dark Matter model.

Being decoupled from baryon matter, the OHe gas does not follow the formation of baryon astrophysical objects (stars, planets, molecular clouds...) and forms dark matter halos of galaxies. It can be easily seen that O-helium gas is collisionless for its number density, saturating galactic dark matter. Taking the average density of baryon matter one can also find that the Galaxy as a whole is transparent for O-helium in spite of its nuclear interaction. Only individual baryon objects like stars and planets are opaque for it.

O-helium atoms can be destroyed in astrophysical processes, giving rise to acceleration of free O^{--} in the Galaxy. If the mechanisms of O^{--} acceleration are effective, the anomalous low Z/A component of -2 charged O^{--} can be present in cosmic rays at the level $O/p \sim n_O/n_g \sim 10^{-9} S_3^{-1}$, and be within the reach for PAMELA and AMS02 cosmic ray experiments [16, 28].

Collisions in the galactic bulge can lead to excitation of O-helium. If 2S level is excited, pair production dominates over two-photon channel and positron production is sufficient to explain the excess in positron annihilation line from bulge[29], measured by INTEGRAL [30].

It should be noted that the nuclear cross section of the O-helium interaction with matter escapes the severe constraints on strongly interacting dark matter particles (SIMPs) [23, 24] imposed by the XQC experiment [31]. Therefore, a special strategy of direct O-helium search is needed, as it was proposed in [32].

3. O-helium effects in underground detectors

3.1. O-helium in the terrestrial matter

The evident consequence of the O-helium dark matter is its inevitable presence in the terrestrial matter, which appears opaque to O-helium and stores all its in-falling flux.

After they fall down terrestrial surface, the in-falling OHe particles are effectively slowed down due to elastic collisions with matter. Then they drift, sinking down towards the center of the Earth with velocity

$$V = \frac{g}{n\sigma v} \approx 80 S_3 A^{1/2} \,\mathrm{cm/s}. \tag{5}$$

Here $A \sim 30$ is the average atomic weight in terrestrial surface matter, $n = 2.4 \cdot 10^{24}/A\,\mathrm{cm}^{-3}$ is the number density of terrestrial atomic nuclei, σv is the rate of nuclear collisions, $m_o \approx M_X + 4m_p = S_3\,\mathrm{TeV}$ is the mass of O-helium, M_O is the mass of the O^{--} component of O-helium, m_p is the mass of proton and $g = 980\ \mathrm{cm/s^2}$.

Near the Earth's surface, the O-helium abundance is determined by the equilibrium between the in-falling and down-drifting fluxes.

The in-falling O-helium flux from dark matter halo is

$$F = \frac{n_0}{8\pi} \cdot |\overline{V_h} + \overline{V_E}|,$$

where V_h-speed of Solar System (220 km/s), V_E-speed of Earth (29.5 km/s) and $n_0 = 3 \cdot 10^{-4} S_3^{-1}\,\mathrm{cm}^{-3}$ is the local density of O-helium dark matter. For qualitative estimation we don't take into account here velocity dispersion and distribution of particles in the incoming flux that can lead to significant effect.

At a depth L below the Earth's surface, the drift timescale is $t_{dr} \sim L/V$, where $V \sim 400 S_3\,\mathrm{cm/s}$ is given by Eq. (5). It means that the change of the incoming flux, caused by the motion of the Earth along its orbit, should lead at the depth $L \sim 10^5\,\mathrm{cm}$ to the corresponding change in the equilibrium underground concentration of OHe on the timescale $t_{dr} \approx 2.5{\cdot}10^2 S_3^{-1}\,\mathrm{s}$.

The equilibrium concentration, which is established in the matter of underground detectors at this timescale, is given by

$$n_{oE} = \frac{2\pi \cdot F}{V} = n_0 \frac{n\sigma v}{4g} \cdot |\overline{V_h} + \overline{V_E}|, \tag{6}$$

where, with account for $V_h > V_E$, relative velocity can be expressed as

$$|\overline{V_o}| = \sqrt{(\overline{V_h} + \overline{V_E})^2} = \sqrt{V_h^2 + V_E^2 + V_h V_E sin(\theta)} \simeq$$

$$\simeq V_h \sqrt{1 + \frac{V_E}{V_h} sin(\theta)} \sim V_h (1 + \frac{1}{2} \frac{V_E}{V_h} sin(\theta)).$$

Here $\theta = \omega(t - t_0)$ with $\omega = 2\pi/T$, $T = 1yr$ and t_0 is the phase. Then the concentration takes the form

$$n_{oE} = n_{oE}^{(1)} + n_{oE}^{(2)} \cdot sin(\omega(t - t_0)) \tag{7}$$

So, if OHe reacts with nuclei in the matter of underground detector, there are two parts of the signal of such reaction: constant and annual modulation, as it is expected in the strategy of dark matter search in DAMA experiments [33, 34].

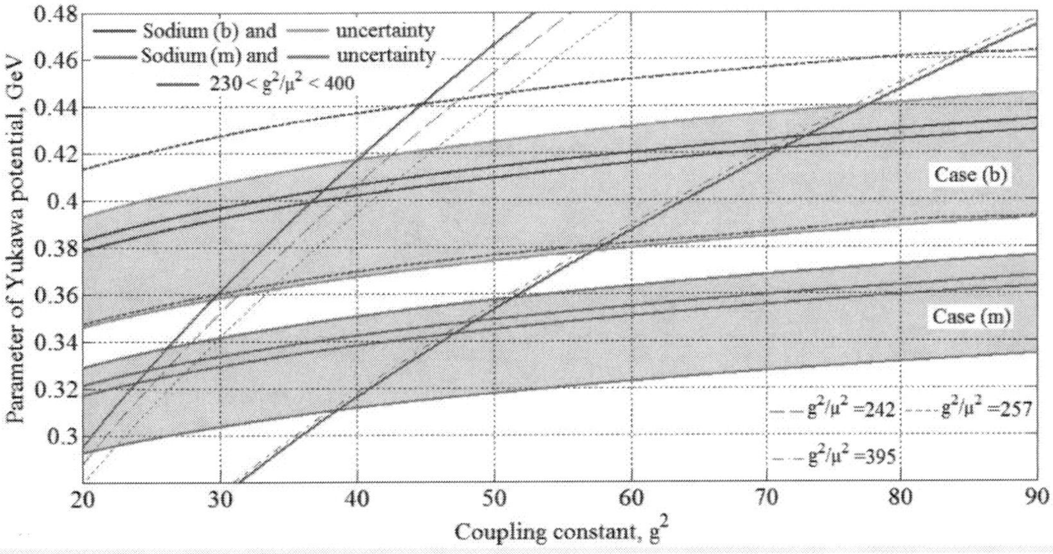

Figure 2. The regions of parameters μ and g^2, for which Na-OHe system has a level in the interval 4 keV for the cases (m) and (b). Two lines determine at $d_o = 1.2/(200\,\mathrm{MeV})$ the region of parameters, at which the bound system of this element with OHe has a 4 keV level. In the region between the two strips the energy of level is below 4 keV. There are also indicated the range of g^2/μ^2 (dashed lines) as well as their preferred values (thin lines) determined from parametrization of the relativistic $(\sigma - \omega)$ model for nuclear matter. The uncertainty in the determination of parameter $1.15/(200\,\mathrm{MeV}) < d_o < 1.3/(200\,\mathrm{MeV})$ results in the uncertainty of μ and g^2 shown by the shaded regions surrounding the lines.

3.2. Low energy bound state of O-helium with Na nuclei

The explanation [17, 18, 19, 20] of the results of DAMA/NaI [33] and DAMA/LIBRA [34] experiments is based on the idea that OHe, slowed down in the matter of detector, can form a few keV bound state with nuclei, in which OHe is situated **beyond** the nucleus (i.e. in region III on Fig. 1). Therefore the positive result of these experiments is explained by reaction

$$A + (^4He^{++}X^{--}) \rightarrow [A(^4He^{++}X^{--})] + \gamma \quad (8)$$

with nuclei in DAMA detector.

The energy of this bound state and its existence strongly depend on the parameters μ and g^2 of nuclear potential. On the Fig. 2 the regions of these parameters, giving 4 keV energy level in OHe bound state with sodium are presented. Radiative capture to this level can explain results of DAMA/NaI and DAMA/LIBRA experiments with the account for their energy resolution [35]. The lower shaded region on Fig. 2 corresponds to the case (m) of nuclear Yukawa potential averaged over the orbit of He in OHe, while the upper region corresponds to the case (b) of this potential taken at the position of He most close to the nucleus. The result is sensitive to the precise value of d_o, which determines the size of nuclei $R = d_o A^{1/3}$. The range of g^2/μ^2 and their preferred values were determined in [36]. In the calculations [19] the mass of OHe was taken equal to $m_o = 1TeV$, however the results weakly depend on the value of $m_o > 1TeV$. It is interesting that the values of μ on Fig. 2 are compatible with the results of recent experimental measurements of mass of sigma meson [37].

The important qualitative feature of this solution is the restricted range of intermediate

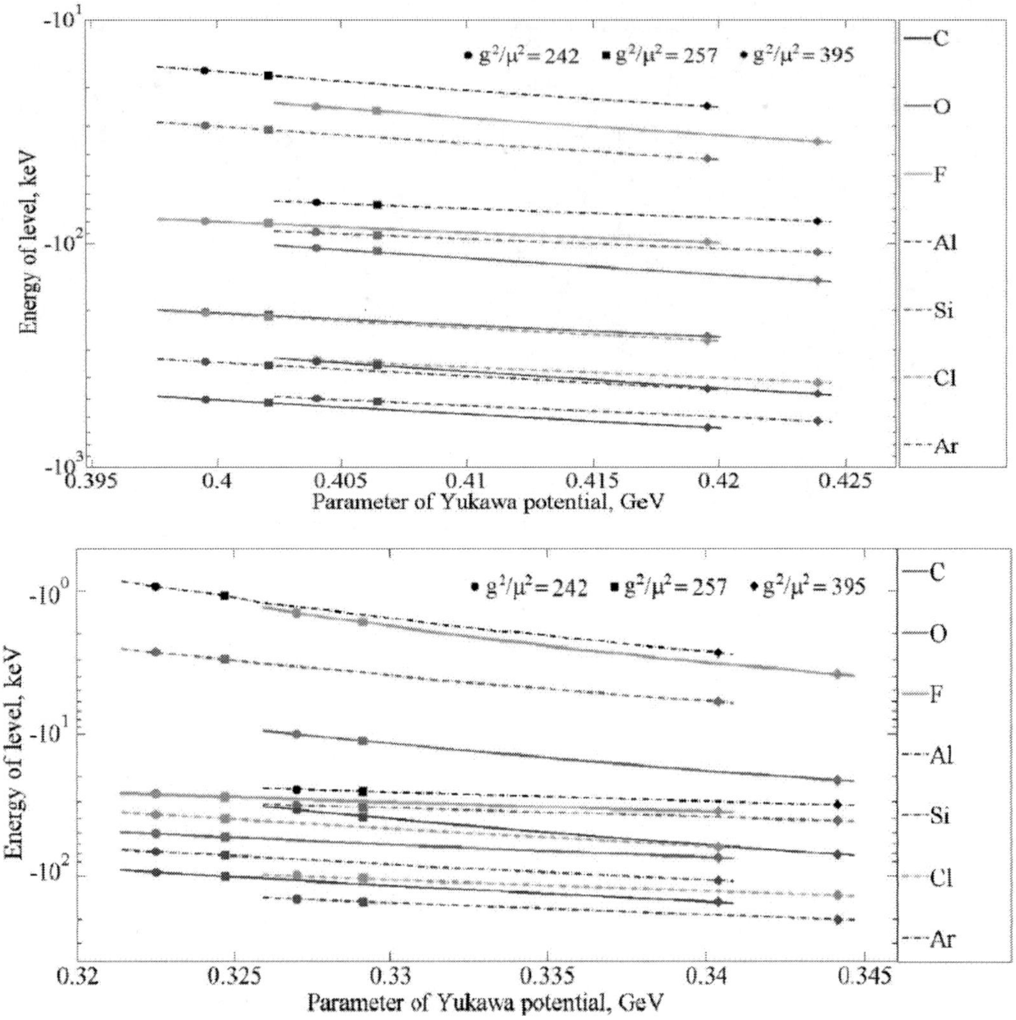

Figure 3. Energy levels in OHe bound system with carbon, oxygen, fluorine, argon, silicon, aluminium and chlorine for the case of the nuclear Yukawa potentials (b - upper plot) and (m - lower plot). The preferred values of g^2/μ^2 are indicated by the corresponding marks (squares or circles).

nuclei, in which the OHe-nucleus bound state is possible. On the Fig. 3 the energy levels for OHe bound states with carbon, oxygen, fluorine, argon, silicon, aluminium and chlorine are plotted for the cases of the nuclear Yukawa potentials (b - upper plot) and (m - lower plot). The predictions are given for the range of g^2/μ^2 determined in [36] from parametrization of the relativistic $(\sigma - \omega)$ model for nuclear matter. The energy range for the signals of radiative capture of OHe in detectors with the chemical content different from the one in DAMA/NaI and DAMA/LIBRA is predicted to differ strongly from the signal in DAMA experiments.

The range of nuclei with bound states with OHe corresponds to the part of periodic table between B and Ti. The results are very sensitive to the numerical factors of calculations and the existence of OHe-Ge and OHe-Ga bound states at a narrow window of parameters μ and g^2 turns to be strongly dependent on these factors so that change in numbers smaller than 1% can give qualitatively different result for Ge and Ga (see [19] for more details).

3.3. Radiative capture of O-helium

In the essence, the explanation [17, 18, 19, 20] of the results of experiments DAMA/NaI and DAMA/LIBRA is based on the idea that OHe concentration in the matter of DAMA detectors possess annual modulation and its radiative capture to a few keV bound state with sodium nuclei leads to the corresponding energy release and ionization signal, detected in DAMA experiments.

The rate of radiative capture of OHe by nuclei was calculated [17, 18, 20] with the use of the analogy with the radiative capture of neutron by proton with the account for: i) absence of M1 transition that follows from conservation of orbital momentum and ii) suppression of E1 transition in the case of OHe. Since OHe is isoscalar, isovector E1 transition can take place in OHe-nucleus system only due to effect of isospin nonconservation, which can be measured by the factor $f = (m_n - m_p)/m_N \approx 1.4 \cdot 10^{-3}$, corresponding to the difference of mass of neutron,m_n, and proton,m_p, relative to the mass of nucleon, m_N. In the result the rate of OHe radiative capture by nucleus with atomic number A and charge Z to the energy level E in the medium with temperature T is given by

$$\sigma v = \frac{f \pi \alpha}{m_p^2} \frac{3}{\sqrt{2}} (\frac{Z}{A})^2 \frac{T}{\sqrt{A m_p E}}. \tag{9}$$

Formation of OHe-nucleus bound system leads to energy release of its binding energy, detected as ionization signal. In the context of the approach [17, 18, 19, 20, 21] the existence of annual modulations of this signal in the range 2-6 keV and absence of such effect at energies above 6 keV means that binding energy of Na-OHe system in DAMA experiment should not exceed 6 keV, being in the range 2-4 keV. The amplitude of annual modulation of ionization signal (measured in counts per day per kg, cpd/kg) is given by

$$\zeta = \frac{3\pi\alpha \cdot n_o N_A V_E t Q}{640\sqrt{2} A_{med}^{1/2} (A_I + A_{Na})} \frac{f}{S_3 m_p^2} (\frac{Z_i}{A_i})^2 \frac{T}{\sqrt{A_i m_p E_i}} = a_i \frac{f}{S_3^2} (\frac{Z_i}{A_i})^2 \frac{T}{\sqrt{A_i m_p E_i}}. \tag{10}$$

Here N_A is Avogadro number, i denotes Na, for which numerical factor $a_i = 4.3 \cdot 10^{10}$, $Q = 10^3$ (corresponding to 1kg of the matter of detector), $t = 86400\,\mathrm{s}$, E_i is the binding energy of Na-OHe system and $n_0 = 3 \cdot 10^{-4} S_3^{-1}\,\mathrm{cm}^{-3}$ is the local density of O-helium dark matter near the Earth. The value of ζ should be compared with the integrated over energy bins signals in DAMA/NaI and DAMA/LIBRA experiments and the result of these experiments can be reproduced for $E_{Na} = 3\,\mathrm{keV}$. The account for energy resolution in DAMA experiments [35] can explain the observed energy distribution of the signal from monochromatic photon (with $E_{Na} = 3\,\mathrm{keV}$) emitted in OHe radiative capture.

At the corresponding values of μ and g^2 there is no binding of OHe with iodine and thallium [19].

It should be noted that the results of DAMA experiment exhibit also absence of annual modulations at the energy of MeV-tens MeV. Energy release in this range should take place, if OHe-nucleus system comes to the deep level inside the nucleus. This transition implies tunneling through dipole Coulomb barrier and is suppressed below the experimental limits.

Since OHe capture rate is proportional to the temperature, it looks like it is suppressed in cryogenic detectors by a factor of order 10^{-4}. However, for the size of cryogenic devices less, than few tens meters, OHe gas in them has the thermal velocity of the surrounding matter and the suppression relative to room temperature is only $\sim m_A/m_o$. Then the rate of OHe radiative capture in cryogenic detectors, containing nuclei with atomic number A is given by

Eq.(9), in which room temperature T is multiplied by factor m_A/m_o, and the ionization signal (measured in counts per day per kg, cpd/kg) is given by Eq.(10) with the same correction for T supplemented by additional factors $2V_h/V_E$ and $(A_I + A_{Na})/A$.

Effects of OHe-nucleus binding should strongly differ in detectors with the content, different from NaI. For the chosen range of nuclear parameters, reproducing the results of DAMA/NaI and DAMA/LIBRA, there are no levels in the OHe-nucleus systems for heavy and very light (e.g. ^3He) nuclei [19]. In particular, there are no such levels in Xe and most probably in Ge, what causes problems for direct comparison with DAMA results in CDMS experiment [38] and seem to prevent such comparison in XENON100 [39] experiments. Therefore test of results of DAMA/NaI and DAMA/LIBRA experiments by other experimental groups can become a very difficult task and the puzzles of dark matter search can reflect the nontrivial properties of composite dark matter.

4. Conclusions

The existence of heavy stable charged particles may not only be compatible with the experimental constraints but can even lead to composite dark matter scenario of nuclear interacting Warmer than Cold Dark Matter. This new form of dark matter can provide explanation of excess of positron annihilation line radiation, observed by INTEGRAL in the galactic bulge. The search for stable -2 charge component of cosmic rays is challenging for PAMELA and AMS02 experiments. Decays of heavy charged constituents of composite dark matter can provide explanation for anomalies in spectra of cosmic high energy positrons and electrons, observed by PAMELA and FERMI. In the context of our approach search for heavy stable charged quarks and leptons at LHC acquires the significance of experimental probe for components of cosmological composite dark matter.

The results of dark matter search in experiments DAMA/NaI and DAMA/LIBRA can be explained in the framework of our scenario without contradiction with the results of other groups. Our approach contains distinct features, by which the present explanation can be distinguished from other recent approaches to this problem [40] (see also for review and more references in [41]).

OHe concentration in the matter of underground detectors is determined by the equilibrium between the incoming cosmic flux of OHe and diffusion towards the center of Earth. It is rapidly adjusted and follows the change in this flux with the relaxation time of few minutes. Therefore the rate of radiative capture of OHe should experience annual modulations reflected in annual modulations of the ionization signal from these reactions. Within the uncertainty of the allowed nuclear physics parameters there exists a range at which OHe binding energy with sodium is in the interval 2-4 keV. Radiative capture of OHe to this bound state leads to the corresponding energy release observed as an ionization signal in DAMA detector. The rate of OHe radiative capture by nuclei is sufficiently low so that there is no significant absorption of OHe by the matter in the course of OHe diffusion to the center of Earth.

An inevitable consequence of the proposed explanation is appearance in the matter of DAMA/NaI or DAMA/LIBRA detector anomalous superheavy isotopes of sodium, having the mass roughly by 1 TeV larger, than ordinary isotopes of this element.

Since in the framework of our approach there should be no OHe radiative capture in detectors, containing heavy or light nuclei, positive result of experimental search for WIMPs by effect of their nuclear recoil would be a signature for a multicomponent nature of dark matter. Such OHe+WIMPs multicomponent dark matter scenarios naturally follow from AC model [8] and can be realized in models of Walking technicolor [16].

The presented approach sheds new light on the physical nature of dark matter. In this context positive result of DAMA/NaI and DAMA/LIBRA experiments may be a signature for exciting phenomena of O-helium nuclear physics.

4.1. Acknowledgments
We are grateful to P. Colas and I. Giomataris for kind invitation to contribute the program of 5th symposium on Large TPCs for Low Energy Rare Events and to J.-R. Cudell and A.S.Romanyuk for discussions.

References
[1] M.Yu. Khlopov *Cosmoparticle physics* (World Scientific, Singapore, 1999).
[2] M.Yu. Khlopov in *Cosmion-94*, Eds. M.Yu.Khlopov et al. (Editions frontieres, 1996) P. 67; M. Y. Khlopov in hep-ph/0612250 , p 51.
[3] M. Y. Khlopov, *Bled Workshops in Physics* **8**, 114 (2007); in arXiv:0711.4681, p. 114; M. Y. Khlopov and N. S. Mankoc-Borstnik, ibid, p. 195.
[4] S. L. Glashow, arXiv:hep-ph/0504287.
[5] D. Fargion and M. Khlopov, arXiv:hep-ph/0507087.
[6] M. Y. Khlopov, arXiv:astro-ph/0607048.
[7] C. A. Stephan, arXiv:hep-th/0509213.
[8] D. Fargion *et al*, *Class. Quantum Grav.* **23**, 7305 (2006); M. Y. Khlopov and C. A. Stephan, arXiv:astro-ph/0603187.
[9] A. Connes *Noncommutative Geometry* (Academic Press, London and San Diego, 1994).
[10] M. Y. Khlopov and C. Kouvaris, *Phys. Rev.* D **77**, 065002 (2008).
[11] F. Sannino and K. Tuominen, *Phys. Rev.* D **71**, 051901 (2005); D. K. Hong *et al*, *Phys. Lett.* B **597**, 89 (2004); D. D. Dietrich *et al*, *Phys. Rev.* D **72**, 055001 (2005); D. D. Dietrich *et al*, arXiv:hep-ph/0510217; S. B. Gudnason *et al*, *Phys. Rev.* D **73**, 115003 (2006); S. B. Gudnason *et al*, *Phys. Rev.* D **74**, 095008 (2006).
[12] K.M.Belotsky *et al*, *Gravitation and Cosmology* **11**, 3 (2005)
[13] M.Yu. Khlopov, *JETP Lett.* **83**, 1 (2006).
[14] K. Belotsky *et al*, arXiv:astro-ph/0602261. K. Belotsky *et al*, *Gravitation and Cosmology* **12**, 1 (2006); K. Belotsky *et al*, arXiv:0806.1067 [astro-ph].
[15] N.S. Mankoč Borštnik, This Volume; A. Borštnik Bračič, N.S. Mankoč Borštnik, *Phys. Rev.* D **74**, 073013 (2006); N.S. Mankoč Borštnik, *Mod. Phys. Lett.* A **10**, 587 (1995); N.S. Mankoč Borštnik, *Int. J. Theor. Phys.* **40**, 315 (2001); G. Bregar, M. Breskvar, D. Lukman, N.S. Mankoč Borštnik, *New J. of Phys.* **10**, 093002 (2008).
[16] M. Y. Khlopov and C. Kouvaris, *Phys. Rev.* D **78**, 065040 (2008)
[17] M. Y. Khlopov, AIP Conf. Proc. **1241** (2010) 388 [arXiv:0911.5685 [astro-ph.CO]].
[18] M. Y. Khlopov, A. G. Mayorov and E. Y. Soldatov, Int. J. Mod. Phys. D **19** (2010) 1385 [arXiv:1003.1144 [astro-ph.CO]].
[19] M. Y. Khlopov, A. G. Mayorov and E. Y. Soldatov, *Bled Workshops in Physics* **11**, 73 (2010) arXiv:1011.4586
[20] M. Y. Khlopov, A. G. Mayorov and E. Y. Soldatov, *Bled Workshops in Physics* **11**, 186 (2010) arXiv:1011.4587
[21] M. Y. Khlopov, A. G. Mayorov and E. Y. Soldatov, Prespace. J. **1** (2010) 1403 [arXiv:1012.0934 [astro-ph.CO]].
[22] M. Y. Khlopov, arXiv:0806.3581 [astro-ph].
[23] C. B. Dover *et al*, *Phys. Rev. Lett.* **42**, 1117 (1979); S. Wolfram, *Phys. Lett.* B **82**, 65 (1979); G. D. Starkman *et al*, *Phys. Rev.* D **41**, 3594 (1990); D. Javorsek *et al*, *Phys. Rev. Lett.* **87**, 231804 (2001); S. Mitra, *Phys. Rev.* D **70**, 103517 (2004); G. D. Mack *et al*, *Phys. Rev.* D **76**, 043523 (2007);
[24] B. D. Wandelt et al., arXiv:astro-ph/0006344; P. C. McGuire and P. J. Steinhardt, arXiv:astro-ph/0105567; G. Zaharijas and G. R. Farrar, *Phys. Rev.* D **72**, 083502 (2005)
[25] R. N. Cahn and S. L. Glashow, *Science* **213**, 607 (1981); M. Pospelov, *Phys. Rev. Lett.* **98**, 231301 (2007); K. Kohri and F. Takayama, *Phys. Rev.* D **76**, 063507 (2007).
[26] M. Y. Khlopov, arXiv:0801.0167 [astro-ph]; M. Y. Khlopov, arXiv:0801.0169 [astro-ph]; M. Yu. Khlopov, A. G. Mayorov, E.Yu. Soldatov, *Bled Workshops in Physics* **10**, 79 (2009); arXiv:0911.5606; M. Yu. Khlopov, A. G. Mayorov, E.Yu. Soldatov, *Bled Workshops in Physics* **9**, 24 (2008)
[27] L. D. Landau, E. M. Lifshitz *Quantum Mechanics: Non-Relativistic Theory* (Fizmatlit, Moscow, 2004).
[28] K.M.Belotsky, A.G.Mayorov, M.Yu.Khlopov. Charged particles of dark matter in cosmic rays. ISBN 978-5-7262-1280-7, Scientific Session NRNU MEPhI-2010, V.4, P.127
[29] D. P. Finkbeiner and N. Weiner, *Phys. Rev.* D **76**, 083519 (2007)
[30] B. J. Teegarden *et al*, *Astrophys. J.* **621**, 296 (2005)
[31] D. McCammon *et al*, *Nucl. Instrum. Methods* A **370**, 266 (1996); D. McCammon *et al*, *Astrophys. J.* **576**, 188 (2002).
[32] K. Belotsky *et al*, arXiv:astro-ph/0606350.

[33] R. Bernabei *et al.*, *Rivista Nuovo Cimento* **26**, 1 (2003)
[34] R. Bernabei *et al.* [DAMA Collaboration], Eur. Phys. J. C **56** (2008) 333 [arXiv:0804.2741 [astro-ph]]; R. Bernabei *et al.*, Eur. Phys. J. C **67** (2010) 39 [arXiv:1002.1028 [astro-ph.GA]]; R. Bernabei *et al.*, arXiv:1007.0595 [astro-ph.CO].
[35] R. Bernabei *et al.* [DAMA Collaboration], Nucl. Instrum. Meth. A **592** (2008) 297 [arXiv:0804.2738 [astro-ph]].
[36] A. b. A. Dadi, Phys. Rev. C **82** (2010) 025203 [arXiv:1005.2030 [nucl-th]].
[37] Yu. A. Troyan, A. V. Beljaev, A. Y. Troyan, E. B. Plekhanov, A. P. Jerusalimov and S. G. Arakelian, arXiv:1011.3124 [hep-ex].
[38] D. S. Akerib *et al.* [CDMS Collaboration], *Phys. Rev. Lett.* **96**, 011302 (2006); Z. Ahmed *et al.* [CDMS Collaboration], arXiv:0802.3530 [astro-ph]; O. Kamaev for the CDMS Collaboration, arXiv:0910.3005 [hep-ex].
[39] E. Aprile *et al.* [XENON100 Collaboration], Phys. Rev. Lett. **105** (2010) 131302 [arXiv:1005.0380 [astro-ph.CO]].
[40] F. Petriello and K. M. Zurek, *JHEP* **0809**, 047 (2008); R. Foot, *Phys. Rev.* D **78**, 043529 (2008); J. L. Feng, J. Kumar and L. E. Strigari, arXiv:0806.3746 [hep-ph]; J. L. Feng, J. Kumar, J. Learned and L. E. Strigari, arXiv:0808.4151 [hep-ph]; E. M. Drobyshevski, arXiv:0811.0151 [astro-ph]; B. Feldstein, A. L. Fitzpatrick and E. Katz, JCAP **1001** (2010) 020 [arXiv:0908.2991 [hep-ph]]; Y. Bai and P. J. Fox, JHEP **0911** (2009) 052 [arXiv:0909.2900 [hep-ph]]; B. Feldstein, A. L. Fitzpatrick, E. Katz and B. Tweedie, JCAP **1003** (2010) 029 [arXiv:0910.0007 [hep-ph]]; A. L. Fitzpatrick, D. Hooper and K. M. Zurek, Phys. Rev. D **81** (2010) 115005 [arXiv:1003.0014 [hep-ph]]; S. Andreas, C. Arina, T. Hambye, F. S. Ling and M. H. G. Tytgat, Phys. Rev. D **82** (2010) 043522 [arXiv:1003.2595 [hep-ph]]; D. S. M. Alves, S. R. Behbahani, P. Schuster and J. G. Wacker, JHEP **1006** (2010) 113 [arXiv:1003.4729 [hep-ph]]; V. Barger, M. McCaskey and G. Shaughnessy, Phys. Rev. D **82** (2010) 035019 [arXiv:1005.3328 [hep-ph]]; C. Savage, G. Gelmini, P. Gondolo and K. Freese, arXiv:1006.0972 [astro-ph.CO]; D. Hooper, J. I. Collar, J. Hall and D. McKinsey, arXiv:1007.1005 [hep-ph]; S. Chang, R. F. Lang and N. Weiner, arXiv:1007.2688 [hep-ph]; S. Chang, N. Weiner and I. Yavin, arXiv:1007.4200 [hep-ph]; V. Barger, W. Y. Keung and D. Marfatia, arXiv:1007.4345 [hep-ph]. A. L. Fitzpatrick and K. M. Zurek, arXiv:1007.5325 [hep-ph]; T. Banks, J. F. Fortin and S. Thomas, arXiv:1007.5515 [hep-ph]; B. Feldstein, P. W. Graham and S. Rajendran, arXiv:1008.1988 [hep-ph].
[41] G. B. Gelmini, Int. J. Mod. Phys. A **23** (2008) 4273 [arXiv:0810.3733 [hep-ph]]; E. Aprile, S. Profumo, New J. Phys. **11** (2009) 105002. J. L. Feng, arXiv:1003.0904 [astro-ph.CO].

Journal of Physics: Conference Series **309** (2011) 012014 doi:10.1088/1742-6596/309/1/012014

MIMAC: A micro-tpc matrix for directional detection of dark matter

D. Santos, J. Billard, G. Bosson, J.L. Bouly, O. Bourrion, Ch. Fourel, C. Grignon, O. Guillaudin, F. Mayet, J.P. Richer
LPSC, Universite Joseph Fourier Grenoble 1, CNRS/IN2P3, Institut Polytechnique de Grenoble

A. Delbart, E. Ferrer, I. Giomataris, F.J. Iguaz, J.P. Mols
IRFU,CEA Saclay, 91191 Gif-sur-Yvette cedex

C. Golabek, L. Lebreton
LMDN, IRSN Cadarache, 13115 Saint-Paul-Lez-Durance

E-mail: Daniel.Santos@lpsc.in2p3.fr

Abstract.
Directional detection of non-baryonic Dark Matter is a promising search strategy for discriminating WIMP events from background. However, this strategy requires both a precise measurement of the energy down to a few keV and 3D reconstruction of tracks down to a few mm. To achieve this goal, the MIMAC project has been developed. It is based on a gaseous micro-TPC matrix, filled with CF_4 and CHF_3. The first results on low energy nuclear recoils (^1H and ^{19}F) obtained with a low mono-energetic neutron field are presented. The discovery potential of this search strategy is discussed and illustrated by a realistic case accessible to MIMAC.

1. Introduction

Directional detection of Dark Matter is based on the fact that the solar system moves with respect to the center of our galaxy with a mean velocity of roughly 220 km/s [1]. Taking into account the hypothesis of the existence of a galactic halo of DM formed by WIMPs (Weakly Interacting Particles) with a negligible rotation velocity, we can expect a privileged direction for the nuclear recoils in our detector, coming out from elastic collision with those WIMPs.

The MIMAC (MIcro-tpc MAtrix of Chambers) detector project tries to get these elusive events by a double detection: ionization and track, at low gas pressure with low mass target nuclei (H, ^{19}F or ^3He). In order to have a significant cross section we explore the axial, spin dependent, interaction on odd nuclei. The very weak correlation between the neutralino-nucleon scalar cross section and the axial one, as it was shown in [2], makes this research, at the same time, complementary to the massive target experiments.

Journal of Physics: Conference Series **309** (2011) 012014 doi:10.1088/1742-6596/309/1/012014

2. MIMAC prototype

The MIMAC prototype consists of one of the chamber of the matrix allowing to show the ionization and track measurement performances needed to achieve the directional detection strategy. The primary electron-ion pairs produced by a nuclear recoil in one chamber of the matrix are detected by driving the electrons to the grid of a bulk micromegas[3] and producing the avalanche in a very thin gap (128 or 256μm).

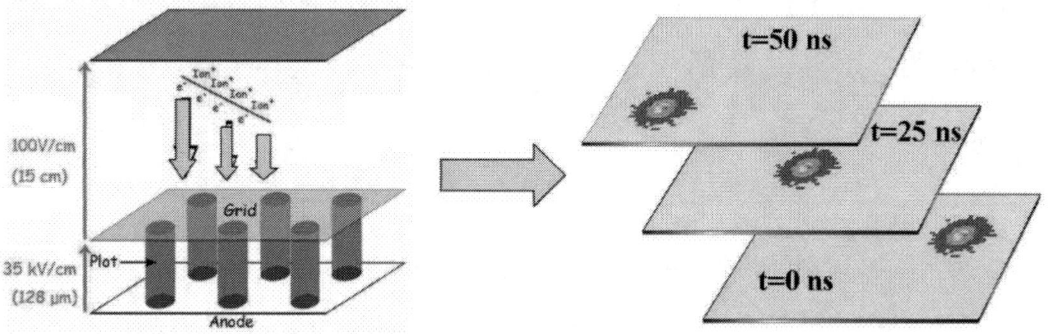

Figure 1. Track reconstruction in MIMAC. The anode is read-out every 25 ns and the 3D track is reconstructed, knowing the drift velocity of primary electrons, from the consecutive number of images, defining the event, from the anode.

As pictured on figure 1, the electrons move towards the grid in the drift space and are projected on the anode thus allowing to get information on X and Y coordinates. To access the X and Y dimensions with a 200 μm spatial resolution, a bulk micromegas with a 4 by 4 cm active area, segmented in pixels with a pitch of 350 μm was used as 2D readout in the first prototype. In order to reconstruct the third dimension Z of the recoil, the LPSC developed a self-triggered electronics able to perform the anode sampling at a frequency of 40 MHz. This includes a dedicated 16 channels ASIC [7] associated to a DAQ [8] in its first version now the 64 channels version is installed, as shown in fig. 2 (left).. The 10cmx10cm micromegas coupled to the MIMAC electronics running the 512 channels shown in fig.2 (center), working, is in fact the validation of the feasibility of a large TPC for directional detection. The installation at Modane of the first module bi-chamber, shown in fig.2 (right) has been accepted for June 2011.

In order to get the total recoil energy we need to know the ionization quenching factor (IQF) of the nuclear recoil in the gas used. We have developed at the LPSC a dedicated experimental facility to measure such IQF. A precise assessment of the available ionization energy has been performed by Santos *et al.* [4] in He + 5%C_4H_{10} mixture within the dark matter energy range (between 1 and 50 keV) by a measurement of the IQF. For a given energy, an electron track in a low pressure micro-TPC is an order of magnitude longer than a recoil one. It opens the possibility to discriminate electrons from nuclei recoils by using both energy and track length information, as it was shown in [5]. The 3D tracks are obtained from consecutive read-outs of the anode, every 25 ns, defining the event. To get the length and the orientation of the track, an independent measurement of the drift velocity is needed. Preliminary measurements, with different gases in the MIMAC prototype, of the drift velocity have been performed and they fit well with Magboltz simulations [6].

Figure 2. : (Left): MIMAC electronic card with the 8 ASIC chips (64 channels each) reading-out the 512 channels of the 10cmx10cm MIMAC prototype.(Center): The 10cmx10cm prototype showing the electronic box side of the chamber. (Right): The bi-chamber module to be installed at Modane in June 2011.

3. First experimental results

The first result concerning the ability to detect tracks with the prototype was performed with a ^{55}Fe X-ray source in order to reconstruct the 5.9 keV electron tracks produced by photoelectric effect in the active volume.

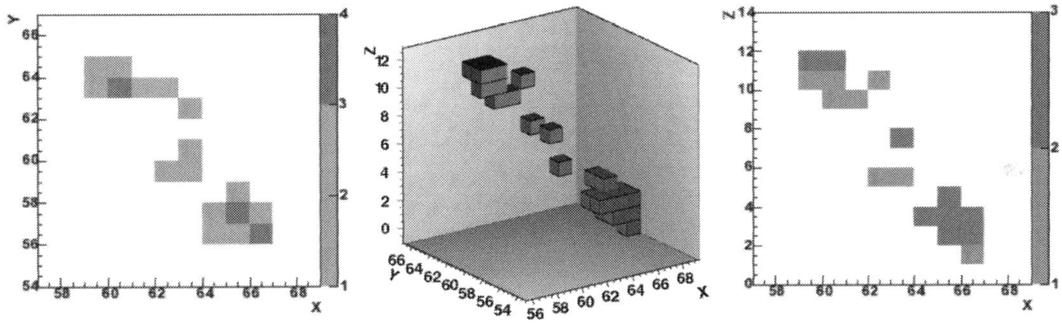

Figure 3. A 5.9 keV electron track in 350 mbar 95% $^4He + C_4H_{10}$. The left panel represents the 2D projection of the recoil seen by the anode, the center panel represents a 3D view of the track after using the reconstruction algorithm and the right panel represents a projection of the 3D track on the XZ plane

Figure 3 presents a typical electron track seen by the anode (X,Y) (left panel), its projection on the XZ plane (right panel) and reconstructed in 3D (center panel). This result shows the MIMAC capability to reconstruct the track of low energy electrons which are the typical background in dark matter experiments.

On fig 4 (left panel), a 3D track reconstruction is achieved for high energy (5.5 MeV) alpha particles issued from the natural radioactivity (^{222}Rn) present in the chamber. However, the final validation concerning the possibility for MIMAC to get directional detection had to be done with neutrons giving nuclear recoils in the range of a few keV. In order to have mono-energetic neutron fields, in the range of a few tens of keV, we perform an experiment at the AMANDE facility (IRSN- Cadarache) allowing to select the energy of the neutrons by the angle with respect to a proton beam producing a neutron resonance on a LiF target.

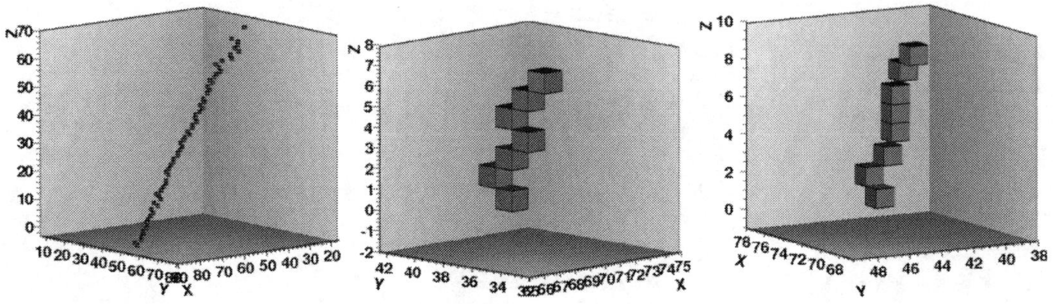

Figure 4. From left to right: recoil of 5.5 MeV He nuclei in 350 mbar $^4He + 5\%C_4H_{10}$, 8 keV hydrogen nucleus in 350 mbar $^4He + 5\%C_4H_{10}$ and fluorine nucleus leaving 50 keV in ionization in 55 mbar 70% CF_4 + 30% CHF_3

On fig 4 (center and right panel), 3D tracks of nuclear recoils following elastic scattering of mono-energetic neutrons are represented. On the center panel, a 8 keV proton recoil leaving a track of 2.4 mm long in 350 mbar $^4He + 5\%C_4H_{10}$ is represented. This event is a typical kind of signal that MIMAC will expect for dark matter search. The right panel presents a 50 keV (in ionization) fluorine recoil of 3 mm long obtained in a 55 mbar mixture of 70% CF_4 + 30% CHF_3. In addition, the electron-recoil discrimination, a very important point for dark matter detection, showing the ability to separate gamma background from nuclear recoils, has been presented in pure Isobutane or $^4He + 5\%C_4H_{10}$ mixture in ref. [5].

4. Directional detection of Dark Matter and its discovery potential

Directional detection of non-baryonic Dark Matter is a promising search strategy for discriminating WIMP events from background offering a complementarity to massive detectors [9, 10]. The directionality gives a solid and unambiguous signature of the correlation between the galactic halo and a signal in our detector. This is achieved by searching for a correlation of the WIMP signal with the solar motion around the galactic center, observed as a direction dependence of the WIMP flux coming roughly from the direction of the Cygnus constellation [1]. This is generally referred to as directional detection of Dark Matter and several projects are being developed for this goal [11, 12, 13, 14, 15].

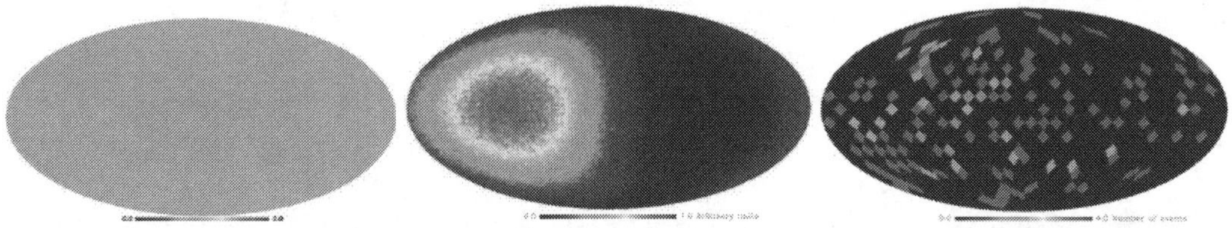

Figure 5. From left to right : isotropic background distribution, WIMP-induced recoil distribution in the case of an isothermal spherical halo and a typical simulated measurement : 100 WIMP-induced recoils and 100 background events with a low angular resolution. Recoils maps are produced for a ^{19}F target, a 100 GeV.c^{-2} WIMP and considering recoil energies in the range 5 keV $\leq E_R \leq$ 50 keV. Figures from [16].

The main asset of directional detection is the fact that the WIMP angular distribution is pointing towards the Cygnus constellation while the background one is isotropic (fig 5). The right panel of figure 5 presents a typical recoil distribution observed by a directional detector : 100 WIMP-induced events and 100 background events generated isotropically. For an elastic axial cross-section on nucleon $\sigma_{\rm n} = 1.5 \times 10^{-3}$ pb and a 100 GeV.c^{-2} WIMP mass, this corresponds to an exposure of $\sim 7 \times 10^3$ kg.day in ^{3}He and $\sim 1.6 \times 10^3$ kg.day in CF_4, on their equivalent energy ranges as discussed in ref. [16]. Low resolution maps are used in this case ($N_{\rm pixels} = 768$) which is sufficient for the low angular resolution, $\sim 15°$ (FWHM), expected for this type of detector. In this case, 3D read-out and sense recognition are considered, while background rejection is based on electron/recoil discrimination by track length and energy selection [5]. It is not straightforward to conclude from the recoil map of figure 5 (right) that it does contain a fraction of WIMP events pointing towards the direction of the solar motion.

To extract information from this example of a measured map, a likelihood analysis has been developed. The likelihood value is estimated using a binned map of the overall sky with Poisson statistics, as shown in Billard *et al.* [16]. This is a four parameter likelihood analysis with m_χ, $\lambda = S/(B+S)$ the WIMP fraction (B is the background spatial distribution taken as isotropic and S is the WIMP-induced recoil distribution) and the coordinates (ℓ, b) referring to the maximum of the WIMP event angular distribution.

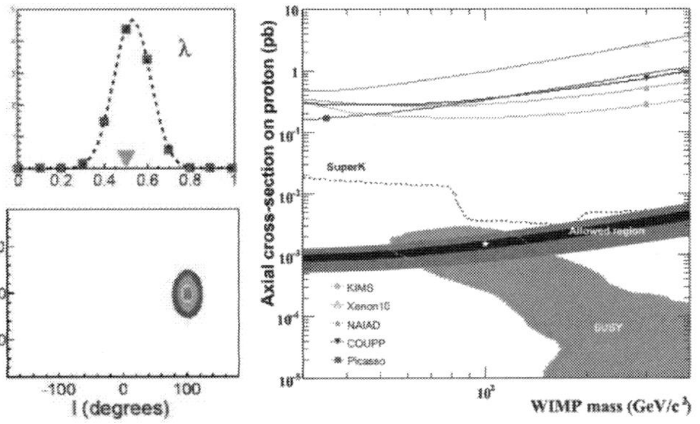

Figure 6. On the left, marginalized probability density functions of λ (top), ℓ, b (bottom) after the likelihood analysis of the simulated recoil map shown in fig. 5. On the right, allowed regions presented in the plane of the spin dependent cross-section on proton (pb) as a function of the WIMP mass (GeV/c^2). Input value for the simulation is shown with a star.

The result of this map-based likelihood method is that the main recoil direction is recovered and it is pointing towards ($\ell = 95° \pm 10°, b = -6° \pm 10°$) at 68 % CL, corresponding to a non-ambiguous detection of particles from the galactic halo. This is indeed the discovery proof of this detection strategy (left panel of fig. 6) [16]. Furthermore, the method allows to constrain the WIMP fraction in the observed recoil map leading to a constraint in the (σ_n, m_χ) plane (right panel of fig. 6). As emphasized in ref. [16], a directional detector could allow for a high significance discovery of galactic Dark Matter even with a sizeable background contamination. For very low exposures, competitive exclusion limits may also be imposed [17].

5. Conclusions

Directional detection is a promising search strategy to discover galactic dark matter. The MIMAC detector provides the energy of a recoiling nucleus and the reconstruction of its 3D track. The first 3D tracks observed with the MIMAC prototype were shown: 5.9 keV electrons (typical background) and low energy proton and fluorine recoils (typical signal). The next step will be to build a demonstrator of 1 m^3 to show that the large micro-tpc matrix for directional detection of dark matter search is accessible.

Acknowledgments

The MIMAC collaboration acknowledges the ANR-07-BLANC-0255-03 funding.

6. References

[1] D. N. Spergel, Phys. Rev. D **37** (1988) 1353.
[2] E. Moulin, F. Mayet et D. Santos. Phys. Lett. B **614** (2005) 143-154.
[3] I. Giomataris *et al.*, Nucl. Instrum. Methods A **560** (2006) 405.
[4] D. Santos *et al.*, arXiv:0810.1137.
[5] C. Grignon *et al.*, JINST 4 (2009) P11003, arXiv:0909.0654.
[6] S.F. Biagi, Nucl. Instrum. and Meth. A**421** (1999) 234-240.
[7] J.P. Richer *et al.*, Nucl. Instrum. Methods A **620** (2010) 470.
[8] O. Bourrion *et al.*,Nucl. Instrum. and Meth. A**662**(2010) 207.
[9] E. Aprile *et al.*, arXiv:1005.0380.
[10] E. Armengaud *et al.*, Phys. Lett. B **687** (2010) 294-298.
[11] D. Santos *et al.*, J. Phys. Conf. Ser. **65** (2007) 012012.
[12] G. J. Alner *et al.*, Nucl. Instr. Meth. A **555** (2005) 173.
[13] S. Ahlen *et al.*, arXiv:1006.2928.
[14] K. Miuchi *et al.*, Phys. Lett. B **654** (2007) 58.
[15] S. Ahlen *et al.*, Int. J. Mod. Phys. A **25** (2010) 1.
[16] J. Billard *et al.*, Phys. Lett. B **691** (2010) 156-162.
[17] J. Billard *et al.*, Phys. Rev. D **82** (2010) 55011.

TPC2010 IOP Publishing
Journal of Physics: Conference Series **309** (2011) 012015 doi:10.1088/1742-6596/309/1/012015

Directional detection of Dark Matter with MIMAC: WIMP identification and track reconstruction

J. Billard, F. Mayet, C. Grignon and D. Santos

Laboratoire de Physique Subatomique et de Cosmologie, Université Joseph Fourier Grenoble 1, CNRS/IN2P3, Institut Polytechnique de Grenoble, Grenoble, France

E-mail: `billard@lpsc.in2p3.fr`

Abstract. Directional detection is a promising Dark Matter search strategy. Indeed, WIMP-induced recoils present a direction dependence toward the Cygnus constellation, while background-induced recoils exhibit an isotropic distribution in the galactic rest frame. Taking advantage on these characteristic features and even in the presence of a sizeable background, we show for the first time the possibility to constrain the WIMP properties, both from particle and galactic halo physics, leading to an identification of non-baryonic Dark Matter. However, such results need highly accurate track reconstruction which should be reachable by the MIMAC detector using a dedicated readout combined with a likelihood analysis of recoiling nuclei.

1. Introduction

Taking advantage of the astrophysical framework, directional detection of Dark Matter is an interesting strategy in order to distinguish WIMP events from background ones [1]. Indeed, like most spiral galaxies, the Milky Way is supposed to be immersed in a halo of WIMPs which outweighs the luminous component by at least one order of magnitude. As the Solar System rotates around the galactic center through this Dark Matter halo, WIMPs should mainly come from the direction to which points the Sun velocity vector and which happens to be roughly in the direction of the Cygnus constellation ($\ell_\odot = 90^\circ, b_\odot = 0^\circ$). Hence, we argue that a clear and unambigous signature of a Dark Matter detection could be done by showing the correlation of the measured signal with the direction of the solar motion.

Several project of directional detectors are being developed [2, 3] and some of them are already taking data [4, 5, 6]. The first step when analysing directional data should be to look for a signal pointing toward the Cygnus Constellation with a sufficiently high significance [7]. If no evidence in favor of a Galactic origin of the signal is deduced from the previous analysis, then an exclusion limit should be derived [8]. On figure 1 we present the projected discovery regions and exclusion limits for a forthcoming directional detector proposed by the MIMAC collaboration. We consider a 10 kg CF_4 detector operated during ~ 3 years, allowing 3D track reconstruction, with a 10° angular resolution, a recoil energy range 5-50 keV and with a conservative background rate of 10 evts/kg/year. From figure 1, two different scenarii may be distinguished:

- The dark and light-grey shaded areas represent the contours where a directional detection of Dark Matter would have a significance greater than 3σ and 5σ. These contours are deduced

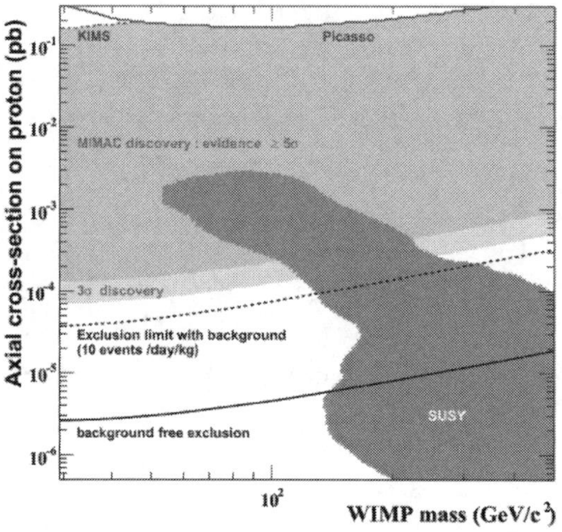

Figure 1. Spin dependent cross-section on proton (pb) as a function of the WIMP mass (GeV/c^2). Exclusion limits from some direct detection experiments are presented, KIMS [9] and Picasso [10] as well as the theoretical region, obtained within the framework of the constrained minimal supersymmetric model from [11]. Contours corresponding to a significance greater than 3σ and 5σ are presented in dark and light grey. The exclusion limit corresponding to pure background data is presented as the black dashed line and the detector sensitivity as the black solid line.

from the map-based likelihood method [7]. As an illustration, if the WIMP-nucleon cross-section is about 10^{-4} pb with a WIMP mass of 100 GeV.c^{-2}, the detector would have a Dark Matter detection with a significance greater than 3σ.

- If the WIMP-nucleon cross-section is lower than 10^{-5} pb, then an exclusion limit is deduced using the extended likelihood method (black dashed line) presented in [8]. As a benchmark and to illustrate the effect of background on exclusion limits, the detector sensitivity (no event) is presented (black solid line).

Figure 1 also presents exclusion limits from direct detection experiments, KIMS [9] and Picasso [10] as well as the theoretical region, obtained within the framework of the constrained minimal supersymmetric model taken from [11]. We can conclude that a directional detector like MIMAC will cover an important region of interest worth being investigated.

Constraining Dark Matter parameters both from particle physics (mass m_χ and cross section σ_n) and galactic halo (WIMP velocity distribution) with upcoming Dark Matter experiments is a main concern of current phenomenological studies. In the context of upcoming experiments which might give WIMP positive detection ($\sigma_n \sim 10^{-3}$ pb) instead of background rejection, it is of particular interest to develop a model-independent formalism to constrain Dark Matter properties. A new approach has been presented in [12]. The idea is to constrain Dark Matter properties using a high dimensional multivariate analysis within the framework of a general halo model. This strategy is referred to as *quasi model-independent* method as all parameters are constrained directly from the data of a single experiment.
Indeed, directional detection presents a high identification potential thanks to the combined measurement of the recoil energy E_R and the recoil direction (ℓ_R, b_R). It allows to achieve

multivariate analysis in order to obviate systematic bias in the determination of the WIMP properties, due to wrong halo model assumptions [14].

2. Identification of Dark Matter

Within the framework of a quasi model-independent method and by using a Markov Chain Monte Carlo analysis of recoil events, we show for the first time the possibility to contrain the WIMP properties, both from particle physics (m_χ, σ_n) and galactic Dark Matter halo physics (velocity dispersions). This leads to an identification of non-baryonic Dark Matter, which could be reached within few years by upcoming directional detectors [2].

2.1. Directional framework

Directional detection depends crucially on the WIMP velocity distribution. The isothermal sphere halo model is often considered but it is worth going beyond this standard paradigm in the case of a model-independent analysis. Indeed, recent results from N-body simulations are in favor of triaxial Dark Matter halos with anisotropic velocity distributions [13]. Moreover, recent observations of Sagittarius stellar tidal stream have shown evidence for a triaxial Milky Way Dark Matter halo [15].

The multivariate Gaussian WIMP velocity distribution [16] corresponds to the generalization of the standard isothermal sphere with a density profile $\rho(r) \propto 1/r^2$, leading to a smooth WIMP velocity distribution, a flat rotation curve and no substructure. The WIMP velocity distribution in the laboratory frame is given by,

$$f(\vec{v}) = \frac{1}{(8\pi^3 \det \boldsymbol{\sigma}_v^2)^{1/2}} \exp\left[-\frac{1}{2}(\vec{v} - \vec{v}_\odot)^T \boldsymbol{\sigma}_v^{-2}(\vec{v} - \vec{v}_\odot)\right] \tag{1}$$

where $\boldsymbol{\sigma}_v = \text{diag}[\sigma_x, \sigma_y, \sigma_z]$ is the velocity dispersion tensor assumed to be diagonal in the Galactic rest frame ($\hat{x}$, $\hat{y}$, $\hat{z}$) and $\vec{v}_\odot$ is the Sun motion with respect to the Galactic rest frame. When neglecting the Sun peculiar velocity and the Earth orbital velocity about the Sun, $\vec{v}_\odot$ corresponds to the detector velocity in the Galactic rest frame and is taken to be $v_\odot = 220$ km.s^{-1} along the $\hat{y}$ axis pointing toward the Cygnus constellation at ($\ell_\odot = 90°$, $b_\odot = 0°$). The velocity anisotropy $\beta(r)$, is then defined [17] as

$$\beta(r) = 1 - \frac{\sigma_y^2 + \sigma_z^2}{2\sigma_x^2} \tag{2}$$

According to N-body simulations, the β parameter at the Solar radius spans the range $0 - 0.4$, corresponding to radial anistropy.
In the following, the input halo model used to generate simulated data, is chosen according to two models : a standard isotropic halo ($\beta = 0$) in which case the velocity dispersions are related to the local circular velocity $v_0 = 220$ km/s as $\sigma_x = \sigma_y = \sigma_z = v_0/\sqrt{2}$; and an extremely anisotropic halo ($\beta = 0.4$), with the following velocity dispersions $\{\sigma_x = 200$ km/s; $\sigma_z = 169$ km/s; $\sigma_y = 140$ km/s$\}$. The latter case corresponds to the logarithmic ellipsoidal halo model from [16] with the Sun located on the major axis of the halo with the axis ratios p and q equal to 0.9 and 0.8 respectively.

The directional recoil rate is given by [18] :

$$\frac{\mathrm{d}^2 R}{\mathrm{d}E_R \mathrm{d}\Omega_R} = \frac{\rho_0 \sigma_0}{4\pi m_\chi m_r^2} F^2(E_R) \hat{f}(v_{\min}, \hat{q}), \tag{3}$$

with m_χ the WIMP mass, m_r the WIMP-nucleus reduced mass, $\rho_0 = 0.3$ GeV/c^2/cm^3 the local Dark Matter density, σ_0 the WIMP-nucleus elastic scattering cross section, $F(E_R)$ the form factor (using the axial expression from [19]) and $v_{\rm min}$ the minimal WIMP velocity required to produce a nuclear recoil of energy E_R. Finally, $\hat{f}(v_{\rm min}, \hat{q})$ is the three-dimensional Radon transform of the WIMP velocity distribution $f(\vec{v})$. Using the Fourier slice theorem [18], the Radon transform of the multivariate Gaussian is,

$$\hat{f}(v_{\rm min}, \hat{q}) = \frac{1}{(2\pi\hat{q}^T\boldsymbol{\sigma}_v^2\hat{q})^{1/2}} \exp\left[-\frac{[v_{\rm min} - \hat{q}.\vec{v}_\odot]^2}{2\hat{q}^T\boldsymbol{\sigma}_v^2\hat{q}}\right]. \tag{4}$$

In the following, the model is characterized by 8 free parameters which are $\{m_\chi, \log_{10}(\sigma_n), l_\odot, b_\odot, \sigma_x, \sigma_y, \sigma_z, R_b\}$, where the direction $(l_\odot, b_\odot)$ refers to the main direction of the recorded events [7], σ_n is the WIMP-nucleon cross section directly related to σ_0 in the pure proton approximation and R_b is the background rate. We have considered flat prior for each parameter. In such case, the Bayes' theorem is simplified and the target distribution reduces to the 8 dimensional likelihood function $\mathscr{L}(\vec{\theta})$ dedicated to unbinned data as,

$$\mathscr{L}(\vec{\theta}) = \frac{(\mu_s + \mu_b)^N}{N!} e^{-(\mu_s+\mu_b)} \times \prod_{n=1}^{N_{\rm event}} \left[\frac{\mu_s}{\mu_s + \mu_b} S(\vec{R}_n) + \frac{\mu_b}{\mu_s + \mu_b} B(\vec{R}_n)\right] \tag{5}$$

where μ_s and $\mu_b = R_b \times \xi$ are the expected number of WIMP and background events respectively. $\vec{R}_n$ refers to the energy and direction of each event while the functions S and B are the directional rate of the WIMP and the background events respectively. This MCMC analysis is based on the Metropolis-Hastings algorithm, using chain subsampling according to the burn-in and correlation lengths to deal only with independent samples [12].

2.2. Results from a benchmark model

We first applied this MCMC method on a benchmark model with the following characteristics : isotropic halo model ($\beta = 0$) with a sizeable background contamination (10 kg^{-1}.year^{-1}), a 50 GeV/c^2 WIMP and a WIMP-nucleon axial cross section $\sigma_n = 10^{-3}$ pb, well below current exclusion limits, see figure 1. The input model is used to generate simulated data considering a MIMAC like detector as described above. These data are then analysed with the directional MCMC method. The full MCMC result with 2D correlations may be found in [12]. However, major results are shown on table 1 and the conclusion is threefold:

- the discovery proof is given by the reconstruction of the main incoming direction $(\ell_\odot, b_\odot)$ pointing toward the Cygnus constellation within 2.5°, which is in favor of a positive Dark Matter detection.
- The three velocity dispersions are strongly and consistently constrained according the input values corresponding to an isothermal sphere. Moreover, using equation (2), we can evaluate the posterior Probability Density Function (PDF) of the velocity anisotropy parameter (β) which is in favor of an isotropic halo ($\beta = -0.073^{+0.29}_{-0.18}$).
- As the galactic Dark Matter halo properties are well constrained, constraints on the WIMP properties (m_χ, σ_n) are also strong and consitent according to the input WIMP parameters: ($m_\chi = 51.8^{+5.6}_{-19.4}$ GeV/c^2, $\log_{10}(\sigma_n) = -3.01^{+0.05}_{-0.08}$ pb).

As a conclusion of this benchmark study, we can deduce that this MCMC analysis, combined with directional data, is very well suited to constrain Dark Matter properties in a quasi model-independent way.

Table 1. Comparison of the values of the parameters for the input model (isotropic halo and a flat background) and as extracted after the MCMC analysis from the marginalized distributions. We quote mean value of the PDF distribution and (68 % CL) error bars.

	m_χ (GeV/c^2)	$\log_{10}(\sigma_n)$ (pb)	$\ell_\odot$ (°)	$b_\odot$ (°)	σ_x (km.s^{-1})	σ_y (km.s^{-1})	σ_z (km.s^{-1})	β	R_b (kg^{-1}year^{-1})
Input	50	-3	90	0	150	150	150	0	10
Output	$51.8^{+5.6}_{-19.4}$	$-3.01^{+0.05}_{-0.08}$	$92.2^{+2.5}_{-2.5}$	$2.0^{+2.5}_{-2.5}$	158^{+15}_{-17}	164^{+27}_{-26}	145^{+14}_{-17}	$-0.073^{+0.29}_{-0.18}$	10.97 ± 1.2

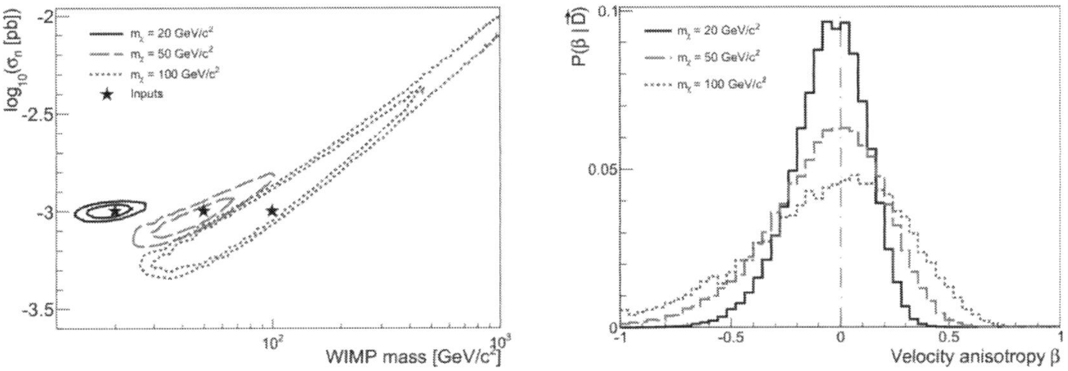

Figure 2. Left panel : 68% and 95% contour level in the (m_χ, σ_n) plan, for the isotropic input model and for a WIMP mass equal to 20, 50 and 100 GeV/c^2. Right panel : posterior PDF distribution of the β parameter for the same models.

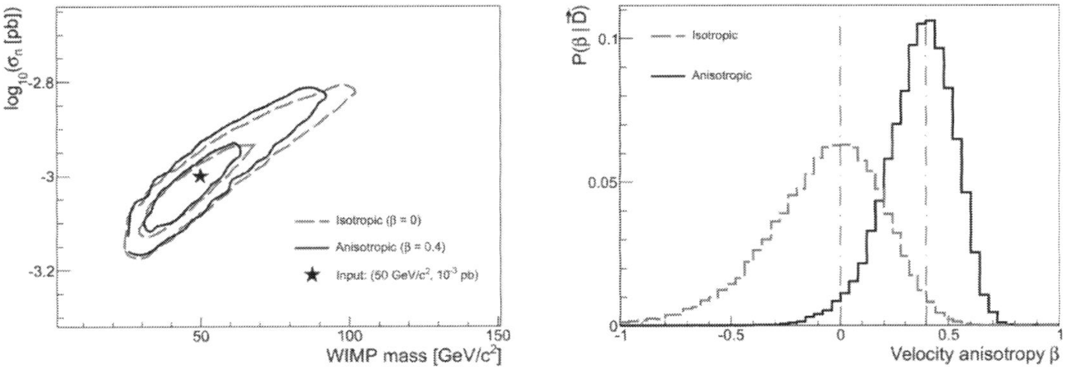

Figure 3. Left panel : 68% and 95% contour level in the (m_χ, σ_n) plan, for a 50 GeV/c^2 WIMP and for two input models : isotropic ($\beta = 0$) and triaxial ($\beta = 0.4$). Right panel : posterior PDF distribution of the β parameter for the same models.

2.3. Varying the input parameters

From the previous section, we have conclude that this MCMC analysis tool is very efficient to constrain Dark Matter properties. However, in this section we will briefly discuss the effect of some of the input parameters on the different constraints in order to estimate the performance

of this analysis tool. We will first focus on the impact of the input WIMP mass. To do so, we have simulated three different sets of directional data corresponding to an input WIMP mass of $m_\chi = 20, 50, 100$ GeV/c^2 with a constant WIMP-nucleon cross-section $\sigma_n = 10^{-3}$ pb and the standard isotropic halo model. The results from the three MCMC runs are illustrated on figure 2. We present for the three WIMP masses, on the left panel, the 68% and 95% CL contours in the $(m_\chi, \log_{10}(\sigma_n))$ plan and on the right panel, the posterior PDF $P(\beta|\vec{D})$ of the anisotropy velocity parameter β.
It can be deduced from figure 2 that the constraints strongly depend on the input WIMP mass, but in each case, the constraints are consistent with the input values. Then, this analysis has been shown to be working for any input WIMP mass although the constraints are stronger for light WIMPs. This is due to the fact that the signal characteristics, *i.e* the slope of the energy distribution and the width of the angular distribution, evolve slowly with the WIMP mass once $m_\chi \geq 100$ GeV/c^2, as shown in [7].

In the following, we investigate the effect of an extremely triaxial halo model with $\beta = 0.4$ on the estimation of the Dark Matter parameters $(m_\chi, \sigma_n, \beta)$. The results from the MCMC run on a simulated dataset corresponding to a WIMP mass of 50 GeV/c^2 with the anisotropic halo model are presented on figure 3. As previously, on the left panel is presented the constraint at 68% and 95% on the $(m_\chi, \log_{10}(\sigma_n))$ plan while on the right panel is given the deduced posterior PDF of the β parameter. For convenience and comparison, the results from the benchmark input model (isothermal sphere with a 50 GeV/c^2 WIMP) are recalled.
From the left panel of figure 3, we can conclude that the two halo models give similar constraints which are both consistent with the input values. In fact, and as foreseen, the fact that the velocity dispersions are set as free parameters in the MCMC analysis allows to avoid induced bias due to wrong model assumption. From the right panel of figure 3 we can deduce that the β parameter is well constrained: $\beta = 0.38^{+0.2}_{-0.1}$ and strongly in favor of an anisotropic Dark Matter halo.

As a conclusion of this study, it should be highlighted that the combination of information from the angular and energy distributions leads to robust allowed regions in the $(m_\chi, \log_{10}(\sigma_n))$ plan, since the halo model is also being constrained with the MCMC analysis from the same dataset of a single directional detection experiment. Moreover, the velocity anisotropy parameter β, *i.e.* the three velocity dispersions, could be sufficiently constrained to discriminate between different halo models with future directional detectors such as the one proposed by the MIMAC collaboration [3].

3. Track reconstruction with MIMAC

3.1. MIMAC prototype

The MIMAC prototype is the elementary chamber of the future large matrix. It allows the possibility to show the ionization and track measurement performance needed to achieve the directional detection strategy. The primary electron-ion pairs produced by a nuclear recoil in one chamber of the matrix are detected by drifting the electrons to the grid of a bulk micromegas [20] and producing the avalanche in a very thin gap (128 or 256μm). The electrons move towards the grid in the drift space and are projected on the anode thus allowing to get information on X and Y coordinates. To access the X and Y dimensions with a 100 μm spatial resolution, a bulk micromegas with a 4 by 4 cm^2 active area, segmented in pixels with a pitch of 350 μm is used as 2D readout. In order to reconstruct the third dimension Z of the recoil, a self-triggered electronics has been developed. It allows to perform the anode sampling at a frequency of 40 MHz. This includes a dedicated 16 channels ASIC [21] associated to a DAQ [22]. The total recoil energy is deduced from the measured ionization quenching factor (IQF) [23].

Table 2. Comparison between the input and the output values of the track characteristics with the $\{S = \text{downward}\}$ hypothesis for a 100 keV proton recoil in 50 mbar of pure isobutane. We quote mean value of the PDF distribution and (68 % CL) error bars.

	X [cm]	Y [cm]	Z [cm]	θ [°]	ϕ [°]
Input	0	0	5	45	0
Output	-0.003 ± 0.009	0.006 ± 0.012	5.068 ± 0.18	45.47 ± 1.7	-0.59 ± 2.6

In order to fully exploit the data from the MIMAC detector and to constrain Dark Matter properties, the recoiling tracks have to be accurately reconstructed: direction, sense and position in the detector volume. To achieve this goal, we first developed a track simulation software using SRIM [24] (simulated tracks), Magboltz [25] (electron drift velocity and dispersions) combined with the detector response simulation (MIMAC DAQ). This way, we can simulate any kind of recoiling track according to the following input parameters: X, Y, Z, θ, ϕ and the sense S (upward or downward). Each track is characterized by three types of topological observables: number of slices, position of the center of gravity X^{bary} and Y^{bary} of the collected charge for each slice and the width ΔX^{bary} and ΔY^{bary} along X and Y of each slice. Hence, the number of topological observables associated to any recoil track is given by:

$$N_{obs} = 1 + 4 \times N_{slice} \tag{6}$$

In this paper, we present a new MIMAC data analysis, based on [26], in order to recover the track parameters: $\{X, Y, Z, \theta, \phi, S\}$. As a working example, we will consider in the following a simulated track of an hydrogen recoil of 100 keV at the position $\{X = 0, Y = 0, Z = 5\}$ cm, in the direction $\{\theta = 45°, \phi = 0°\}$ and going downward. The analysis is based on the computation of the likelihood function $\mathscr{L}(X, Y, Z, \theta, \phi|S)$ defined as,

$$\mathscr{L}(X, Y, Z, \theta, \phi|S) = P(N_{slice}) \prod_{n=1}^{N_{slice}} P(X_n^{bary})P(Y_n^{bary})P(\Delta X_n^{bary})P(\Delta Y_n^{bary}) \tag{7}$$

where the different probability P are estimated using a frequentist approach by simulating a large number of tracks for each step in the parameter space.

In order to compute the likelihood function, we used a Markov Chain Monte Carlo algorithm and the full MCMC result following the $\{S = \text{downward}\}$ hypothesis is given in [26]. Deduced constraints on the five parameters and comparison with their input values are described in table 2. From this single track analysis it can be deduced that the spatial uncertainty is of the order of 100 μm for X and Y and about 2 mm for the absolute height Z; and the angular uncertainty is about 3°. However, even if systematic studies are needed to accurately estimate the different resolutions, results from table 2 are strongly in favor of accurately reconstructed tracks.

So far, we did not discuss the possibility to discriminate between the two hypothesis: downward or upward, the so-called "head-tail effect". To do so with our track reconstruction method, we compute the likelihood function according to the second hypothesis $\{S = \text{upward}\}$. Then, a direct comparison of the maximum Log-Likelihood values of each hypothesis (-3.8 for $\{S = \text{downward}\}$ and -5.1 for $\{S = \text{upward}\}$) allows to conclude in favor of a downwarding track consistent with the input simulated track. Hence, this method seems to be able to achieve the sense recognition of the recoiling tracks, but here again, systematic studies must be done to estimate its efficiency [26].

4. Conclusion

In this paper, we have shown that the directional detector proposed by the MIMAC collaboration should be sensitive to an important region of interest of the Dark Matter parameter space (m_χ, σ_n) motivated by constrained minimal supersymmetric models. In the case of a strong WIMP positive detection with $\sigma_n \sim 10^{-3}$ pb, an identification of Dark Matter could be achieve in a model independent way using the MCMC analysis method presented in section 2. We have shown that constraining the velocity dispersions of the galactic Dark Matter halo allows to obviate bias in the determination of the WIMP properties, due to wrong halo model assumptions. Moreover, we have shown the possibility to discriminate between various halo model characterized by their velocity anisotropy parameter. However, such interesting results requires accurate track reconstruction: direction, position and sense. To reach this goal, we developed a new track reconstruction method based on a likelihood analysis which has shown very promising results, *i.e* high resolutions. Although, systematic studies must be done to accurately estimate the differents resolutions and the sense recognition efficiency.

References

[1] Spergel D N, Phys. Rev. D **37** (1988) 1353.
[2] Ahlen S *et al.*, International Journal of Modern Physics A25 (2010) 1-51.
[3] Santos D *et al.*, arXiv:1012.1166.
[4] Ahlen S *et al.*, Phys. Lett. B **695** (2011) 124-129.
[5] Alner G J *et al.*, Nucl. Instr. Meth. A **555** (2005) 173.
[6] Miuchi K *et al.*, Phys. Lett. B **654** (2007) 58.
[7] Billard J, Mayet F, Macias-Perez J F and Santos D, Phys. Lett. B **691** (2010) 156-162.
[8] Billard J, Mayet F and Santos D, Phys. Rev. D **82** (2010) 055011.
[9] Lee H S *et al.*, Phys. Rev. Lett. **99** (2007) 091301.
[10] Archambault S *et al.*, Phys. Lett. B **682** (2009) 185.
[11] Trotta R *et al.*, JHEP **0812** (2008) 024.
[12] Billard J, Mayet F and Santos D, arXiv:1012.3960.
[13] Ling F S *et al.*, JCAP **1002** (2010) 012.
[14] Green A M, JCAP **0708** (2007) 022.
[15] Law D R, Majewski S R and Johnston K V, Astrophys. J. **703** (2009) L67.
[16] Evans N W, Carollo C M and de Zeeuw P T, Mon. Not. Roy. Astron. Soc. **318** (2000) 1131.
[17] Binney J and Tremaine S, Galactic dynamics, Princeton University Press (2008).
[18] Gondolo P, Phys. Rev. D **66** (2002) 103513.
[19] Lewin J D & Smith P F, Astropart. Phys. **6** (1996) 87.
[20] Giomataris I *et al.*, Nucl. Instrum. Methods A **560** (2006) 405.
[21] Richer J P *et al.*, Nucl. Instrum. Methods A **620** (2010) 470.
[22] Bourrion O *et al.*, Nucl. Instrum. and Meth. A**662**(2010) 207.
[23] Santos D *et al.*, arXiv:0810.1137.
[24] Ziegler J F, Biersack J P and Littmark U, Pergamon Press New York (1985), www.srim.org.
[25] Biagi S F, Nucl. Instrum. and Meth. A**421** (1999) 234-240.
[26] Billard J *et al.*, *in preparation.*

Journal of Physics: Conference Series **309** (2011) 012016 doi:10.1088/1742-6596/309/1/012016

Single electron sensitive GridPix TPCs and their application in Dark Matter search and ν-less Double Beta Decay Experiments

Harry van der Graaf
NIKHEF, Amsterdam

Email: vdgraaf@nikhef.nl

Abstract. The industrial production of monolithic GridPix pixel chips for the digital readout of primary electrons in a TPC is under study at IZM-Fraunhofer, Berlin. With future low-cost chips, new applications are foreseen in TPCs for bi-phase Ar/Xe dark matter experiments, and for ν-less double beta decay experiments.

1. Introduction and status of the GridPix TPC

With the GridPix detector [1], each MPGD hole is read out individually by the circuitry in a pixel of a CMOS chip (see fig.1). For this, we applied the Timepix chip [2]. Since the source capacity at the pixel preamp input can be as low as 10 fF, the pixel circuitry is sensitive to avalanches initiated by one single electron. Therefore, if the granularity is sufficient, the GridPix device detects each individual electron liberated in the ionization process in the gaseous (drift) volume.

With 'Micro Electro Mechanical Systems' MEMS technology, we constructed a Micromegas onto the pixel chip, forming the integrated grid 'InGrid' [3]. This guaranties perfect alignment of the grid holes with the pixel input pads, and a constant avalanche gap thickness. The insulating support pillars could be made narrow such that they could be positioned between active pixel input pads, eliminating dead regions. With a GridPix detector, in which InGrid is combined with a time-resolved pixel readout, all the information of all individual primary electrons, due to ionisation radiation, can be collected. Diffusion in the gas during the drift of the electrons is the only limit on the intrinsic information (see fig.2).

The 'Micro Electro Mechanical Systems' (MEMS) technology, developed at MESA+ at the University of Twente, for producing the protection layer and InGrid, is being transferred to IZM-Fraunhofer in Berlin. Once this transfer is complete, wafers with pixel chips are expected to be processed commercially, at acceptable costs.

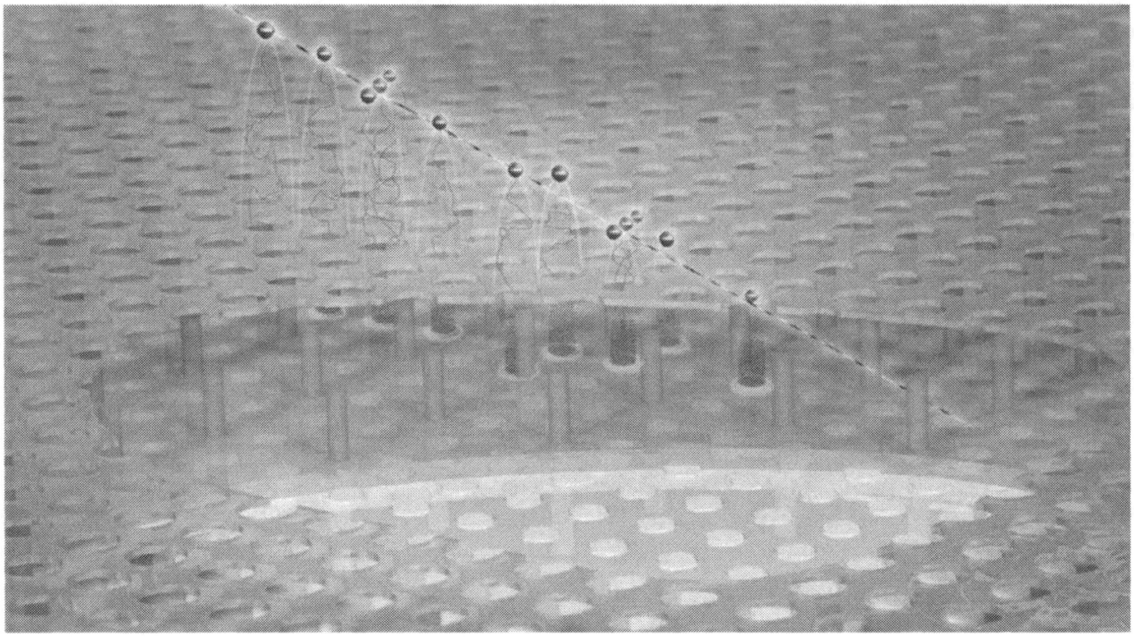

Figure 1. The GridPix detector. Electrons from the drift volume initiate an electron avalanche in the gap between the pixel chip and the grid.

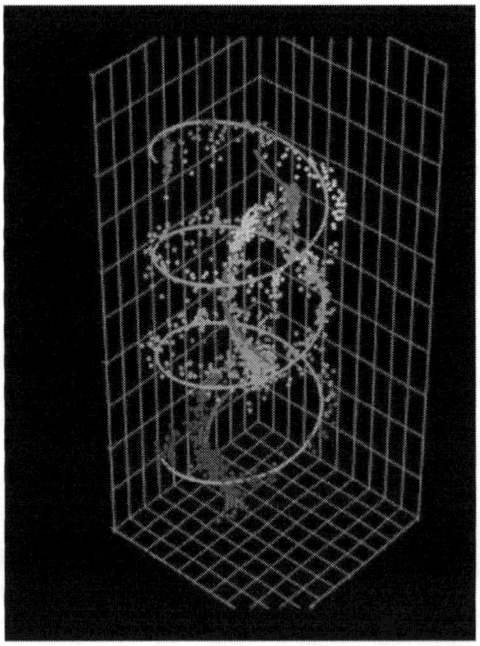

Figure 2. Two ß's from a 90Sr source, recorded with a Timepix based GridPix detector. Drift length: 30 mm. Gas: He/i-butane 80/20, with a magnetic field of 0.2 T oriented parallel to the (vertical) drift field. The bottom plane represents the Timepix chip (256 x 256 pixels; square pixel pitch 55 μm), fiducial surface 14 mm x 14 mm.

2. Applications in Dark Matter and Double Beta decay experiments

Operating Micro Pattern Gas Detectors (MPGD) in pure and cold Ar gas has been successfully demonstrated [4]. The readout sensors of bi-phase xenon or argon based WIMP search experiments are usually photomultipliers, registering the photons due to ionization and scintillation. We are studying the possibility of replacing the top array of photomultipliers with GridPix detectors. Here, the grids are facing downwards, positioned in the gas phase (see fig.3). With a sufficiently strong drift field, primary electrons, created by an ionizing event in the liquid, are drifting upwards, crossing the phase transition, and continue drifting in the gas. A GridPix detector can individually detect these electrons. Events with 1, 2, or n electrons, generated in a small volume by a WIMP event, could be identified. If the pixel chip is placed some 50 μm above the liquid surface plane, then the InGrid could be omitted if cohesion does not bring the liquid in contact with the pixel chip: gas amplification would occur in the gap between the liquid and the pixel chip. If the grid of the GridPix detector is made sensitive for photons (CsI grid, see above), then the photons created in WIMP events would also be recorded in the gaseous GridPix detector.

GridPix TPCs could be well employed in neutrinoless double-beta decay experiments: see fig.4.

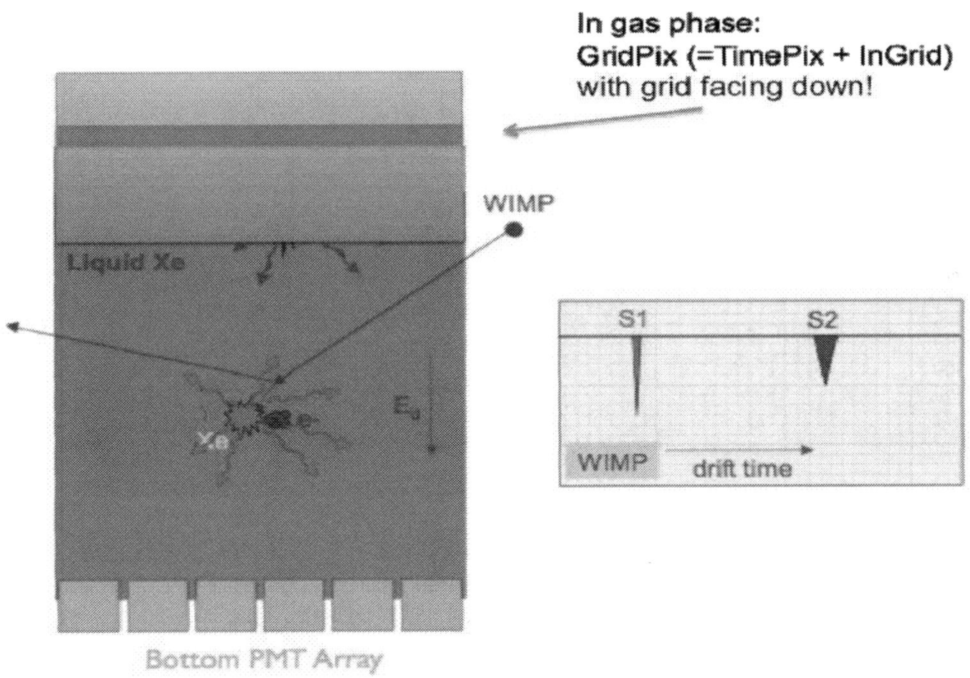

Figure 3. A WIMP search experiment with GridPix as single electron detector, located in the gaseous top section [5].

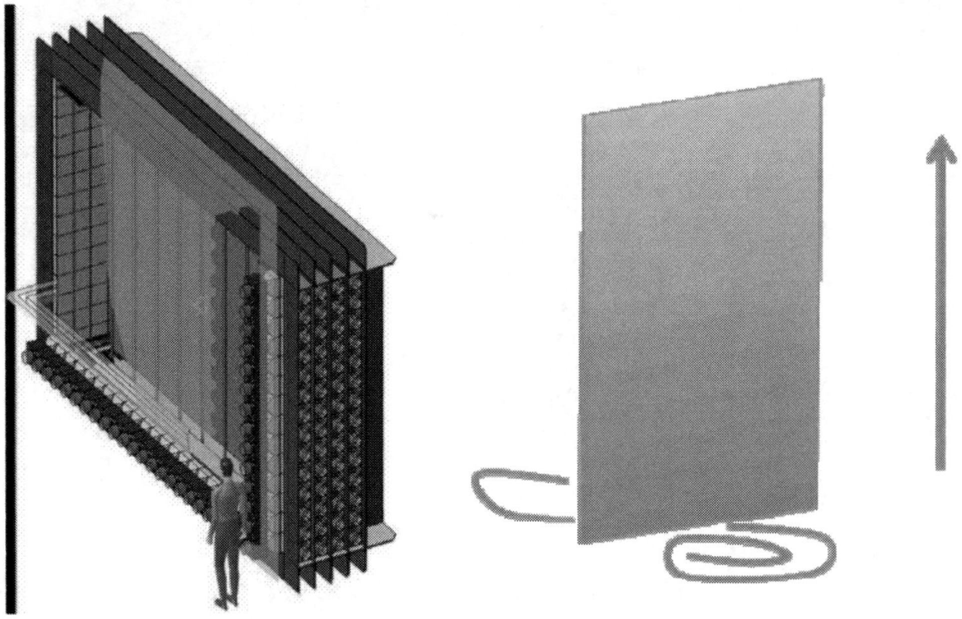

Figure 4. Left: the neutrinoless double beta decay superNemo experiment. Right: A large TPC with a GridPix readout at the bottom. The beta-emission foil must be segmented in horizontal strips in order to create a homogeneous electric drift field. The beta energies are measured from the total charge deposited at (total) energy absorption in the gas, and from the initial radius of curvature.

3. Other applications

With a GridPix (or thin gas layer Gossip) detector, a track vector is measured, whereas Si trackers measure a track point in space. This is relevant for future tracking detectors in particle physics experiments where Si pixel and Si strip tracking detectors could be replaced by rad-hard Gossip detectors. In the upgraded sLHC experiments, level-1 triggers consisting of bi-Si layers are being considered, measuring the track momentum from the local track angle. The problematic inter-communication between these two layers is not required for a GridPix miniPC where the projected track length (see fig. 5) is a direct measure for the track momentum and particle charge sign [6].

Fig. 6 shows an image from a Gridpix detector, equipped with a quartz window, irradiated with UV light from a low-pressure deuterium arc lamp. The image is the difference between a clean window and a window with fingerprint. The efficiency of this photon detector is improved by the deposition of a CsI layer on top of the aluminium grid, reaching the theoretical maximum [7].

With the time-resolved Timepix chip, GridPix detectors provide 3D images of the primary ionization in the drift gap (see fig.7). Using this information, the origin, and the initial direction and the energy of the electron created in the interaction of a photon with the gas, can be measured. The polarization of photons can be measured to a standard that is relevant for astro particle physics. An experiment to test GridPix ('PolaPix') in a beam of polarised photos at the University of Erlangen-Nürnberg is in preparation [8].

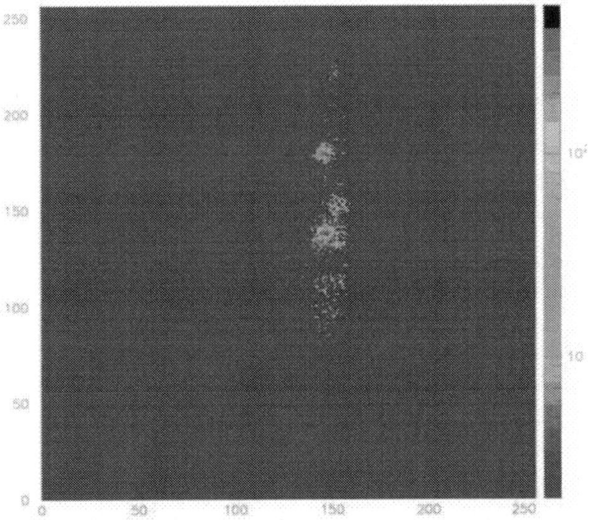

Figure 5. Response of tilted GridPix detector to a 5 GeV electron, viewing the projected track. Note the diffusion of the drifting primary electrons, largest for the largest drift length (20 mm). A transition radiator was placed in fron of the detector; the TR-quantas are clearly visible. The projected track length is a direct measure for the angle between the detector plane and the track. With a known attitude of the detector, the track angle can be obtained in a rapid way, and therefore the particles' momentum.

Figure 6 Image of Martin Fransen's right-hand index fingerprint.

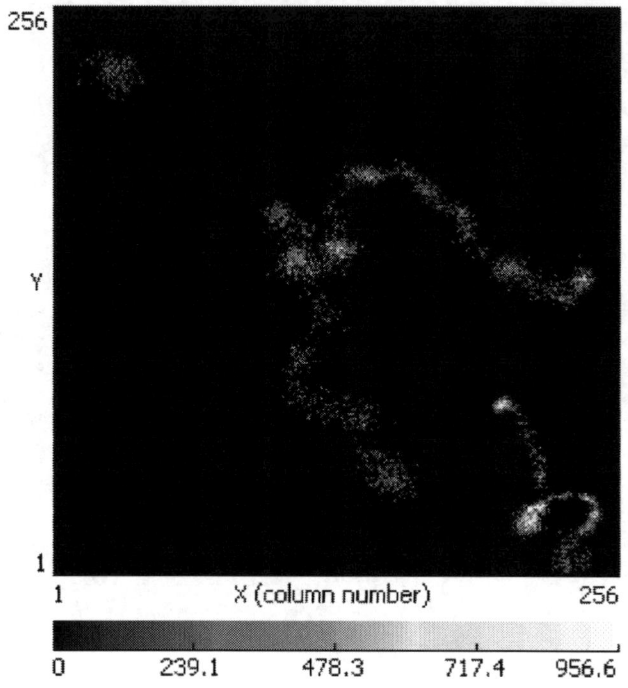

Figure 7. Image of GridPix TPC irradiated with 10 – 50 keV X-ray quanta. By a reconstruction of the path of the photo electron, information concerning the position, direction, and the energy of the incident photon can be obtained. From a group of photons, their degree of polarization can be measured.

4. Infrastructure: a new gas system, the ReLaXd DAQ system and miniHV

A new generation of miniTPCs requires new gas systems, and new safety rules. The quantity of flammable gas is often very small, and the applied mixtures are often below certain levels of flammability. At Nikhef, a gas mixing station has been made, providing 12 liter light-weight bottles containing up to 20 bars of pre-mix gas. By controlling and monitoring the bottle pressure, the gas flow out of the bottle, and the gas flow at the detector' exhaust, the intrinsic leak rate of the complete system can be measured. This system is easy to use, but the existing safety rules need to be reviewed.

With the ReLaXd DAQ systems, a set of Medipix and TimePix chips can be read out. A ReLaXd board has dimensions of 46 mm x 90 mm, and has Ethernet as communication/data output [9].

For GridPix detectors, HV supplies are required for the grid, the cathode and the guard. The current is usually less than 1 μA. Commercial supplies are expensive, voluminous and can deliver much more current, which is only a disadvantage: there is less risk of damage if the maximum power is limited intrinsically. In addition, for our applications in MPGDs, fast and high-resolution monitoring of the actual current is relevant for guarding the health of the detector, and for counting discharges. For GEM detectors, a larger number of these low-cost HV channels is needed.

We have developed a miniature Cockroft-Walton circuit with its input supply and monitoring/control connected to the bottom stage only. In future versions, the volume of this supply will be reduced to a few cubic centimeters. In another development, the resistor chain of the field-shaping electrodes of TPCs is replaced by a chain of diodes and capacitors forming the local Cockroft-Walton chain.

Journal of Physics: Conference Series **309** (2011) 012016 doi:10.1088/1742-6596/309/1/012016

References

[1] H. van der Graaf *et al.*: Nucl. Instr & Methods. A 566 (2006) 62-65.
[2] X. Llopart *et al.*: Nucl. Instr. & Meth. A 581 (2007) 485-494
[3] H. van der Graaf, Nucl. Instr. and Meth. A607 (2009) 78–80
[4] A.Rubbia et al: Operation of a double-phase pure argon Large Electron Multiplier Time Projection Chamber: comparison of single and double phase operation. 11th Pisa meeting, Isola d'Elba, Italy May 2009. Proceedings: A.Badertscher *et al.*: Nucl. Instr. & Meth. A 617 (2010) 188-192
[5] E.Aprile, T.Doke, Rev.Mod.Phys.82:2053-2097,2010 [arXiv:0910.4956].
[6] H. van der Graaf: Data Compression and LVL1 track triggering by means of Digital Signal Processing in GridPix/Gossip FE pixel chips. Workshop on Intelligent Trackers WIT2010, Feb 2010, Berkeley, CA, USA.
[7] J. Melai *et al.*: Nucl. Instr. & Meth. A, http://dx.doi.org/10.1016/j.nima.2010.06.301.
[8] Ulrike Gebert: PhD Thesis. ECAP/FA Univ. Erlangen-Nurnberg, DE
[9] J. Visser, et al., Nucl. Instr. and Meth. A (2010), doi:10.1016/j.nima.2010.06.110

Journal of Physics: Conference Series **309** (2011) 012017 doi:10.1088/1742-6596/309/1/012017

RD51, a world-wide collaboration for the development of Micro Pattern Gaseous Detectors

M. Chefdeville, on behalf of the RD51 collaboration

Laboratoire d'Annecy-le-Vieux de Physique des Particules, 9 Chemin de Bellevue - BP 110, 74941 Annecy-le-Vieux CEDEX - FRANCE

E-mail: `chefdevi@lapp.in2p3.fr`

Abstract. Originally introduced to improve the rate capability of traditional wire chambers, Micro Pattern Gaseous Detectors (MPGD) actually demonstrate many more benefits. Be it for medical and industry imaging, collider experiments or more interestingly in the framework of this conference for the search of rare events, they are the subject of constant research and development in several laboratories over the world. The RD51 collaboration has been coordinating this work since April 2008 and is meant to advance the technological development and application of MPGD. The collaboration is presented and emphasis is put on its latest achievements which do make these devices an attractive option for the detection of low energy rare events: the possibility to instrument large area and to detect UV photons.

1. Introduction

1.1. Micro Pattern Gaseous Detectors

An experimental fact familiar to gaseous detector physicists is the drop of gas gain of wire chambers above some rate. This limit rate primarily depends on the collection time of the avalanche ions, the space charge of which reduces the applied field. As a result of the rapidly decreasing radial field around the wire, the intrinsically low mobility of ions and the relatively large wire to cathode distance (1–10 mm), the maximum sustainable rate hardly exceeds 10 kHz/mm^2. With thin alternating cathode and anode strips printed on a substrate at a few hundred micron pitch, the Micro Strip Gas Counter [2] is considered as the first Micro Pattern Gaseous Detector. In spite of a poor robustness against discharges its invention in 1988 is a major breakthrough in the field of gaseous detectors, paving the way for higher rate operation. The fast collection of ions in MSGC indeed pushes the rate capability by two order of magnitude with respect to wire chambers, up to roughly 1 MHz/mm^2 [3].

Stimulated by the progress in printed circuit board (PCB) technology, the invention of the Micro Mesh Gaseous Structure (Micromegas [4]) and the Gas Electron Multiplier (GEM [5]) followed in 1996 and 1997 respectively. Compared to MSGCs, they show even higher rate capability [6]. Their functioning is easily grasped in Figure 1 where drawings of the electrodes and electric field lines are shown. Based on Micromegas and GEM geometry, other detectors with specific advantages have been developed. Non exhaustively, they are the micro-Bulk, Thick GEM (THGEM [7]), Resistive Electrode Thick GEM (RETGEM [8]), Resistive mesh detectors [9], Micro Hole Strip Plate (MHSP [10]) and the GridPix [11]. Stacking GEMs or combining different technologies is possible, for instance to increase the total gain [12, 13] or reduce ion feedback to the drift space or top electrodes [14].

Optimised MPGD show great performance: gas gains close to the breakdown limit (10^6–10^7 for single electrons), spatial resolution of a few tens of microns [15, 16], timing capability with MIPs of a few ns [18] and below the ns as photodetector [17], energy resolution close to the statistical limit (11.7 % and 1 % FWHM to 5.9 keV X-rays and 5.5 MeV alphas respectively [19, 20]), ion backflow fractions (IBF) of a few 10^{-3} [14, 21]. Finally, MPGD can be made sensitive to photons in the UV range with coatings of the electrodes with photoconverting materials like CsI [17, 22].

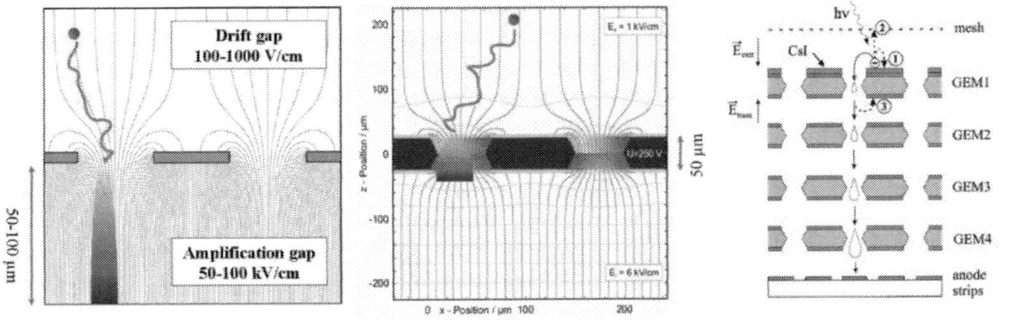

Figure 1. Illustration of the working principle of Micromegas (left), single (center) and cascaded (right) GEM. The electron multiplication occurs in the high field region between the mesh and the anode plane in the case of Micromegas and inside the GEM holes.

1.2. The RD51 collaboration

The RD51 collaboration goal is to advance the technological development and application of MPGD [23]. It coordinates the efforts of several universities and research laboratories from 25 countries in Europe, America, Asia and Africa. It is organised in working groups, each with objectives, tasks and regular meetings for sharing information, results and experience. The collaboration tries to optimize the cost of R&D projects by sharing resources, creating common projects and providing common infrastructures. The main objectives of the collaboration can be summarized as follows:

- the development of fabrication techniques and production facility of large area detectors which also includes industrial partnership to increase mass production capability;
- the understanding of the physical phenomena limiting the performance of current MPGD and the definition of test standards for comparison between different detector technology;
- the development of common software, test infrastructure and readout systems.

For a more detailed description of the working groups the reader is referred to [24].

1.3. Detection of rare events with MPGDs

Micro Pattern Gaseous Detectors can be applied as the readout of calorimetric or tracking devices for the search of rare events like neutrino scattering off nuclei, dark matter interactions and neutrino-less double-beta decays. On-going projects using gas/liquid media as well as some proposals for future experiments are based on large Time Projection Chambers (TPCs) filled with liquefied noble gases (see [25, 26] for reviews of experiments, projects and proposals). Argon and xenon are targets of choice for the detection of dark matter because of their high density. They do moreover scintillate with a high photon yield allowing the discrimination of backgrounds by measuring the ionisation density through the ratio of scintillation light to charge.

This dark detector has to be sensitive to UV light and to provide good light resolution over large area. Charge resolution may also be required if the primary ionisation is to be directly detected rather than converted into light through scintillation in moderate electric fields. Another TPC application is the search for the neutrino-less double-beta decay of ^{136}Xe. Here, the TPC is used to precisely measure the energy of the two emitted electrons and equipped so as to trap and identify the daughter ^{136}Ba ion by laser spectroscopy. Clearly, excellent energy resolution is mandatory to distinguish the neutrino-less decay peak in the tail of the standard double beta decay energy spectrum.

The capability to cover large area with uniform response and an excellent energy resolution are essential ingredients which are partly determined by the fabrication techniques of MPGD. These techniques are detailed in the next section with examples of running experiments or current projects within the RD51 collaboration. Also related to large area detectors is the development of a readout system that can be applied to different MPGD and easily scales with the number of channels. This RD51 development is motivated in section 3 where its recent successful implementation with a detector is presented. Finally, the detection of UV light with MPGD-based photomultipliers relevant to the detection of rare events is discussed.

2. Fabrication of large area detectors

2.1. Micromegas

First Micromegas were made of a few micron thin electroformed nickel mesh subsequently stretched and glued on an insulating frame which is then placed on a flat PCB dressed with photolithographically printed pillars and readout electrodes. Possible wrinkles of the mesh would disappear upon application of some voltage difference between the mesh and the anode as the electric force pulls the mesh toward the PCB. This assembly process is not straightforward and a simpler procedure called Bulk has been proposed [27].

The Bulk Micromegas consists of a woven wire mesh. Available in rolls of 24×40 m^2, the mesh is solidly mounted on the PCB in a single process. The anode plane, a photoresistive film, the mesh and a second film are laminated at high temperature. The pillars are then formed by photolithography. Facts and figures about standard Bulk are: stainless steel 30 μm diameter wires woven at a pitch of 80 μm, an anode to mesh distance of 128 μm, a pillar diameter and pitch of 300 μm and 2 mm respectively and similar gas gain property as electroformed Micromegas.

The principal production facility is the CERN workshop where detector sizes up to 150×50 cm^2 are currently obtained. The CERN management is supporting future upgrade of some LHC sub-detectors with large area MPGD. Accordingly it has approved the construction of a new workshop and its equipment with new machines of larger size processing capability. Eventually, the fabrication of Bulk Micromegas with active areas up to 2×1 m^2 will be possible.

2.2. Large-area Micromegas experiments and projects

Bulk Micromegas has been chosen by the T2K collaboration to equip the 3 TPC of the off beam axis detector [28]. Each TPC has a central cathode plane of 2×0.7 m^2 and two end-plates of 12 modules (fig. 2 left). Individual modules are equipped with 1726 readout channels and have a size of 34×36 cm^2, 95 % of which is active. The distance between adjacent modules is such that 85 % of the end-plate is active. Pad by pad X-ray characterisation of the modules in a dedicated test chamber revealed a uniformity of the photopeak position of 3 % and an RMS dispersion of the mean peak position of all 72 modules of about 8 %. Since January 2010, the three TPCs are operational and successfully took data during the first T2K physics run.

A modular approach to the construction of large area Bulk chambers is also adopted by a group of the Calorimeter for linear collider experiment collaboration (CALICE) [29]. The mid-term goal of the project is the construction of a 1 m^3 sampling hadronic calorimeter with thin Micromegas chambers as active media interleaved with steel absorbers. Six modules of 48x32 cm^2

with square pad of 1 cm^2 are assembled in a same gas volume. With 1 mm between each module and 2 mm wide inactive borders, dead regions amount to less than 3 % of the chamber area. A first prototype was constructed and tested in beam in 2010. A technical challenge here is the embedding of the front-end electronics on the Bulk PCB inside the chamber while keeping the total thickness below 12 mm.

An example of large area coverage with a single Bulk mesh is the ATLAS muon chamber upgrade project undertaken by the MAMMA collaboration [30]. The increase of luminosity and thus of rate foreseen at super LHC will impose the replacement of the Cathode Strip Chambers (CSC) in the forward ATLAS muon system. Micromegas chambers, thanks to their high rate capability have been chosen for this upgrade and first prototypes of the size of half a CSC have been constructed (wedge-shaped and of roughly 100×50 cm^2, fig. 2 center). When voltage is applied to the mesh, a large energy can be released in the front-end electronics upon discharges. That makes spark protections essential for the stable operation of such large area single mesh detectors. Within the RD51 collaboration, protections using resistive strips have been successfully tested on small prototypes and are now applied on full scale chambers.

Figure 2. From left to right: photograph of a TPC for T2K showing an endplate (0.7×2 m^2) with Micromegas modules and the field cage; 1 m long Bulk mesh with strip readout for Super ATLAS; triple GEM chamber for the TOTEM tracking system (0.2 m^2).

2.3. GEM

Standard GEM foils are fabricated using a double-mask photolithography process. The raw material is usually 50 μm thick kapton with a 5 μm copper cladding on both sides. The hole pattern (50–100 μm diameter at 100–200 μm pitch) is transferred from flexible masks to the top and bottom copper surfaces by photolithography and the kapton is removed by wet etching. The alignment between the top and bottom masks is crucial to insure a good uniformity of the hole geometry across the GEM surface. It becomes however awkward for area larger than 40×40 cm^2. Moreover, the width of the raw material is limited to roughly 60 cm which is another obstacle to fabricate larger area GEMs. These two limitations have recently been overcome with the introduction of a single-mask process and a new technique to splice foils with minimum dead zone at the junction. The key of the process is to use the top copper layer (patterned by photolithography) as a mask to remove the kapton; the patterned kapton then acts as a mask to etch the bottom copper layer [31]. At the moment of writing, the size of single GEM foils produced at the CERN workshop is 70×40 cm^2, limited by the machine dimensions. After the upgrade of the workshop (mentioned in the previous section), the fabrication of GEM foils of 200×50 cm^2 will be possible.

2.4. Large area GEM projects

Single-mask GEMs are foreseen for the upgrade of the tracking stations of the TOTEM experiment at LHC [32]. A prototype based on 66×66 cm^2 foils spliced together demonstrated the possibility to instrument large area, in that case 0.2 m^2 (fig. 2 right). It consists of a stack of three GEMs, each electrode being connected to a voltage divider chain. Triple GEMs are also considered to extend the forward region of the CMS muon spectrometer in the rapidity region above 1.6 [33]. In 2010, a first prototype was fabricated and tested in a high energy particle beam. Of trapezoidal shape, it is 1 m long with a total area of about one third of a m^2. In order to avoid possible damage of the GEM foils by discharges, they are segmented in several sectors so as to reduce the electric energy stored in the foils.

The relative fragility of GEM against discharges can be overcome by segmentation of the foils. It also motivated the development of the so-called Thick-GEM [7] which is a GEM with dimensions (thickness, hole diameter and pitch) scaled up by a factor of ten roughly. Based on relatively cheap and simple production techniques they are robust and suitable for the fabrication of large detectors. Single THGEMs do show high gas gains (10^5) and even higher when stacked (10^7) [34]. They are an alternative to GEM for applications where the requirements on spatial resolution are not too stringent (sub-mm).

3. A multi-channel scalable readout system of MPGD

3.1. Concept of the Scalable Readout System

In parallel to the fabrication of large area MPGD, an important mission of the RD51 collaboration is the development of a portable multi-channel readout system of scalable architecture. This so-called Scalable Readout System (SRS) can be applied to small test systems as well as large LHC like systems and should insure protection of the front-end electronics against discharges. It is also intended for facilitating the access of RD51 groups to electronics and for avoiding parallel developments of very similar systems.

The SRS architecture is composed of a detector specific part (the ASIC), a front-end concentrator card (FEC) which can also perform signal digitization, a Scalable Readout Unit (SRU) for multiplexing data when more than one FEC is used and a PC with DAQ and slow-control software and analysis framework (fig. 3 left). The SRU multiplexes data from up to 36 FEC which is about several tens of thousand channels. Even larger systems can be implemented with multiple SRS.

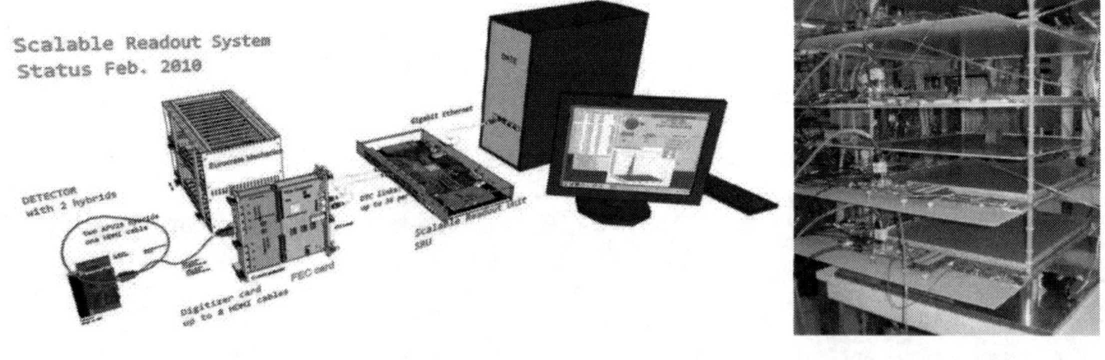

Figure 3. Sketch of a complete Scalable Readout System from front-end chip to computer (left). Photograph of a minimal muon tomography station consisting of GEM chambers of 30×30 cm^2.

3.2. First implementation of the SRS in a GEM detector

The first complete small SRS has become available in fall 2010 and was successfully applied to the readout of GEM tracking stations for muon tomography [35]. The project aims at detecting and imaging heavily shielded high-Z nuclear materials such as enriched uranium by exploiting multiple scattering of cosmic muons [36]. A minimal tomography station based on several Triple GEM chambers of 30×30 cm^2 equipped with two sets of 768 strips running in both X and Y directions was realised. Tests performed over a small active area with preliminary electronics demonstrated the capability to image medium and high-Z targets of small volumes [37].

In a second step, entire Triple GEM chambers (1536 channels/chamber, fig. 3 right) were read out, one by one, with a small SRS. The system consists of 12 front-end hybrid boards based on an analogue amplifier chip, which are connected via HDMI cables to a pair of ADC/FEC electronics cards. A Gb-ethernet link connects the small-system SRS electronics to a Linux PC running DATE on-line data acquisition software developed for the ALICE experiment. The ROOT-based ALICE/AMORE package is used for on-line and off-line data analysis. The full chain from GEM detector to off-line analysis was made operational within one day and enabled to take data continuously during several days. The group is now working on the production of the full SRS electronics so as to equip 10 tracking stations (≈16 thousand channels). This first successful implementation of the SRS is very promising and should benefit to other large area MPGD projects in the future.

4. Photomultipliers based on MPGDs

4.1. Introduction

Gaseous detectors have some unique advantages over vacuum-based and solid-state devices (PMT, MCP and SiPM) for the detection of single UV photons. They can operate in a magnetic field and as far as large area coverage is concerned, they are the most cost effective solution and involve little material budget (see [38] and references therein). Three generations of gaseous PM can be distinguished: the first is based on the addition of a photoconverting vapour (TMAE, TEA) to the detector atmosphere where photons enter through a transparent window. The second generation consists of a wire chamber with CsI coated cathode pads. The most popular photocathode material is CsI. Because of a relatively high work function it is more robust than others used in vacuum-based detectors. The production of CsI photocathodes has been optimised and presently quantum efficiency of 27 % at 170 nm are routinely obtained. Maintaining the efficiency to its original value, however, is difficult above a collected charge of some mC/mm^2. This ageing phenomenon is caused by the impact of avalanche ions and photons onto the cathode surface and prevents operation at high gas gain and/or rate. The new generation of gaseous PMs combine MPGD with photosensitive coatings to reduce feedback mechanisms and improve the rate capability. Their functioning is illustrated with two leading projects of the RD51 collaboration: the development of Ring Imaging Cerenkov and cryogenic detectors.

4.2. Ring imaging Cerenkov detectors

A typical arrangement of GEM-based photomultiplier taken from [14] is showed in Figure 1 (right). CsI is evaporated on the top metal layer of the first foil. With optimised field settings and gas composition photoelectrons are efficiently transferred to the first GEM holes where they are multiplied. Multi-staged structures have the advantage that the photocathode is fully concealed from avalanche photons. Distribution of the gain over the various GEMs allows very large total gains of 10^6–10^7 to be achieved [34]. Moreover, most of the ions are produced at the bottom GEM and have a high probability to be captured on some electrodes before reaching the photocathode. The exact ion backflow fraction is a complex function of the fields and foil geometry. Typical values are of the order of some 10^{-3} at a total gain of 10^4 and decrease further at higher gains.

Photosensitive GEM stacks can operate in pure CF_4 as a window-less Cerenkov detector; the gas acting both as a radiator and a charge amplification medium. A conductive mesh placed above the stack and set at the same voltage as that of the top electrode collects the primary charge from charged particles traversing the gas. As a result, only particles emitting Cerenkov radiation in the gas atmosphere are detected. These so-called Hadron Blind Detectors (HBD) find application for instance in the PHENIX experiment at RHIC where they are used at identify low momentum electrons in a high density hadron environment [39].

Cascaded GEM with CsI coatings are also considered for extending the particle identification capability of the ALICE RICH to higher momenta. Several optimisation studies are being carried out in view of this application [40]. For instance: the use of THGEM to achieve very high gas gain, the introduction of new foil geometry with ion-trapping electrodes ([10, 41]) to reduce the IBF below 10^{-3} and the measurement of photoelectron extraction efficiency versus electric field in various gas mixtures to maximize the single photon sensitivity.

4.3. Cryogenic application in double phase TPC

Liquefied noble gas filled TPCs have been proposed for the direct detection of dark matter. This technique makes use of both charge and scintillation signals produced by interacting particles in the liquid (fig. 4 left). CsI-coated multi-stage MPGD could be applied as the readout of double-phase TPCs with or without separation of the two phases by a window. In both cases, cryogenic temperatures result in high molecular density which can possibly lower the photoelectron extraction efficiency. Moreover, the window-less design implies that the amplifying structures are standing in a quencher-free atmosphere of a high-purity noble gas where high gains are difficult to achieve. The operation of MPGD at low temperature is thus a real challenge undertaken by several groups who study the performance of GEM-based prototypes in various noble gases at low temperature or high pressure [42, 43]. A successful example of this research is the detection, a few years ago, of both primary scintillation and ionisation signals produced by β-electrons in liquid argon in a triple GEM chamber [44]. The experimental set-up and scintillation and charge signals are shown in fig. 4 center and right respectively.

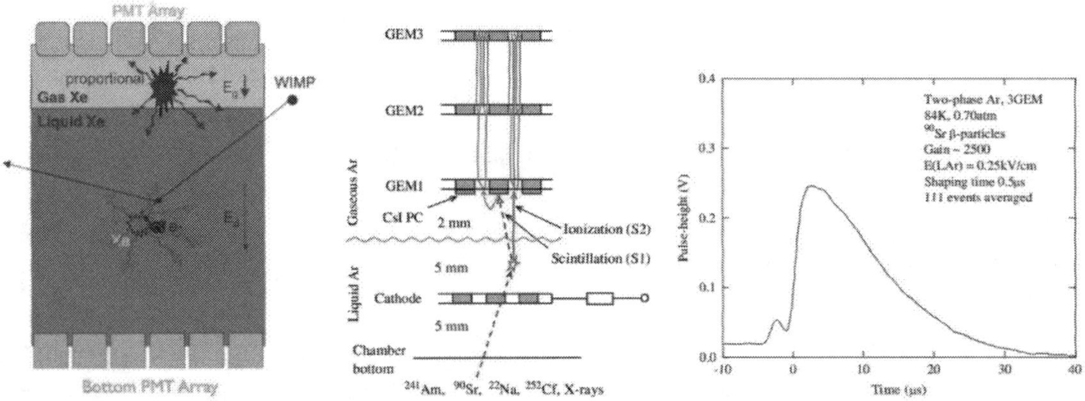

Figure 4. From left to right: concept of the detection of WIMP with double phase TPC; schematic view of an experimental set-up to study the performance of a small two-phase argon detector with CsI coating; top GEM signals (averaged over 111 events) induced by ^{90}Sr β-electrons, the scintillation signal is the small peak, prior to the higher ionisation signal. The center and right pictures are taken from [44].

5. Conclusion

Some fifteen years after their invention, Micro Pattern Gaseous Detectors have become inescapable in the landscape of particle detection. Precise, fast and versatile, they are the subject of constant research to understand their basic functioning and to reinforce and extend their application in various fields of science and industry. The RD51 collaboration is coordinating world-wide efforts in this direction such as the development of large area detector fabrication techniques and of a common multi-channel scalable readout system. Besides this, the capability of MPGD to detect UV photons is a topic of constant interest. It has been underlined in these proceedings as it paves the way for the search of rare events with MPGDs.

References

[1] A. Oed *Nucl. Instr. Meth. and Meth.* A **471** (2001) 109.
[2] A. Oed *Nucl. Instr. Meth. and Meth.* A **263** (1988) 351.
[3] F. Sauli, A. Sharma, *Annu. Rev. Nucl. Part. Sci.* **49** (1999) 341.
[4] Y. Giomataris et al. *Nucl. Instr. Meth. and Meth.* A **376** (1996) 29.
[5] F. Sauli *Nucl. Instr. Meth. and Meth.* A **386** (1997) 531.
[6] G. Charpak et al. *Nucl. Instr. Meth. and Meth.* A **412** (1998) 47.
[7] R. Chechick et al. *Nucl. Instr. Meth. and Meth.* A **535** (2004) 303.
[8] G. Agòcs et al. *JINST* **3** (2008) 02012.
[9] V. Peskov et al. *Nucl. Instr. Meth. and Meth.* A *In Press*.
[10] A. Breskin et al. *Nucl. Instr. Meth. and Meth.* A **553** (2005) 46.
[11] H. van der Graaf *Nucl. Instr. Meth. and Meth.* A **580** (2007) 1023.
[12] S. Bachmann et al. *Nucl. Instr. Meth. and Meth.* A **471** (2001) 115.
[13] P. Fonte, V. Peskov and B.D. Ramsey *Nucl. Instr. Meth. and Meth.* A **419** (1998) 405.
[14] A. Bondar et al. *Nucl. Instr. Meth. and Meth.* A **496** (2003) 325.
[15] J. Derré and I. Giomataris *Nucl. Instr. Meth. and Meth.* A **477** (2002) 23.
[16] B. Ketzer et al. *Nucl. Instr. Meth. and Meth.* A **535** (2004) 314.
[17] J. Derré et al. *Nucl. Instr. Meth. and Meth.* A **449** (2000) 314.
[18] G. Bencivenni et al. *Nucl. Instr. Meth. and Meth.* A **478** (2002) 245.
[19] A. Delbart et al. *Nucl. Instr. Meth. and Meth.* A **461** (2001) 84.
[20] T. Dafni et al. *Nucl. Instr. Meth. and Meth.* A **608** (2009) 259.
[21] I. Giomataris, P. Colas and V. Lepeltier *Nucl. Instr. Meth. and Meth.* A **535** (2004) 226.
[22] D. Mörmann et al. *Nucl. Instr. Meth. and Meth.* A **478** (2002) 230.
[23] Collaboration webpage: http://rd51-public.web.cern.ch/RD51-Public/
[24] S. Duarte Pinto, *Proc. of the XLVII Int. Winter Meeting on Nucl. Phys.*, Bormio (2009) 26.
[25] D. Akimov *Nucl. Instr. Meth. and Meth.* A **628** (2011) 50.
[26] F.T. Avignone, S.R. Elliott and J. Engel *Rev. Mod. Phys.* A **80** (2008) 481.
[27] I. Giomataris et al. *Nucl. Instr. Meth. and Meth.* A **560** (2006) 405.
[28] A. Delbart *Nucl. Instr. Meth. and Meth.* A **623** (2010) 105.
[29] C. Adloff et al. *JINST* A **4** (2009) 11023.
[30] T. Alexopoulos et al. *Nucl. Instr. Meth. and Meth.* A **617** (2010) 161.
[31] M. Villa et al. *Nucl. Instr. Meth. and Meth.* A **628** (2011) 182.
[32] G. Anelli et al. *JINST* A **3** (2008) 08007.
[33] D. Abbaneo et al. RD51 internal note RD51-Note-2010-005.
[34] A. Breskin et al. *Nucl. Instr. Meth. and Meth.* A **623** (2010) 132.
[35] L.J. Schultz et al. *Nucl. Instr. Meth. and Meth.* A **519** (2004) 687.
[36] K. Gnanvo et al. RD51 internal note RD51-Note-2010-004.
[37] K. Gnanvo et al., Submitted for publication in *Nucl. Instr. Meth. and Meth.* A, arXiv:1007:0256.
[38] S. Dalla Torre *Nucl. Instr. Meth. and Meth.* A *In Press*
[39] A. Kozlov et al. *Nucl. Instr. Meth. and Meth.* A **523** (2004) 345.
[40] M. Alexeev et al. *Nucl. Instr. Meth. and Meth.* A **623** (2010) 129.
[41] J.F.C.A. Veloso et al. *Nucl. Instr. Meth. and Meth.* A **581** (2007) 261.
[42] A. Bondar et al. *Nucl. Instr. Meth. and Meth.* A **556** (2006) 273.
[43] A. Breskin et al. *Nucl. Instr. Meth. and Meth.* A *In Press*
[44] A. Bondar et al. *Nucl. Instr. Meth. and Meth.* A **598** (2009) 121.

Journal of Physics: Conference Series **309** (2011) 012018 doi:10.1088/1742-6596/309/1/012018

The T2K Experiment

Kate Scholberg, for the T2K collaboration
Department of Physics, Duke University, Durham, NC, 27708
E-mail: `schol@phy.duke.edu`

Abstract. This talk will describe briefly the physics goals of the Tokai to Kamioka (T2K) experiment, and its current status and prospects.

1. The Quest for θ_{13}

In the current standard three-flavor picture of neutrino oscillations, the neutrino flavor states are related to the mass eigenstates by a 3×3 unitary mixing matrix, the Maki-Nakagawa-Sakata (MNS) matrix. A consequence of this framework is that neutrinos oscillate in flavor as they propagate, with transition probabilities dependent on the differences between the squares of the masses of the states (Δm^2_{ij}), the MNS matrix elements, and the energy of the neutrino. The MNS matrix can be parameterized using three mixing angles, θ_{12}, θ_{23} and θ_{13}, and a CP-violating phase δ (and two Majorana phases, unobservable in oscillation experiments). The first two mixing angles have been measured by solar/reactor and atmospheric/beam oscillation measurements, respectively; the third, θ_{13}, is known to be small from the CHOOZ reactor experiment results[1], which are consistent with atmospheric neutrino oscillation[2] and beam[3, 4] studies. The two mass-squared differences are also known; however the pattern of masses (as well as the absolute mass scale) are unknown. The CP phase is also unknown. The next step for oscillation physics is the measurement of the unknown mixing angle θ_{13}.

Globally, there are two experimental strategies for determining θ_{13}. First, one may look for disappearance of few-MeV reactor $\bar{\nu}_e$ on a ~km distance scale: several experimental programs are taking this approach[6, 7, 9]. Alternatively, one may look for an appearance signal of ν_e in a GeV ν_μ beam on a few-hundred km distance scale. The first-generation long-baseline beam experiments, K2K[3] and MINOS[4, 5], have searched for non-zero θ_{13}; to go further requires a higher intensity beam. The next experiments to tackle ν_e appearance are T2K, NOνA[10], and LBNE[8].

2. The Long Baseline Beam Approach

The ν_e appearance probability is approximated by $P(\nu_\mu \to \nu_e) \sim \sin^2 2\theta_{13} \sin^2 \theta_{23} \sin^2 \left(\frac{\Delta m^2_{23} L}{4E_\nu}\right)$, where E_ν is the energy of the neutrino; this expression holds in vacuum for $\Delta m^2_{23} >> \Delta m^2_{12}$, and $E_\nu \sim L\Delta m^2_{23}$, $\delta = 0$. The oscillation is driven by the "atmospheric" Δm^2; from the known limit on θ_{13}, the appearance amplitude cannot be more than about 7%. Therefore good statistics and a clean sample are both needed to observe an appearance signal of non-zero θ_{13}. For a baseline of ~300 km, the first oscillation maximum is at around 600 MeV.

To make a ~GeV neutrino beam, one accelerates protons, collides them with a target, and focuses the produced mesons forward with magnetic horns; the mesons (primarily pions) are then allowed to decay in a long pipe via $\pi \rightarrow \mu + \nu_\mu$. The neutrino beam is sampled at a near detector before propagating to a far detector. Two on-axis beam experiments, K2K and MINOS, have already observed long-baseline ν_μ disappearance oscillations[12, 13]. Two next-generation long baseline oscillation experiments, T2K and NOνA, will improve their sensitivity to oscillations via clever configuration of beam and detector: they will site their detectors slightly off beam axis. According to two-body decay kinematics, at locations few degrees off beam-axis, neutrino energy becomes relatively independent of pion energy. The neutrino spectrum is then more sharply peaked, which allows enhanced flux at the oscillation maximum and reduction of backgrounds from off-peak tails of the spectrum.

3. The T2K Experiment

The T2K experiment[11] is the first off-axis long-baseline neutrino oscillation experiment, employing a high-intensity beam from the J-PARC facility in Tokai (designed eventually to achieve 750 kW beam power) sent to the 22.5 kton fiducial mass Super-Kamiokande (Super-K) water Cherenkov detector 295 km away in Kamioka. The experiment also includes a number of components deployed near the neutrino source for beam characterization: a muon monitor, an on-axis near detector (the INGRID), and an off-axis suite of detectors (the ND280).

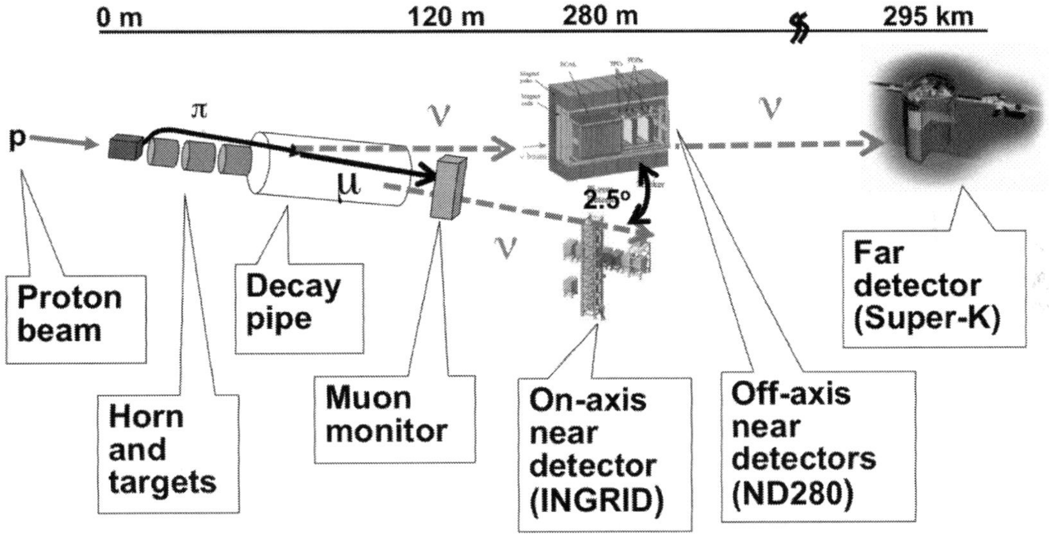

Figure 1. Components of the T2K experiment.

3.1. Neutrino Beamline

Protons for T2K are extracted from the J-PARC 30 GeV main ring proton synchrotron. The neutrino beam target station includes a graphite target and three magnetic horns for parent meson focusing. A muon detector monitors muon flux at 120 m from the target. For the first running period, there were 6 bunches per spill, with spills every 3.5 seconds; the rate has been increased to 8 bunches per 5 μs spill, and 3.2 seconds between spills. Spills are matched to events at Super-K using GPS timing. The beam energy spectrum peaks at about 600 MeV, which is optimized for oscillation sensitivity.

3.2. Near Detectors
The on-axis detector, the INGRID, comprises an iron target and scintillator active elements, and is used for beam profile monitoring; the first run's results indicated good beam alignment.

The off-axis detector complex at 280 m employs fine-grained trackers inside a 0.2 T magnetic field (provided by the relocated UA1/NOMAD magnet). The off-axis detectors are used for flux normalization, background characterization for oscillation physics, and cross-section studies. Three time projection chambers instrumented with Micromegas are interleaved with two mixed plastic-scintillator/water fine-grained detectors read out by multi-pixel photon counters. (The TPCs were described in more detail in a separate contribution to this symposium.) These tracking detectors are designed to determine ν_e and ν_μ flux and spectrum for extrapolation to Super-K. A "P0D" module and downstream and barrel electromagnetic calorimeters are also installed. The P0D consists of water target and scintillator planes and is designed for measurement of neutral current π^0 production. Additionally the magnet yoke is instrumented with scintillator detectors. Installation of all near detector components was complete by summer 2010. Please see references [14, 15] for more detailed description of the near detectors and their capabilities.

3.3. Far Detector
The far detector, Super-K[16], has been running as "Super-K-IV" since 2008, with a full complement of 11,000 photomultiplier tubes and completely refurbished electronics. In Super-K, the signature of ν_e appearance is an excess of charged-current quasi-elastic events, $\nu_e + n \to p + e^-$, in the expected energy range; these interactions manifest themselves in a water Cherenkov detector as single fuzzy rings (electron showers appear fuzzy because electrons scatter and shower, in contrast to ν_μ-induced muon rings which appear sharp). Understanding of backgrounds is critical to the measurement: dominant backgrounds are intrinsic beam ν_e contamination, misidentified μ-like events, and neutral-current π^0 events for which the double gammas from $\pi^0 \to \gamma\gamma$ are not distinguished from single electron showers.

4. First Data
First beam arrived in 2009, and production running began in early 2010. Over the first running period, through June 2010, stable operation at 50 kW beam power was achieved, and a total of 33×10^{18} protons on target (pot) were delivered (see Fig. 2). The second running period started in November 2010; at the time of this conference contribution, 114 kW had been achieved. The aim is for 150 kW $\times10^7$ seconds by mid-2011.

The first long-baseline T2K beam event observed in Super-K occurred in February 2010. Beam neutrinos at SK are divided into several subsamples: fully-contained events have light in SK's inner detector (ID) only; outer detector (OD) events have light in SK's outer veto (they can be entering, exiting, or contained in the OD); low energy fully-contained events are selected to search for neutral current nuclear excitation events. During the first running period, 33 fully-contained events were observed in the beam window. The time structure of the observed events is consistent with the beam bunch structure: see Fig. 3. The updated data sample for the second running period has an SK event rate per pot consistent with the first running period.

5. Status and Prospects
T2K construction is complete, and near and far detectors are running well.

The eventual aim of the T2K experiment, for running of 750 kW times (5×10^7 s), is for a sensitivity to $\sin^2 2\theta_{13}$ between $2 - 15 \times 10^{-3}$ (depending on δ and the mass hierarchy) at the current best measured value of Δm^2_{23}. See Fig. 4.

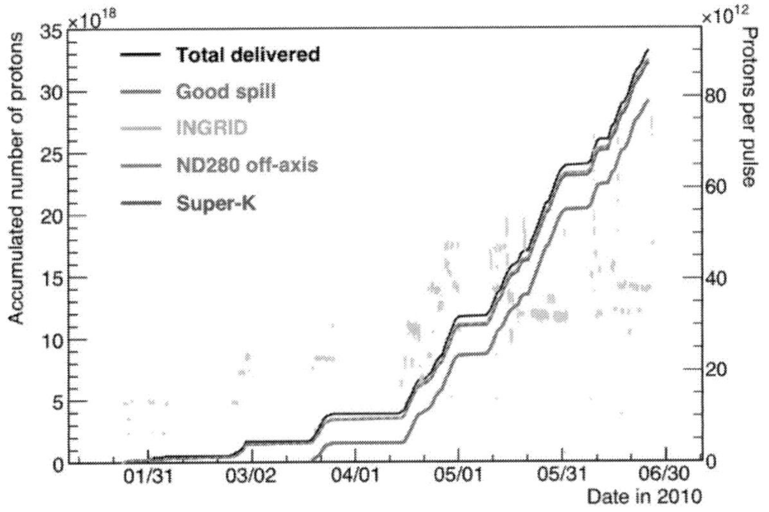

Figure 2. Delivered protons on target for the first T2K running period.

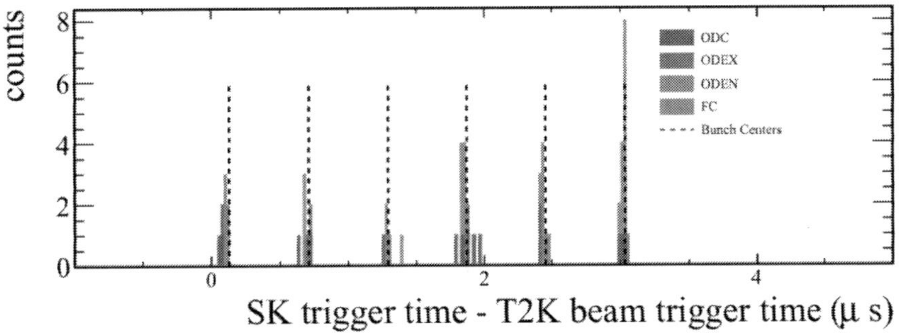

Figure 3. Time difference between the Super-K event trigger and the T2K beam trigger for the first T2K running period for beam events passing neutrino selection cuts. The plot includes different categories of events with light in the OD as well as FC events.

References

[1] M. Apollonio *et al.* [CHOOZ Collaboration], Phys. Lett. **B466**, 415-430 (1999). [hep-ex/9907037].
[2] R. Wendell *et al.* [Super-Kamiokande Collaboration], Phys. Rev. **D81**, 092004 (2010). [arXiv:1002.3471 [hep-ex]].
[3] S. Yamamoto *et al.* [K2K Collaboration], Phys. Rev. Lett. **96**, 181801 (2006) [arXiv:hep-ex/0603004].
[4] P. Adamson *et al.* [MINOS Collaboration], Phys. Rev. Lett. **103**, 261802 (2009) [arXiv:0909.4996 [hep-ex]].
[5] P. Adamson *et al.* [The MINOS Collaboration], Phys. Rev. D **82**, 051102 (2010) [arXiv:1006.0996 [hep-ex]].
[6] F. Ardellier *et al.* [Double Chooz Collaboration], [hep-ex/0606025].
[7] W. Wang [Daya Bay Collaboration], AIP Conf. Proc. **1222**, 494-497 (2010). [arXiv:0910.4605 [hep-ex]].
[8] http://lbne.fnal.gov/index.shtml
[9] J. K. Ahn *et al.* [RENO Collaboration], [arXiv:1003.1391 [hep-ex]].
[10] R. E. Ray, J. Phys. Conf. Ser. **136**, 022019 (2008).
[11] Y. Itow *et al.* [hep-ex/0106019].
[12] M. H. Ahn *et al.* [K2K Collaboration], Phys. Rev. **D74**, 072003 (2006). [hep-ex/0606032].
[13] P. Adamson *et al.* [MINOS Collaboration], Phys. Rev. Lett. **101**, 131802 (2008) [arXiv:0806.2237 [hep-ex]].
[14] http://www.nd280.org/info

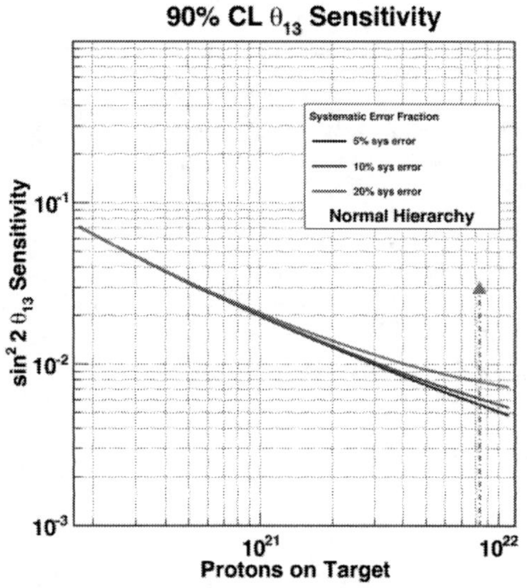

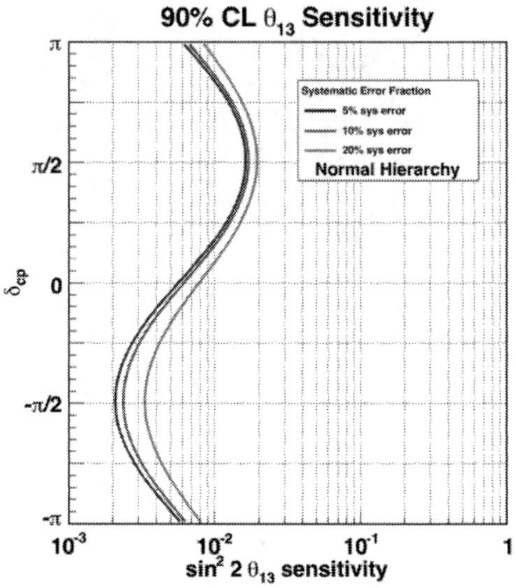

Figure 4. Left: T2K sensitivity to θ_{13} at the 90% confidence level as a function of protons on target. 5%, 10% and 20% systematic error fractions are plotted. The dashed arrow indicates a 5-year run. The following oscillation parameters are assumed: $\sin^2 2\theta_{12} = 0.8704$, $\sin^2 2\theta_{23} = 1.0$, $\Delta m^2_{12} = 7.6 \times 10^{-5}\text{eV}^2$, $\Delta m^2_{23} = 2.4 \times 10^{-3}\text{eV}^2$, $\delta_{CP} = 0$, normal hierarchy.
Right: T2K sensitivity to θ_{13} at the 90% confidence level as a function of δ_{CP}. Beam is assumed to be running at 750 kW for 5 years, using the 22.5 kton fiducial volume SK detector. 5%, 10% and 20% systematic error fractions are plotted. The following oscillation parameters are assumed: $\sin^2 2\theta_{12} = 0.8704$, $\sin^2 2\theta_{23} = 1.0$, $\Delta m^2_{12} = 7.6 \times 10^{-5}\text{eV}^2$, $\Delta m^2_{23} = 2.4 \times 10^{-3}\text{eV}^2$, normal hierarchy.

[15] C. Hearty and E. Mazzucato [T2K/TPC Collaboration], J. Phys. Conf. Ser. **136**, 042036 (2008).
[16] Y. Fukuda *et al.*, Nucl. Instrum. Meth. A **501**, 418 (2003).

The T2K Time Projection Chambers: first year of operation

Edoardo Mazzucato, for the T2K/TPC group
CEA Saclay, Irfu-SPP, 91191 Gif-sur-Yvette, France
E-mail: edoardo.mazzucato@cea.fr

Abstract. After a 3-year construction period and a commissioning phase at the end of 2009, the T2K Time Projection Chambers (TPCs) have been operated successfully during the first physics run that took place in 2010 at the J-PARC neutrino facility. This talk presents the main performance achieved by the T2K TPCs in the course of their first year of operation.

1. Introduction

The T2K Time Projection Chambers (TPCs), as part of the Near Detector (ND280) Tracker, are key elements of the experiment. The ND280 Tracker plays indeed an important role in the characterization of the neutrino beam before oscillation. Since a sizeable fraction ($\sim 40\,\%$) of neutrino interactions at the 700 MeV peak energy are Charged Current Quasi Elastic ones ($\nu_\mu + n \rightarrow p + \mu^-$), the ND280 Tracker provides precise information about the energy spectrum of the neutrino beam through the measurement of the charged lepton momentum vector. The particle identification capability of the Tracker allows also the ν_e contamination in the beam to be determined giving thus essential information on the irreducible background for the $\nu_\mu \rightarrow \nu_e$ search[1].

The required performance of the T2K TPCs are a dE/dx resolution better than 10 % for a good $e - \mu$ separation, a relative momentum resolution better than 10 % at 1 GeV/c and an absolute momentum scale known to better than 2 %. The latter requirement is particularly important for the Δm^2_{23} parameter measurement in the ν_μ disappearance channel.

The following sections give a brief description of the T2K TPCS and present their main performance. More detailed information can be found elswehere[2].

2. The T2K Time Projection Chambers

The ND280 Tracker is made of three TPCs interleaved with two scintillator-based Fine Grained Detectors which are used as active targets. It is surrounded by the former UA1/Nomad magnet which produces a horizontal magnetic field of 0.2 T. Each TPC has a rectangular shape and is made of an inner drift volume surrounded by an external box connected to ground (Fig. 1). The inner box is divided into two parts by a central cathode maintained at a voltage of -25 kV. Each end of the inner box contains an ensemble of twelve Micro Pattern Gaseous Detectors (MPGDs) based on the bulk micromegas technology[3]. This ensemble of detectors forms a plane parallel to the cathode. The drift E-field, along the magnetic field direction, is shaped by copper strips located on the inner and outer faces of the inner box. The active tracking length in each TPC is

Journal of Physics: Conference Series **309** (2011) 012019 doi:10.1088/1742-6596/309/1/012019

720 mm along the beam direction while the maximum drift distance on each side of the cathode is 897 mm. In order to minimize E-field distortions, the central cathode was built with a flatness to about 0.1 mm, and the frame that houses the MPGDs, with a planarity to about 0.2 mm.

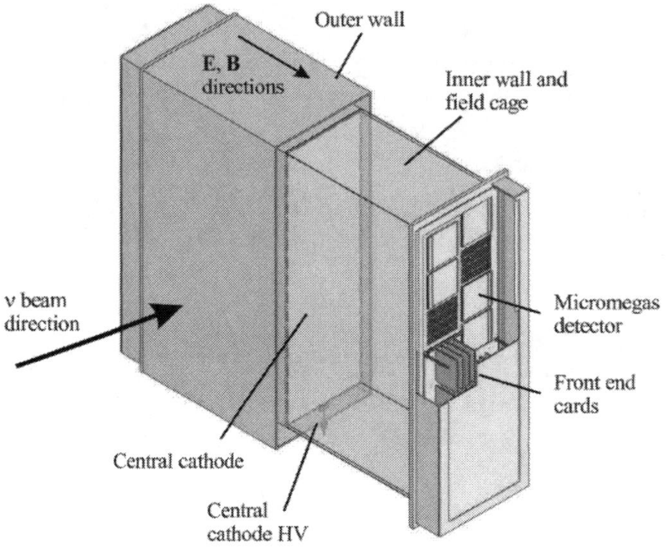

Figure 1. Schematic drawing of a T2K TPC.

The drift and amplification regions are filled with an Ar(95%)/CF_4(3%)/iC_4H_{10}(2%) gas mixture while the gap between the inner and outer volumes is filled with CO_2 for electrical insulation purposes. Spacers behind the micromegas detector planes are used for services: cooling, readout electronics, high-voltage (HV) and low-voltage (LV) connections, temperature probes and calibration system. The outer dimensions of each TPC are approximately 2.3 m × 2.4 m × 1 m.

2.1. Bulk micromegas detectors

The MPGDs which are used as charge amplification devices are based on the bulk micromegas technology. This technique allows large detection areas to be built and combines detector robustness and operation reliability. Each of the 72 detectors or modules instrumenting the three TPCs has dimensions of $36 \times 35\,\mathrm{cm}^2$ and contains 1726 active pads of $9.7 \times 6.9\,\mathrm{mm}^2$ area. The amplification gap between the conductive mesh and the grounded anode pad plane is 128 μm.

The choice of Ar(95%)/CF_4(3%)/iC_4H_{10}(2%) as gaseous medium for the TPCs allows sufficiently high gains to be obtained at low operation voltage ($\sim$1500 at -350V). This gas mixture has a small transverse diffusion coefficient and the maximum drift velocity for electrons is 7.8 cm/μs, almost saturated at the operation drift field of 279 V/cm. Moreover, electron attachment in the gas is small and good space-point resolution can be obtained with pads of several mm^2 area.

Given the rather large size of one module, a uniform response at the level of a few percent over the entire detection surface is desirable. In the case of the T2K TPCs, the rms value for the dispersion of the collected charge on a single module is typically 3%. The energy resolution at 5.9 keV, measured with a ^{55}Fe source, is about 8%, almost independent on the illumination point. Over the 72 modules instrumenting the three TPCs, the rms values for the dispersions of the average gain and resolution per module are 8% and 3%, respectively.

The TPCs are operated with a drift-to-amplification E-field ratio of about 100 allowing maximum electron transparency between the drift and amplification regions to be reached. Finally, cross-talk between neighbouring pads is typically 1%, mainly due to parasitic a capacitance from the routing strips inside the dectector printed circuit board. A picture of a bulk micromegas detector mounted on its stiffening frame is shown in Fig. 2.

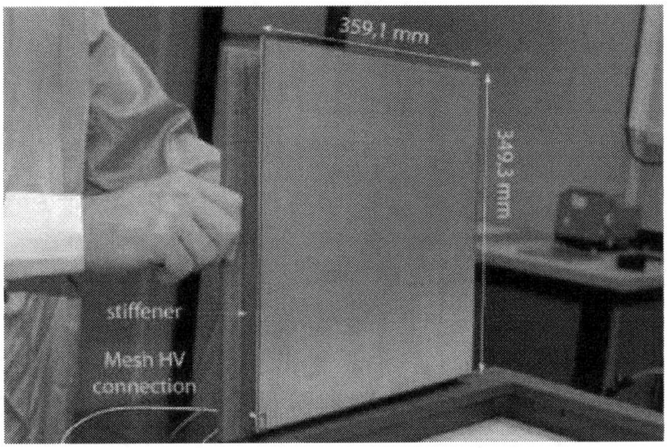

Figure 2. A bulk micromegas detector for the T2K TPC.

2.2. Readout electronics

The readout electronics has to cope with a total of 124,416 channels for the three TPCs. The Front End Electronics (FEE)[4] has a modular structure with six Front End Cards (FECs) and one Front End Mezzanine (FEM) mounted at the back of each micromegas detector. The readout system is based on an ASIC chip called AFTER[5]. Each AFTER chip reads 72 channels and has 511 analog memory cells provided by a switched capacitor array. The AFTER ASIC combines important features like programmable maximum charge in the 120 to 600 fC ranges, signal peaking time from 0.1 to 2 μs and a sampling frequency up to 100 MHz.

Each FEC houses four AFTER chips and their associated ADCs. The FEM board is used to collect the data from the six FECs. It also performs zero-suppression as well as slow-control functions. The data from the 72 FEM boards are collected by 18 Data Concentrator Cards[6] and are then sent to the data acquisition system (DAQ). A picture of TPC modules, instrumented with their FEE, associated shielding and cooling system, is shown in Fig. 3.

2.3. Calibration system

Two small micromegas chambers are used in the experiment to monitor the supply and return gas of the TPCs. Each of these monitoring chambers is operated with one ^{55}Fe source to control the gas gain, and with two ^{90}Sr sources, positioned at precisely known locations inside the chambers, to measure the drift velocity. In addition, a laser calibration system that is continuously running during data taking periods allows E-field and B-field distortions, detector gain variations as well as electron drift velocity to be studied. This calibration system consists of well-defined patterns of thin aluminum discs and strips glued on the cathode, illuminated by an external UV 266 nm (ND-YAG) laser through a set of optical fibers.

Figure 3. Micromegas detectors instrumented with their Front End Electronics, associated shielding and cooling system.

3. TPC operation

All 3 TPCS of the ND280 Tracker were operational during the first T2K physics run which took place between January and June 2010. The ND280 magnet provided a B-field intensity of 0.188 T. The drift E-field inside the TPCs was set to 279 V/cm and the bulk micromegas modules were operated at the nominal mesh voltage of -350 V corresponding to a gas gain of about 1500. The latter was found to be stable within 1 % after correcting for temperature and atmospheric pressure variations which accounted for the largest part of the gain changes. At this gas gain, the average spark rate per module was measured to be below 0.1 per hour, inducing thus negligible detection dead time.

The stability of the gas mixture was found to be better than 0.01 % for each of the three gas components while impurities of O_2, H_2O and CO_2 were measured to be below 2 ppm, 5 ppm and 120 ppm, respectively. By looking at charge variations of track clusters as a function of drift distance, no evidence for electron attachment was found. The micromegas detectors gave stable and uniform responses except for two modules which showed significantly lower average charge collection due to faulty HV filters.

The FEE electronics were operated with the following parameters of the AFTER chip: a charge measurement range from 0 to 120 fC, a peaking time of 200 ns and a sampling frequency of 25 MHz. The electronic noise was measured to be about 800 e^- corresponding to a signal to noise ratio of about 100 for minimum ionizing particles. The total power dissipation in the FEE and the LV cables amounted to 2.8 kW for the three TPCs. The temperature at the level of the FEE and near the gas input to the TPC inner volumes was found to be stable within the 24-26 °C and 20-22 °C ranges, respectively.

After a zero-suppression by a factor of about 1900, the average size of beam and cosmic-ray events was typically 60 kB. Pedestal and laser events were continuously recorded during data taking for noise and calibration purposes. The TPC to DAQ throughput at the nominal 20 Hz trigger rate was below 2 MB/s.

An event display of a neutrino interaction in the ND280 Tracker is shown in Fig. 4.

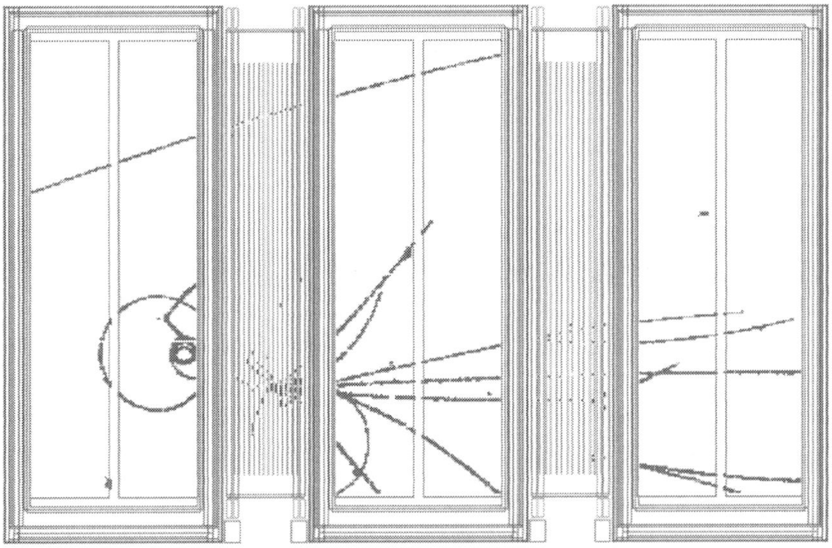

Figure 4. Example of an event reconstructed in the ND280 Tracker. The track crossing the first two TPCs (top left) is due to a neutrino interaction in front of the Tracker. A second more complex neutrino interaction (deep inelastic scattering) occurs in the first Fine Grained Detector with production of charged particles detected in all three TPCs.

4. TPC performance

4.1. Spatial resolution

Space point resolution has been investigated for tracks reconstructed inside each single TPC. A track is defined by an ensemble of charge clusters formed by neighboring pads, within a column for roughly horizontal tracks, or within a row for vertical ones. The track parameters and the ionization width are then determined by maximizing the likelihood of the observed charge sharing between pads. A typical value of $247\,\mu\text{m}/\sqrt{\text{cm}}$ for the transverse diffusion coefficient was obtained for B=0.188 T, to be compared with $288\,\mu\text{m}/\sqrt{\text{cm}}$ when the magnet was off. These values are known, however, to be underestimated by about 10%[7].

To determine the space-point resolution, the transverse coordinate from the track fit is compared with the result from a single cluster fit. For the single cluster fit, all other track parameters were fixed. Fig. 5 shows the variation of the spatial resolution as a function of drift distance. For drift lengths above 150 mm, the resolution is below 0.8 mm but increases above 1 mm at short distances. Such a behaviour is mainly due to contributions from 1-pad clusters which are more likely to be present on a track located near the anode plane.

The variation of the spatial resolution as a function of tan(ϕ), where ϕ is the track angle in the vertical plane, is presented in Fig. 6. A strong dependence on angle due to the ionization fluctuations along the track is observed owing to the fact that the variance of charge sharing in a cluster increases when tracks have an angle with respect to pad boundaries. This variation is well reproduced by a Monte Carlo simulation that incorporates most of the important detector effects, including tranverse and longitudinal diffusion as well as a parametrization of the electronics response.

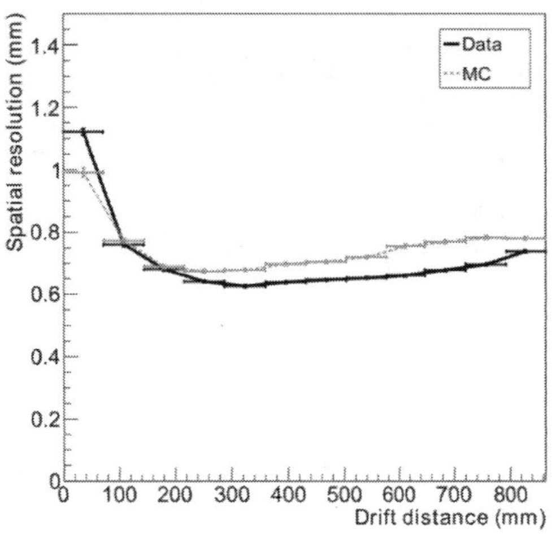

Figure 5. Spatial resolution as a function of drift distance. The rise at low drift distances is due to 1-pad cluster contributions.

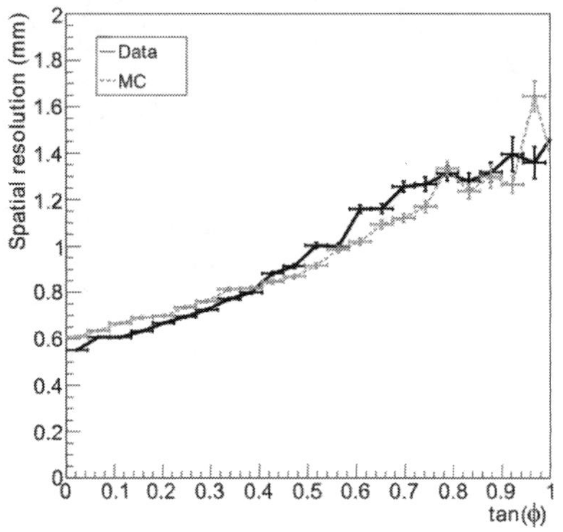

Figure 6. Spatial resolution as a function of tan(ϕ) where ϕ is the track angle in the vertical plane.

4.2. Field distortions

Distortions in the electron drift due to non-homogeneous and misaligned electric and magnetic fields can affect the quality of the reconstructed track parameters. Such effects can be precisely determined using laser calibration data.

E-field distortions are obtained when the ND280 magnet is turned off by comparing the measured position offsets of the aluminum discs on the cathode with respect to survey. Although work is still in progress, present measurements indicate that the displacement rms values do not exceed the space-point resolution.

B-field distortions have been investigated using laser calibration data collected with and without magnetic field. Offsets are found to be typically below 1 mm but can reach as much as 5 mm in some regions of the most downstream TPC where field non-homogeneities are stronger.

4.3. Momentum resolution

The track momentum resolution obtained for a single TPC was estimated with a Monte Carlo sample of muons obtained from simulated neutrino interactions. Fig. 7 shows the relative resolution of the transverse momentum component with respect to the B-field direction. The result from the simulation indicates that for transverse momenta above 400 MeV/c, the momentum resolution goal of 0.1 p_T [GeV/c] is achieved.

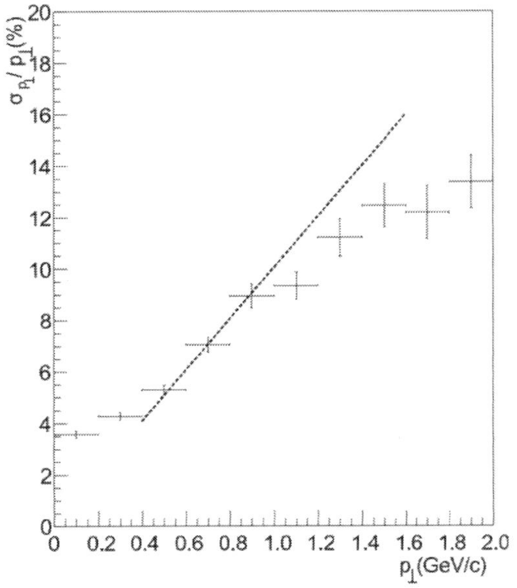

Figure 7. Relative transverse momentum resolution for a single TPC as a function of transverse momentum, perpendicular to the magnetic field. The dashed line represents the momentum resolution goal.

4.4. Particle Identification

Particle identification (PID) in the TPCs is performed through dE/dx measurement. To determine the energy loss, a truncated mean method is used. For each cluster belonging to a track, the linear charge density is estimated after taking into account corrections for temperature and pressure variations. The mean energy loss (C_T) is then computed by keeping only 70% of the clusters corresponding to the lowest charge density values. This choice allows fluctuations from the high-energy loss tails to be minimized, improving thus the resolution on the dE/dx measurement. For minimum ionizing particles with momenta between 0.4 and 0.5 GeV/c, the relative dE/dx resolution is found to be 7.8 %, better than the design goal of 10 %.

Particle identification is then performed by computing the "pull" variable $\delta_E(i)$ for each particle hypothesis ($i = e, \mu, \pi, K, p$):

$$\delta_E(i) = \frac{C_T - C_E}{\sigma(i)} \tag{1}$$

where C_E is the expected energy loss value and $\sigma(i)$ the associated uncertainty. As an example of the PID capability of the T2K TPCs, a $-1 < \delta_E(e) < 2$ cut applied to a sample of through-going muons with momenta below 1 GeV/c gives only a 0.2 % probability of identifying particles as electrons.

The measured energy loss distributions as a function of particle momentum for through-going muons and neutrino interaction in the ND280 detector are shown in Figs. 8 and 9 for negative and positive tracks, respectively. The data are well reproduced by the expected mean energy loss curves for different particle types. For negatively charged tracks, most tracks are identified as muons or electrons while for positive tracks, protons, pions and positrons are clearly observed.

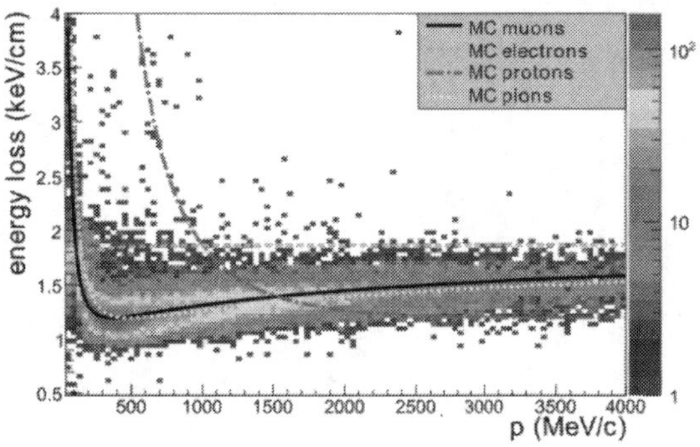

Figure 8. Energy loss distribution as a function of momentum for negative tracks.

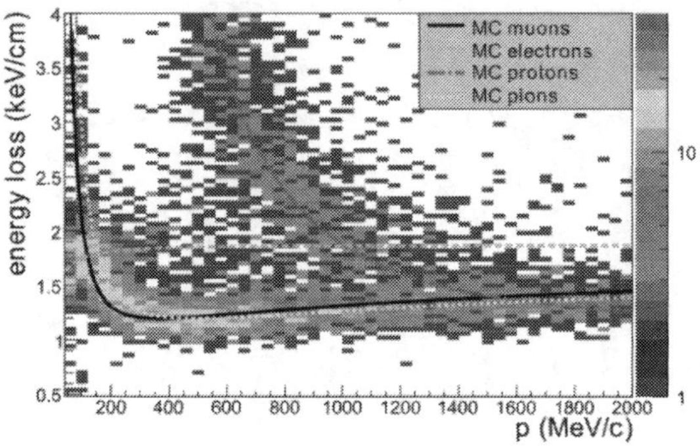

Figure 9. Energy loss distribution as a function of momentum for positive tracks.

5. Conclusion

Since January 2010, the T2K Time Projection Chambers have been successfully taking data. They are performing as expected and first studies indicate that the design requirements are met. More detailed analyses to better understand their performance are in progress. The TPCs have demonstrated to provide an essential contribution to the T2K physics programme.

References

[1] K. Scholberg, "The T2K experiment", contribution to this Symposium.
[2] N. Abgrall et al., T2K ND280 TPC collaboration, to be published in Nucl. Instr. and Meth. A, arxiv:1012.0865.
[3] I. Giomataris et al., Nucl. Instr. and Meth. A560 (2006) 405-408.
[4] P. Baron et al., IEEE Trans. Nucl. Sci. 57 (2009) 406-411.
[5] P. Baron et al., IEEE Trans. Nucl. Sci. 55 (2008) 1744-1752.
[6] D. Calvet et al., Proc. IEEE 17th Real Time Conference, Lisboa, Portugal, 24-28 May 2010.
[7] D. Karlen et al., Nucl. Instr. and Meth. A555 (2005) 80-92.

Search for sterile neutrinos of the νMSM

F Vannucci
Université Paris 7, Paris, France

E-mail: vannucci@in2p3.fr

Abstract. Sterile neutrinos arise in several extensions of the Standard Model to accommodate massive neutrinos. In particular the recently proposed νMSM predicts 3 right-handed neutrinos, one of them could explain the dark matter puzzle.

1. The νMSM model [1]

Singlet (sterile) neutrino states arise in models which try to implement massive (light) neutrinos in extensions of the Standard Model. In particular the recently developed νMSM model considers 3 singlet states N_1, N_2 and N_3 associated with the 3 active neutrinos.

N_1 having a mass around 10 keV has a lifetime very long compared to the age of the Universe. Because of its small mass, it is essentially stable on cosmological times. It could account for the missing mass in the form of Warm Dark Matter.

The other two states, N_2 and N_3, if they are almost degenerate could solve the problem of the matter/antimatter asymmetry in the Universe. Their masses are expected to be in the range 100 MeV-few GeV. N_2 and N_3 can be searched for in the laboratory through their mixings with light neutrinos.

2. Production and decay of N_2 and N_3

If heavy neutrinos exist, they mix with active neutrinos through a unitary transformation. Any neutrino beam will contain a fraction of heavy neutrinos at the level U_{Nl}^2 where U denotes the mixing matrix element between the heavy state N and l, either e or μ or τ.

At accelerators, neutrinos are emitted in pion and kaon decays. At higher energies, charm, beauty and W contribute. Kinematically the mass range allowed for a heavy N depends on the emission process. In πμ decays, sterile neutrinos can reach a mass of 30 MeV. In πe, the range is increased to 130 MeV. Kaons allow larger potential masses of up to 450 MeV.

The flux of N's accompanies the flux of known neutrinos at the level of U_{Nl}^2. Corrections to this straightforward result come from helicity conservation. For massless neutrinos, it suppresses πe decays relative to πμ decays. This is not true anymore for $\pi \to eN$.

N's will decay through weak interactions. The decay modes depend on the N mass. The first channel to open is $N \to ee\nu$, as soon as the mass is greater than 1 MeV. With increasing masses, new modes open and one can obtain eμν, πe, μμν, πμ…

The lifetime is given by the formula applying to weak decays, with a suppression factor coming again from the mixing U_{Nl}^2. Phase space factors are also to be taken into account.

The search consists in looking for a decay signature, typically two charged tracks reconstructing a vertex in an empty volume arising in a neutrino beam.

This has been attempted by the low energy experiment PS191 [2] with 5 10^{18} protons of 19 GeV on target. This put strong limits on the existence of sterile neutrinos in the mass range between 20 and 450 MeV.

For masses of a few MeV, a limit is given by the Borexino experiment in the $N \to ee\nu$ mode [3]. This limit could be improved with data from nuclear reactor experiments.

3. Sterile neutrinos as WDM?

From astrophysical measurements we know that the missing mass of the Universe corresponds to a local density of 300 MeV/cm^3. If this is in the form of 10 keV neutrinos, here called N_1, this means that we are surrounded by a density of 3 10^4 N_1/cm^3. It has been argued that the local density of the dark halo could be much higher.

The solar system moves within this halo at a velocity of 200 km/s. This means that a detector is crossed by an average flux of 6 10^{11} N_1/cm^{2}s. These particles are far from relativistic, they have an average β of 200/300000 = 0.7 10^{-3} and $\gamma = 1$.

The earth travels around the sun at the velocity of 30 km/s. This secondary motion gives a yearly modulation of $\pm$ 15% over periods of 6 months. The flux is maximum in June corresponding to 6.9 10^{11} N_1/cm^{2}s, and minimum in December with a value of 5.1 10^{11} N_1/cm^{2}s. The relative velocity β of the N_1 is also modulated, the particles cross the detector faster in June.

4. Radiative decay of N_1

Neutrinos having a mass of 10 keV can decay radiatively. Their decay mode gives a ν_e together with a monoenergetic photon carrying half of the N_1 mass. The lifetime in vacuum for this process is extremely long:

$$t_0 = 7\ 10^{43}\ (1\text{eV}/m)^5\ 1/U^2\quad (\text{s}) \tag{1}$$

It has been noticed that the radiative decay is very efficiently increased in matter(3):

$$t_0/t_m \sim 9\ 10^{23}\ F(v)(N_e/10^{24}\text{cm}^{-3})^2\ (1\text{eV}/m)^4 \tag{2}$$

This means that a 10 keV neutrino will have a lifetime in matter equal to:

$$t_m = 0.8\ 10^{16}\ 1/U^2\ (\text{s}) \tag{3}$$

5. Mixing of N1 and possible signal

If the interpretation of dark matter as composed of 10 keV sterile neutrinos is correct, it is possible to extract the mixing parameter from cosmological considerations.

Within this scenario, the number of N_1 in the Universe is U^2 times the number of ν_e. Furthermore the sum of all N_1 constitutes 25% of the total content of the Universe. We also know that ordinary matter composed of nucleons constitute 4% of the total content of the Universe. The mass of nucleons being 1 GeV, this means that there are (25/4) 10^5 N1 for 1 nucleon.

We know that the ratio between nucleons and photons in the Universe is 3 10^{-10}. The density of photons is 400/cm^3 while for ν_e it is 100/cm^3. We can then derive the number of N_1 per ν_e. This directly gives $U^2 = 7\ 10^{-4}$.

The problem is then entirely soluble. Let us consider a sensitive detector having a volume of $10\times10\times10$ m^3. It is traversed all the time by a flux of 6 10^{17} N_1/s (slightly modulated over the year). With a 10 m long decay path, the number of expected decays is 4 10^{-3} U_2/s. With the U^2 extracted from cosmology this gives a total of 3 10^{-6} photons of 5 keV or 100 per year.

One can solve the same exercise for 1 keV and 100 keV sterile neutrinos. One finds respectively 1000 and 10 events per year.

Journal of Physics: Conference Series **309** (2011) 012020 doi:10.1088/1742-6596/309/1/012020

6. Conclusion

The νMSM model is a natural and attractive extension of the Standard Model. It predicts new states which can be searched for in forthcoming experiments. The fascinating possibility of the existence of warm dark matter as sterile neutrinos of keV mass is a challenging idea.

The search is difficult but could be made easier if the dark matter is locally strongly amplified. Existing data obtained with germanium crystals could be analysed to put limits on local density of such candidates for the dark matter puzzle.

References

[1] Asaka T and Shaposhnikov M 2005 *Phys.Lett.* B **620** 17; Shaposhnikov M 2007 *Nucl.Phys.* B **763** 49; Shaposhnikov M *Preprint* astro-ph/0703673; Dolgov A D *et al.* 2000 *Preprint* astro-ph/0008138 ; Viel M *et al.* astro-ph/0501562

[2] Bernardi G *et al* 1986 *Phys.Lett.* B **166** 479; Bernardi G *et al* 1988 *Phys.Lett.* B **203** 332; Astier P *et al* 2001 *Phys.Lett.* B **506** 27

[3] Back H O *et al.* 2003 *JETP Letter* **78** 5 261

[4] Giunti C *et al.* 1991 *Phys Rev* **D43** 164

Journal of Physics: Conference Series **309** (2011) 012021 doi:10.1088/1742-6596/309/1/012021

The mysterious Sun: a source and a trap of exotica.

K Zioutas[1,2], M Tsagri[2], Y Semertzidis[3], T Papaevangelou[4], E Georgiopoulou[1,6], T Dafni[5] and V Anastassopoulos[1]

[1] University of Patras, Patras, Greece
[2] European Organization for Nuclear Research (CERN), CH-1211 Genève 23, Switzerland
[3] Brookhaven National Laboratory, NY-USA
[4] IRFU, Centre d' Études Nucléaires de Saclay, Gif-sur-Yvette, France
[5] Laboratorio de Física Nuclear y Astropartículas, Universidad de Zaragoza, Zaragoza

E-mail: zioutas@cern.ch

Abstract. A few solar phenomena are given whose origin remains rather mysterious. The claimed observations of anomalous nuclear decays suggest new experiments, whose design should be observationally driven, i.e., going beyond conventional thinking. Also, the non-observation of delayed solar radioactivity after few large flares makes this issue more puzzling, while the solar flare trigger is considered anyhow as one of the biggest solar mysteries.

1. Introduction

In the past, first hints from the Sun for new physics were observationally driven. The solar energy output and the solar neutrino "deficit" problem were two typical examples. At present, similarly, persistent solar mysteries/anomalies might be the otherwise overlooked fingerprints of new solar ingradients like axions, chameleons, paraphotons, or more generally for the dubbed WISPs (Weakly Interacting Slim Particles). In fact, the Sun may emit some ~100 ktons energy equivalent per second in exotica like axions, and this since 4.5 Gyears, without giving rise to visible ageing effects. The unexpected surface photon excess / deficit or some other anomalous solar behaviour could be due to some locally occurring "invisible" processes, resulting thus to the many faces of the paradoxical Sun. Thus, in an inside-out approach moving from the solar core outwards to the solar surface, its atmosphere and further in outer space all along to Earth's orbit and even beyond, unknown solar physics might still be at work at such relatively large scales. We mention below, for example, gravitational anomalies observed within the solar system. Though, it is an open question the actual dark matter distribution within the solar system and the formation / evolution of the solar system on a dark matter solar halo [1].

2. Sun's anomalous behaviour: harbinger of new physics?

Helioseismology does not describe exactly the inner Sun, while the peak-like appearance of an overshooting near the tachocline could be due to some locally occurring additional heating. From the axion or chameleon point-of-view, this place (some 200000 km underneath the solar surface) is widely considered to be magnetized with a magnetic field strength of about 50 T. This is also the roots of the solar surface magnetism reaching few kGauss in sunspots [2]. Note that below $\sim 0.2R_{solar}$ the rotation profile is unknown.

The Sun is a target/source for the indirect detection of dark matter particles being gravitationally trapped and accumulated over cosmic times by the Sun, e.g., the celebrated WIMPs. In addition, other massive exotic particles like Kaluza-Klein (KK) axions created inside the Sun can be gravitational

[6] Diploma Thesis.

self-trapped. The self-trapping efficiency can be small (about 10^{-7} per created particle) [3]. But, their accumulation over the last 4.5 Gyears can result to a total mass orbiting the Sun of M_{KK} ~10-11M_{solar}. This is, however, not sufficient to explain the pioneer anomaly ($\alpha=10^{-7}$cm/s^2), for which one needs a trapped total mass of ~$10^{-4}M_{solar}$. At the atmosphere, there is the so called solar coronal mystery, since the known physics could not provide as yet a viable microscopic heating mechanism [4]. In fact, it was suggested that a whole Sun external self-irradiation by the spontaneous decay of the trapped exotica, e.g., KK-type axions or other WISPs, can occur. This, along with the outwards emitted X-rays from converted axions / WISPs streaming out of the inner Sun (occasionally), can be the main two components behind the unexpected solar spectral shape, which deviates enormously from a thermal distribution expected to be even below ~5800K. In fact, it is remarkable and fascinating that the Sun emits intense X-rays, and this is a mystery [5].

In addition to the striking solar corona problem, which is considered as one of the biggest problems in the whole of astrophysics, we mention here some solar gravitational anomalies. We recall cosmic gravitational anomalies like the flat rotation curves of spiral galaxies and the accelerated expansion of the universe, point at the existence of dark matter and dark energy, which are still the biggest unsolved problems of all physics. Thus, the Sun gives rise somehow to the following observations of as yet unknown explanation:

- **The Pioneer anomalous acceleration** in outer solar system (~100 AU). Various distributions of DM in the solar system have been proposed already to explain the deceleration due to an additional gravitational pull towards the Sun, but the mystery remains.
- **The increase of the [AU]** of about +15 cm/year, which should be decreasing instead due to the escaping solar mass and radiation, though to a much lesser extent to the solar wind.
- **Earth Flyby anomalies** during gravity assist, giving rise to a strange, step-like acceleration change (max. 13 mm/s), and this is also a mystery [6, 7].
- **Anomalous nuclear decay rate signatures**? ? Data taken over long time intervals with radioactive sources in BNL/USA and at the PTB/Germany exhibited frequencies in the range 1-15 / yr. Except the annual modulation, other oscillation frequencies are compatible with inner solar rotation frequencies (e.g., T~33 days), where the nuclear reactions take place. The question arises then as to how the Sun could be affecting (terrestrial) nuclear decays? While neutrinos have been considered as the first potential candidates, there is no reaction mechanism with such a large cross section, which could be behind the claimed nuclear decay rate oscillations [8].

Such an effect, if definitely true, seems not to fit any conventional physics reasoning. Therefore, the solution should be searched beyond known physics. Exotic particles like axions, chameleons, and WISPs, can also not provide a quantitative explanation, i.e., they also cannot change an intrinsic nuclear property like the lifetime. Nevertheless, such particles can be inspiring for novel experimental approaches [9]. In addition, the claimed anomalous nuclear decay (^{54}Mn decay), being correlated with solar flares, makes the whole issue more intriguing, but also stronger (see below) [10]. These results raise questions concerning solar physics and particle physics alike. Not only are such observations a mystery, but also their explanation [11]. Therefore, we face then a real challenging situation, if all this is true.

If it is a new particle or field of solar origin behind this claim, then it is reasonable to assume that its impact on nuclear intrinsic properties should be stronger near to the Sun. While keeping in mind that of potential importance in this context are the various solar mysteries, e.g., the solar corona problem, we also have searched for a similar nuclear behaviour that is taking place near the sun. Indeed, an interesting example is the search for long-lived radioactive nuclei expected from solar flares, due to the nuclear interaction of flare-accelerated ions with the ambient solar atmosphere.

Journal of Physics: Conference Series **309** (2011) 012021 doi:10.1088/1742-6596/309/1/012021

Prompt gamma-ray line emission has been observed in many flares, with the most notable one being the isotope ^{56}Co, which has a half-life of $T_{½}$=78.8 days and two lines at 847 keV and 1238 keV. The long half-life of ^{56}Co, allows the produced radioactive material to accumulate in case of multiple flares [12]. Surprisingly, the delayed 847 keV line should appear after very large flares, but it has not yet been observed with the best performing orbiting telescope (i.e., RHESSI) [13]. Moreover, *"delayed X- and γ-ray line emission from solar flares has been recently searched for with the RHESSI spectrometer but unsuccessfully...."* [14]. This non-observation should be seen along with the claimed anomalous nuclear decay (^{54}Mn) being correlated with solar flares (see above), which makes the whole issue even more interesting. In addition, we recall that the flare trigger is considered to be among the biggest solar mysteries, while its mean temperature resembles that of the Sun's core some 700000 km underneath [15]. Is this just a chance coincidence or does this reflect an exotic coupling between the inner and the outer Sun?

3. Conclusion

The question arises as to what exotic particles like axions, chameleons, paraphotons, and more generally WISPs can do, provided the aforementioned anomalous nuclear observations are real? Given the fact that such phenomena do not obey the law of physics, new unconventional terrestrial / celestial experimental approaches must be invoked. The design of the novel experiments should be observationally driven, going thus beyond conventional thinking [16].

References

[1] De Diego J A, Nunez D, Zavala Z 2005 Pioneer anomaly? Gravitational pull due to the Kuiper belt *Preprint* astro-ph/0503368
[2] Hiremath K M, Lovely MR 2007 *ApJ.* **667** 585
[3] Di Lella L, Zioutas K 2003 *Astroparticle Phys.* **19** 145
[4] Marsch E 2008 http://www.mps.mpg.de/solar-system-school/lectures/sonnensystem_2008/Corona.pdf
[5] Tsuneta S, 2009 *AAPPS Bulletin* **19**(#3) 11 http://www.cospa.ntu.edu.tw/aappsbulletin/data/19-3/11Hinode.pdf
[6] Nieto MM, Anderson JD 2009 **Physics Today 62** (# 10) 76
[7] Anderson JD, Nieto MM 2009 *Proc IAU Symposium* **261**
Klioner S, Seidelmann PK and Soffel M 2009 Astrometric Solar-System anomalies *Preprint* gr-qc/ 0907.2469v2
Turyshev SG, Toth VT 2010 *Living Rev. Relativity* **13**, 4 (*Preprint* gr-qc/1001.3686)
[8] Fishbach E *et al.* 2010 Evidence for solar influences on nuclear decay rates *Preprint* hep-ph/1007.3318v1
[9] Sturrock PA *et al.*v2010 *Solar Physics* **267** 251 (*Preprint* astro-ph/1010.2225)
[10] Fischbach E *et al.* 2009 *Space Sci. Rev.* **145** 285
[11] Sturrock PA *et al.* 2010 Analysis of PTB Decay-Rate Data: Evidence for Solar Rotational Modulation *Preprint* astro-ph/1010.2225v1
[12] Shih AY, Smith DM, Lin RP and Hurford GJ 2007 *AAS* (#68.04) **38** 175
[13] Smith DM 2005 Astronomy with Radioactivities, **http://www.mpe.mpg.de/gamma/science/lines/workshops/clemson05/presentations/smith_d.pdf**
[14] Tatischeff V, 2010 Private Communication.
[15] Orlando S, Peres G, Reale F, 2003 Adv. Space Res. **32** 955
[16] Work in progress in collaboration with Axel Lindner / DESY.

Journal of Physics: Conference Series **309** (2011) 012022 doi:10.1088/1742-6596/309/1/012022

Laguna: Future Megaton Detectors in Europe

T. Patzak[1] **on behalf of the Laguna Collaboration**

[1] AstroParticule et Cosmologie (APC), CNRS, Univ. Paris 7, CEA, Obs. de Paris

E-mail: patzak@apc.univ-paris7.fr

Abstract. The FP7 Design Study LAGUNA (Large Apparatus studying Grand Unification and Neutrino Astrophysics) supports studies of European research infrastructures in deep underground cavities able to host a very large multipurpose next-generation neutrino observatory dedicated to nucleon decay, neutrinos from supernovæ, solar and atmospheric neutrinos, as well as neutrinos from a future Super-Beam or β-Beam to measure the mixing angle θ_{13}, the CP violating phase δ and the mass hierarchy.

1. Introduction

Neutrinos are messengers from astrophysical objects as well as from the early universe and can give us information on processes, which cannot be studied otherwise. Underground experiments, like SuperKamiokande (SK) [1], have made important discoveries. Next-generation very large volume underground experiments will answer fundamental questions on particle and astroparticle physics. They will search for a possible finite lifetime for the proton with a sensitivity one order of magnitude better then the current limit. With a neutrino beam they will measure with unprecedented sensitivity the last unknown mixing angle (θ_{13}) of neutrinos and unveil through neutrino oscillations the existence of CP violation in the leptonic sector, which could provide an explanation of the matter-antimatter asymmetry in the Universe. Moreover they will study astrophysical objects, in particular the Sun and Supernovæ [2]. The construction of a large scale detector devoted to particle and astroparticle physics in Europe is one of the priorities of the ASPERA[1] roadmap (2008). The FP7 Design Study LAGUNA (Large Apparatus studying Grand Unification and Neutrino Astrophysics) [3] support studies of European research infrastructures in deep underground cavities able to host a very large multipurpose next-generation neutrino observatory - GLACIER (Liquid Argon) [4], LENA (Liquid Scintillator) [5], MEMPHYS (Water Cherenkov) [6].

2. Laguna

The FP7 Design Study LAGUNA [3] is a Pan-European effort of 21 beneficiaries, composed of academic institutions from Denmark, Finland, France, Germany, Poland, Spain, Switzerland, United Kingdom, as well as industrial partners specialized in civil and mechanical engineering and rock mechanics, is assessing the feasibility of this Research Infrastructure in Europe.

The LAGUNA consortium is evaluating possible extensions of the existing deep underground laboratories in Europe: Boulby (UK), Canfranc (Spain) and Modane (France) and considers the

[1] ASPERA: http://www.aspera-eu.org

Journal of Physics: Conference Series **309** (2011) 012022 doi:10.1088/1742-6596/309/1/012022

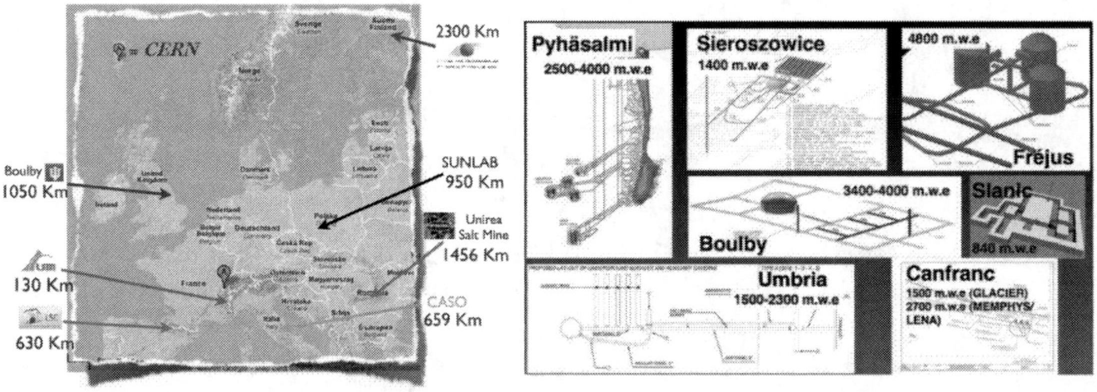

Figure 1. Left: Map of the seven possible underground sites in Europe. Right: Typical example of layouts studied in LAGUNA DS for each site.

creation of new laboratories in the following regions: *Caso* Umbria Region (Italy), *Pyhäsalmi* (Finland), *Sierozsowice* (Poland) and *Slanic* (Romania).
In Europe there are three different proposed detectors, based on different techniques: GLACIER (Liquid Argon), LENA (Liquid Scintillator) and MEMPHYS (Water Cherenkov). For all three detectors there are, in the LAGUNA context, specific studies concerning the construction feasibility, the required depth, the muons and reactor neutrinos flux etc. In figure 1 the seven sites are shown, as well as an example of the construction studies developed in the LAGUNA context by the different beneficiaries.

Location	Type	Envisaged depth m.w.e.	Distance from CERN [km]	Energy 1st osc. max. [GeV]
Fréjus (F)	Road tunnel	$\simeq 4800$	130	0.26
Canfranc (ES)	Road tunnel	$\simeq 2100$	630	1.27
Umbria (IT)[a]	Green field	$\simeq 1500$	665	1.34
Sieroszowice (PL)	Mine	$\simeq 2400$	950	1.92
Boulby (UK)	Mine	$\simeq 2800$	1050	2.12
Slanic (RO)	Salt mine	$\simeq 600$	1570	3.18
Pyhäsalmi (FI)	Mine	up to $\simeq 4000$	2300	4.65

[a] $\simeq 1.0°$ off axis.

Figure 2. Potential sites being studied with the LAGUNA design study [7].

The design study is dedicated to the comprehensive and coordinated technical assessment of each site (rock studies, external activities compatibility, etc...). Moreover the study must produce a coherent cost estimation. The next generation deep underground neutrino detector in Europe should be coupled to advanced neutrino beams from CERN. Each site has specific depth characteristics and is situated at a specific distance from the CERN complex (see figure 2). This offers a large number of possible "detector-site" combinations and must be considered a benefit and not a disadvantage compared to other countries where there is only one possible scenario. The selection of the optimized configuration involves several aspects (physics performances, technical feasibility, safety and legal aspects, socio-econimic and environmental impact, costs, ...) and this implies a particular attention of LAGUNA to interdisciplinary matter, as both physicists and engineers as well as geo-technical experts are directly involved.

2.1. GLACIER

The GLACIER (Giant Liquid Argon Charge Imaging ExpeRiment) [4] detector is based on a new liquid argon detector concept, scalable to a single unit of mass 100 kton: it relies on a cryogenic storage tank developed by the petrochemical industry (LNG technology) and on a novel method of operation called the LAr LEM-TPC. LAr LEM-TPCs operate in double phase with charge extraction and amplification in the vapor phase. The concept has been very successfully demonstrated on small prototypes: ionization electrons, after drifting in the LAr volume, are extracted by a set of grids into the gas phase and driven into the holes of a double stage Large Electron Multiplier (LEM), where charge amplification occurs.

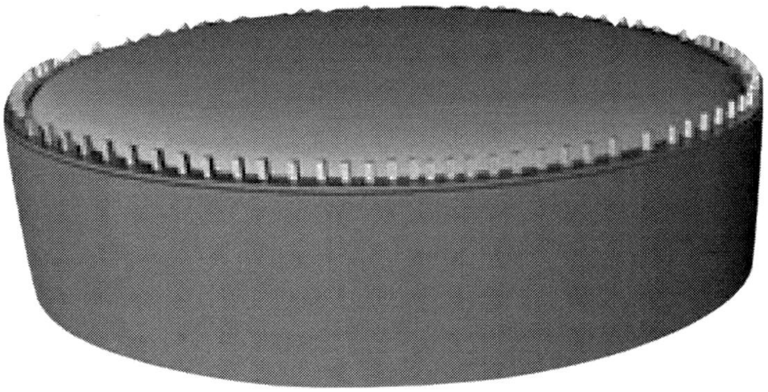

Figure 3. Design of the Glacier detector [4]. The maximum diameter is 70 m; the height is limited to 20 m by the maximum possible drift length.

Effective extrapolation to the required scale needs concrete R&D. A ton-scale LAr LEM-TPC detector is being operated at CERN in Blg 182 within the CERN RE18 experiment (ArDM). In order to prove the performance for neutrino physics, additional dedicated test beam campaigns are being considered, to test and optimize the readout methods and to assess the calorimetric performance of such detectors. A 1 kton detector can be built assuming the GLACIER design with a 12 m diameter and 10 m vertical drift. From the point of view of the drift path, a mere factor 2 will be needed to extrapolate from the prototype to the 100 kton device. Hence, the prototype will be the real demonstrator for the long drifts. At the same time, the rest of the volume scaling from the 1 kon to the 100 kton achieved by increasing the diameter to about 70 m, can be realized noting that (a) large LNG tanks with similar diameters and aspect ratios already exist (b) the LAr LEM-TPC readout above the liquid will be scaled from an area of 80 m^2 (1 kton) to 3800 m^2 (100 kton). This will not require a fundamental extrapolation of the principle, but rather only pose technical challenges of production, which can be solved in collaboration with industry.

In addition, researches into neutrino oscillations point to the need to couple the observatory to intense neutrino beams, for instance from CERN, to address the puzzle of the origin of matter excess over antimatter in the Universe (discovery of CP-violation in the Lepton Sector). In such a detector the level of precision in the measurements of the last unknown oscillation angle θ_{13}, the phase of CP violation and the mass hierarchy when combined with a neutrino beam is unprecedented. Several beams can be considered for the GLACIER detector.

2.2. LENA

LENA (Low Energy Neutrino Astronomy) is a proposed large $\sim$ 50ktons liquid scintillator neutrino (LSc) detector for particle-astrophysics, located in a deep underground laboratory. Because of its low energy threshold, LENA would be sensitive to neutrinos from very different sources: the measurement of the diffuse Supernova neutrino background; the precise determination of thermo-nuclear fusion processes and the matter-effects in solar matter by measuring solar neutrinos with high statistics; a measurement of geo-neutrinos probing Earth's models; in case of an actual galactic type II Supernova an accurate measurement of the time development and flavour content of the emitted neutrino burst. Moreover, LENA can search for proton decay, especially $p \rightarrow K^{+}\bar{\nu}$, thus probing grand unified theories. In addition, LENA will be used as a detector for low energy atmospheric neutrinos and may perform an indirect search for Dark Matter.

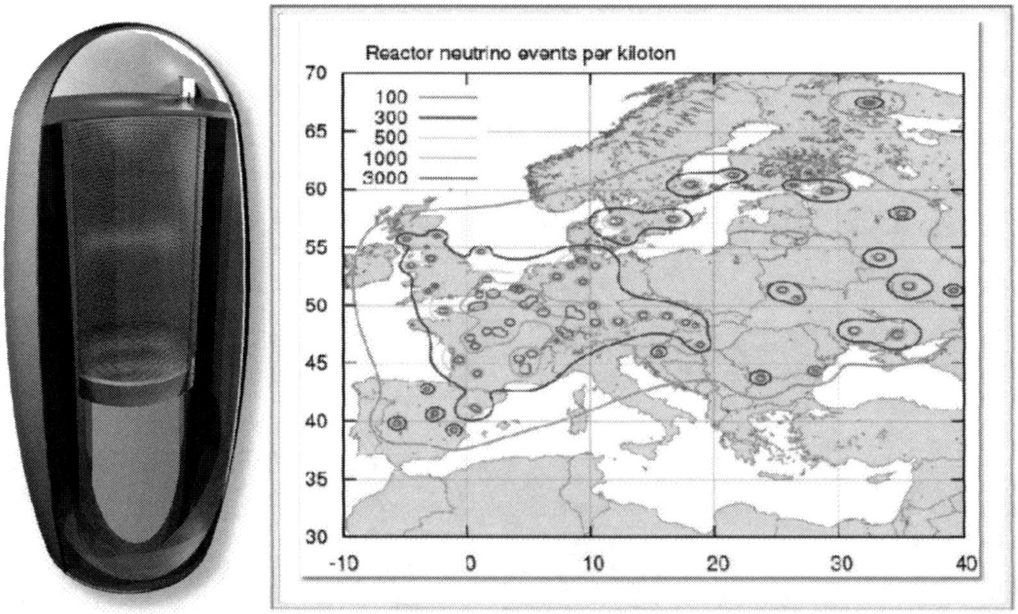

Figure 4. Left: Design of the Lena detector [5]. The tank is 100 m height and includes the water muon veto. Right: Reactor neutrinos background flux in Europe per year and per kiloton.

Recent studies indicate that such a large volume of LSc detector can resolve both momentum and energy of GeV particles [8]. Monte-Carlo simulations have been performed to study the reconstruction capability of a LSc detector for complex event topologies of CC neutrino interactions. If this new approach is feasible it will offer the possibility to complement the already rich low-energy physics program of LENA by long-baseline neutrino oscillation experiments, either from atmospheric neutrinos or an accelerator-produced neutrino beam.

Journal of Physics: Conference Series **309** (2011) 012022 doi:10.1088/1742-6596/309/1/012022

2.3. MEMPHYS

One of the techniques most understood for neutrino detection is based on the Cherenkov light emission in water by the final state particles resulting from neutrino interactions. Therefore the possibility of building a water Cherenkov detector with a fiducial mass of about 20 times larger than SK is currently investigated by different groups around the world, and for different underground sites. The MEMPHYS project is discussed here with particular interest for deployment in an extended Modane Laboratory (LSM -Fréjus), which for a low energy neutrino beam is located at an optimal distance from the CERN accelerator complex [9].

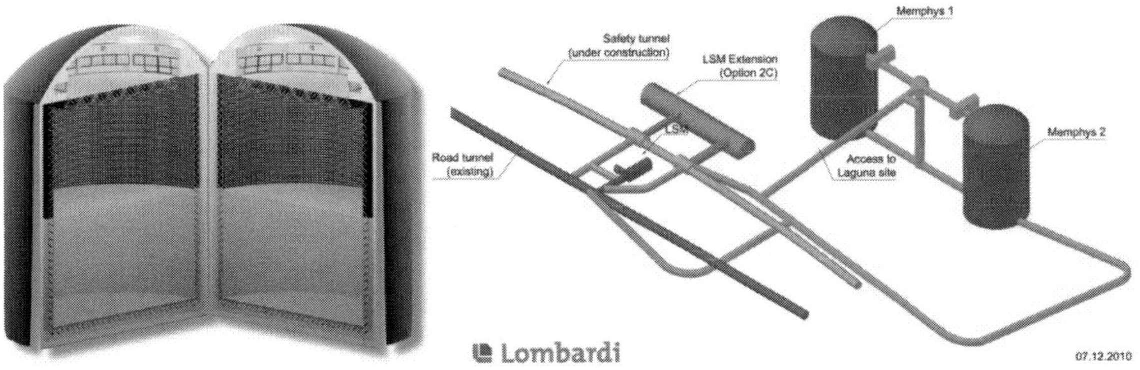

Figure 5. Design of the Memphys detector. On the left the schematic view of a cylinder [10]. On the right one of the possibles option for the MEMPHYS detector: two tanks, 64 m in diameter and 103 m high, separated by 140 m from each other.

The huge size of MEMPHYS and the cost of the light sensors of a such a big experiment require an attentive choice concerning the detection technique and the data acquisition system. The project PMm^2 has realized a new electronics board dedicated to a grouped acquisition in the form of a matrix of 16 PMTs [11]. In the MEMPHYS detector each PMT matrix will have a common board (PARISROC) for the high voltage distribution and for the signal detection. Such a system should be tested with real physical signals and with the same detection technique as MEMPHYS in order to improve and adapt the system to the needs of the megaton scale experiments.

For this purpose a prototype of MEMPHYS, Memphyno, has been constructed at APC in order to make a full test of the complete chain "electronic and acquisition" [10]. The first 16 PMTs matrix of PMm^2 will be placed in the tank and studied first with cosmic muons in water with an option of introducing Gadolinium salt it a later stage. Then Memphyno will be moved to LSM for background tests and later to CERN for electron, pion and kaon beam measurements (an electron beam at the LAL is also possible). The test with electrons will be used to study the collection efficiency of the Cherenkov light from a point-like source and to check the single photoelectron range with the new electronics system.

The Memphyno prototype comprises a tank of $2 \times 2 \times 2$ m^3 filled with water and an Hodoscope made by 4 scintillator planes (2 on the top and 2 on the bottom) to provide the trigger and the direction of crossing cosmic muons.

At the present time, Memphyno has been built at APC and the Hodoscope is operational since May 2010. Whilst waiting for the PMm^2 demonstrator (16 8" Hamamatsu PMTs + PARISROC electronic board) the acquisition system have been performed with 4 8" PMTs (ETL-Electron Tubes Limited) from Borexino.

3. Summary

The Laguna design study allowed to accomplish a very detailed investigation of the different underground locations. It also helped to structure the European neutrino community and define the physics goals of the next generation experiments.

Topics	GLACIER 100 kton	LENA 50 kton	MEMPHYS 440 kton
Proton decay			
$e^+\pi^0$	0.5×10^{35}	—	1.0×10^{35}
$\bar{\nu}K^+$	1.1×10^{35}	0.4×10^{35}	0.2×10^{35}
SN ν (10 kpc)			
CC	$2.5 \times 10^4(\nu_e)$	$9.0 \times 10^3(\bar{\nu}_e)$	$2.0 \times 10^5(\bar{\nu}_e)$
NC	3.0×10^4	3.0×10^3	—
ES	$1.0 \times 10^3(e)$	$7.0 \times 10^3(p)$	$1.0 \times 10^3(e)$
DSNB ν (S/B 5 years)	40-60/30	9-110/7	43-109/47 (*)
Solar ν (Evts. 1 year)			
^{8}B ES	4.5×10^4	1.6×10^4	1.1×10^5
^{8}B CC	—	360	—
^{7}Be	—	2.0×10^6	—
pep	—	7.7×10^4	—
Atmospheric ν (Evts. 1 year)	1.1×10^4	—	4.0×10^4 (1-ring only)
Geo ν (Evts. 1 year)	below threshold	≈ 1000	need 2 MeV threshold
Reactor ν (Evts. 1 year))	—	1.7×10^4	6.0×10^4 (*)
Dark Matter (Evts. 10 years)	3 events ($\sigma_{ES} = 10^{-4}$,$M > 20$ GeV)	—	—

Figure 6. Summary of the physics potential of the proposed detectors for astro-particle. The (*) stands for the case where Gadolinium salt is added in the water of one of the MEMPHYS tanks [3].

In Figure 6 we resume the non-accelerator physics goals of the three detectors. In the framework of the LAGUNA program the detectors have to be considered in their underground sites taking into account the distance from CERN. According to the characteristics of the experiments and of their potential distance from the CERN facility reach different sensitivities to θ_{13} and to the δ CP phase. The synergy with the EUROnu program, that investigates a possible European neutrino beam, is necessary to performed a realistic and competitive experiment.

4. References

[1] S. Fukuda *et al.*, *Nucl. Instrum. Methods* **A501**, 418 (2003).
[2] K.S. Hirata *et al.*, *Phys. Rev.* **D38**, 448 (1988); W.D. Arnett, J.L. Rosner, *Phys. Rev. Lett.* **58**, 1906 (1987).
[3] D. Autiero *et al.*, *JCAP* **11**, 011 (2007) [arXiv:0705.0116v2 [hep-ph]].
[4] A. Rubbia *et al.*, *J. Phys. Conf. Ser.* **171**, (2009).
[5] L. Oberauer, F. von Feilitzsch, W. Potzel, *Nucl. Phys. Proc. Suppl.* **138**, 108 (2005).
[6] A. de Bellefon *et al.*, arXiv:hep-ex/0607026.
[7] A. Rubbia *et al.*, *Acta Physica Polonica B* **41**, (2010).
[8] M. Wurm *et al.*, *Acta Physica Polonica B* **41**, (2010).
[9] J-E. Campagne *et al.*, *JHEP* **04**, (2007).
[10] J.L.. Borne *et al.*, *Acta Physica Polonica B* **41**, (2010).
[11] PMm2 webpage: www.pmm2.in2p3.fr

Journal of Physics: Conference Series **309** (2011) 012023 doi:10.1088/1742-6596/309/1/012023

Low energy investigations and applications with the spherical TPC

E Bougamont[1], P Colas[1], J Derre[1], I Giomataris[1], G Gerbier[1], M Gros[1], P Magnier[1], X F Navick[1], P Salin[3], I Savvidis[2], G Tsiledakis[1*], J D Vergados[4]
[1]CEA-Saclay, France
[2]University of Thessaloniki, Greece
[3]APC University of Paris, France
[4]University of Ioannina, Greece

Abstract. The Spherical Proportional Counter, recently developed, allows to instrument large target masses with good energy resolution and low energy threshold. Ultra low energy results are shown here, leading to an energy threshold as low as 25 eV and a single electron detection sensitivity. The bench mark result is the observation of a well resolved peak at 270 eV due to carbon fluorescence, which is a unique performance for such large massive detector. This very promising feature can fulfill the demands of many challenging projects from dark matter detection to low energy neutrino searches. Details of this study are given in reference [1].

1. Introduction

The development of massive detectors, having a low energy threshold and low noise, remains generally a daunting challenge for present-day and future low-background experiments. The search for WIMP (Weakly Interacting Massive Particles) dark matter is under intense development and relies on the detection of low energy (keV scale) recoils produced by the elastic interaction of WIMP's with the nuclei of the detector [2-5]. The need to go to very low energies may become even more crucial, if the WIMP's turn out to be very light [6], since then, the energy transfer to the nucleus is expected to be smaller. New generation of directional detection of dark matter is under development to perhaps allow for an unambiguous observation even in the presence of backgrounds by observing directional anisotropy of the recoils [7-10].

The question of detecting and exploiting neutrinos from both terrestrial and extra terrestrial sources has become central to physics and astrophysics. Coherent neutrino-nucleus scattering is a famous but as yet untested prediction of the Standard Model [11,12]. The process is mediated by neutral currents (NC), and hence is flavor-blind. Despite having relatively high rates, neutrino-nucleus scattering is difficult to observe because its only signature is a small nuclear recoil of energy ~ keV (for MeV neutrinos). The nucleon energy (T_N) depends on the neutrino energy (E_ν) and the scattering angle (θ).

*georgios.tsiledakis@cea.fr

In the laboratory frame (forward scattering) and for sufficiently small energies, the recoil energy of the nucleon can be approximated as follows [13]:

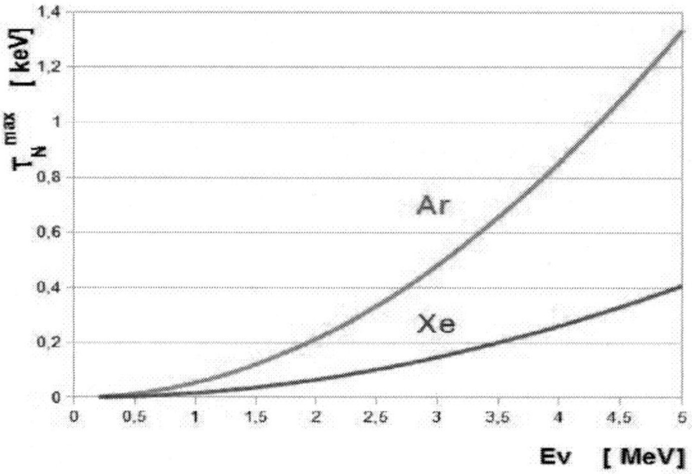

Figure 1. The maximum recoil energy T_N^{max} as a function of the neutrino energy Eν for Ar and Xe respectively (from top to bottom)

$$T_N \approx \frac{2(E_\nu \cos\theta)^2}{m_N} \qquad (1)$$

The maximum recoil energy T_N^{max} as a function of the neutrino energy (Eν) for Ar and Xe as a target nucleus is shown in Figure 1. For a typical neutrino energy of 2 MeV (expected at nuclear reactors) the maximum recoil energy is ~ 66 eV for Xe and ~ 215 eV for Ar. Because the neutrino is light, the nuclear recoil energy is extremely small leading to a signal below threshold for conventional solid or liquid state detectors. Thus, the challenge is to achieve a very low energy threshold (typically below 100 eV).

Several other cases that require a low energy threshold like Supernova neutrinos, the OAK Ridge neutron spallation source, reactor neutrinos, geoneutrinos are discussed in [1]. Ultra low-noise germanium detectors with sub-keV energy capability have been recently developed and are operating at underground laboratories [14,15]. The small excess observed around 1 keV needs to be clarified and verified by detectors having a lower energy threshold.

Now we will report results at low energy obtained by using the novel Spherical Proportional Counter (SPC) which has been recently developed.

2. Detector description

The detector consists of a large spherical copper vessel 1.3 m in diameter and a small metallic ball 16 mm in diameter located at the center of the drift vessel, which is the proportional counter. The ball is maintained in the center of the sphere by a stainless steel rod and is set at high voltage. A second electrode (umbrella-shaped) that is placed 24 mm away from the ball along the rod, is powered with an independent but lower high voltage, serving as electric field corrector. The detector operates in a seal mode: the spherical vessel is first pumped out and then filled with an appropriate gas at a pressure from few tens of mbar up to 5 bar. Electrons originating from ionization of the gas in the volume drift to the region of the central ball where the intense electric field allows gas amplification to occur. The produced signal is amplified through a charge amplifier and a shaper and is read-out by a 14 bit ADC. Detailed description of the detector, its electronics, its operation and its performance could be found in references [16,17,18]. In a previous paper [18] we reported results obtained at higher energy and we pointed out the excellent energy resolution obtained with radon nuclide and its daughters. In this work we will focus our studies to detect very-low energy gamma or *X*-rays emitted by radioactive sources or fluorescence process.

3. Low energy calibration and results

3.1. *Results with radioactive sources and fluorescence X-rays*

For this study (a detailed description can be found in [1]) we are using a gas filling of Argon with 2% admixture of CH_4 at various pressures. The pressure (*P*), the high voltage of the ball (*HV1*), the high voltage of the umbrella field corrector (*HV2*) and the gain (*G*) of the amplifier-shaper in all measurements in the section, are written on each of the following figures.

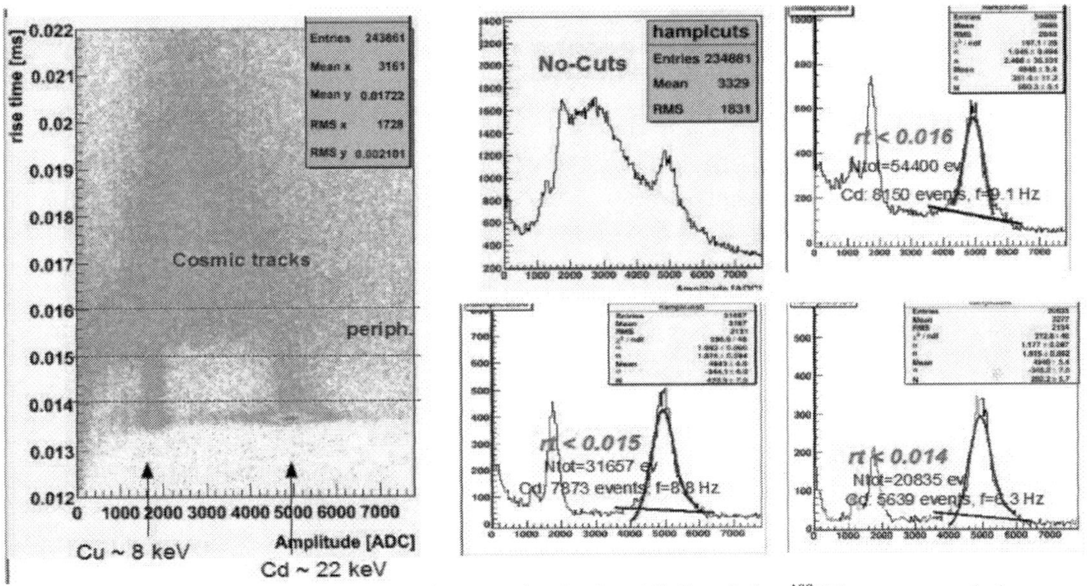

Figure 2. The rise time (rt) versus the amplitude (in ADC) of the ^{109}Cd source and the energy spectrum with the energy lines of 8 and 22 keV respectively, for various cuts in rise time (P = 107 mb, HV1 = 2520 V, HV2 = 1036 V, G = 20).

Calibration of the counter was initially performed using a ^{109}Cd source by irradiating the gaseous volume through a thin 200 μm aluminum window. By applying a pulse shape cut (rejecting long rise time pulses >0.014 ms) we keep 70% of signal in Cd but the effect on the background is spectacular as it is shown in Figure 2. The expected background from cosmic tracks is highly reduced and the energy spectrum with the 22 keV line from the ^{109}Cd source and the 8 keV line, which is an induced fluorescence at the copper vessel, provides a satisfactory energy resolution of 6% and 9% (FWHM) respectively.

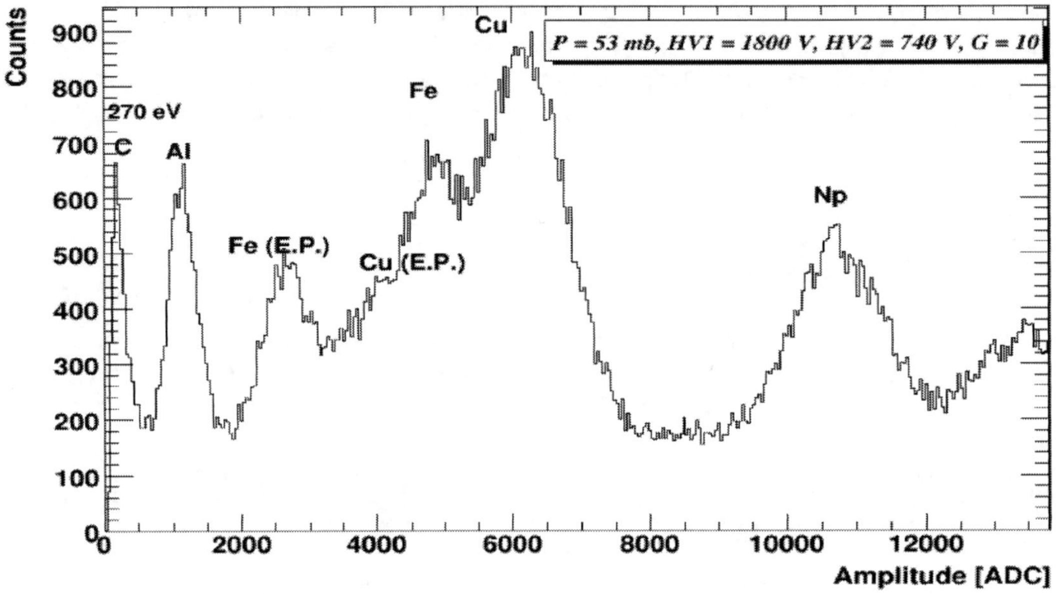

Figure 3. Peaks observed from the ^{241}Am radioactive source through aluminium and polypropylene foil. On the left the Carbon (270 eV) peak is shown, followed by the Aluminium peak (1.45 keV), the escape peak (E.P.) of Iron in Argon (3.3 keV), the escape peak of Copper in Argon (5 keV), the Iron peak (6.4 keV), the Copper peak (8 keV) and the Neptunium peak (13.93 keV) .

Gamma fluorescence is adequate for producing fluorescence lines in the range above several keVas the Cu line at 8 keV. However, it is very difficult to produce low energy calibration lines below a few keV. In order to create lower energy *X*-rays we have used an ^{241}Am source, which decays by the following process: $^{241}\text{Am} \rightarrow (^{237}\text{Np})^{*} + {}^{4}\text{He} + 5.6\ \text{MeV}$. The ^{237}Np nucleus then decays into a lower energy state by emitting a 59.537 keV gamma ray and other L rays , which are used to fluoresce the elements. The source was evaporated to a stainless steel holder and attached to the sensor rod at middle distance; the source then is covered by thin foils with minimal thickness to totally absorb the 5.6 MeV alpha emitted, leaving only gamma rays and fluorescence induced *X*-rays to pass into the gas volume. Covering the source with a 20 μm thick aluminum foil we were able to fluoresce the K *X*-rays ranging from Aluminum whose K electron has a binding energy of 1.56 keV, to Cu whose K electron has a binding energy of 8.98 keV. In order to obtain even lower energy calibration lines we replaced the aluminum foil by a thinner 10 μm one and we attached a 20 μm polypropylene foil. Therefore, the alpha particle that is crossing the aluminum foil, is fully absorbed by the polypropylene foil and induces both aluminum and carbon fluorescence as shown in Figure 3.

High gain combined with low electronic noise can provide energy thresholds clearly below 100 eV. Figure 4 shows the energy spectra at two different gains of the amplifier when no source is used. By applying a cut at the rise time of the signal (which actually provides the depth of the ionized electrons produced in the gas) we can exclude the signal induced by cosmic rays and measure only the Copper energy line of 8 keV (plotted with dashed line in Figure 4). Then, by increasing the gain of the amplifier 20 times (from Gain = 10 to Gain = 200), we keep in the ADC acceptance only the ultra low energy region (plotted with full line in Figure 4). The value of 1000 ADC corresponds to ~ 150 eV and the peak at ~ 50 eV is compatible with single electrons (see section 3.2). Thus, the detection threshold of the Spherical Proportional Counter is as low as 25 eV. The exponential shape of the spectrum has to be confirmed by underground measurements in the LSM laboratory, which are on-going. Preliminary results are shown in Figure 5, where the rise time versus the amplitude is plotted. The 8 keV line from Copper is clearly still present and can be used to calibrate the detector at low energies.

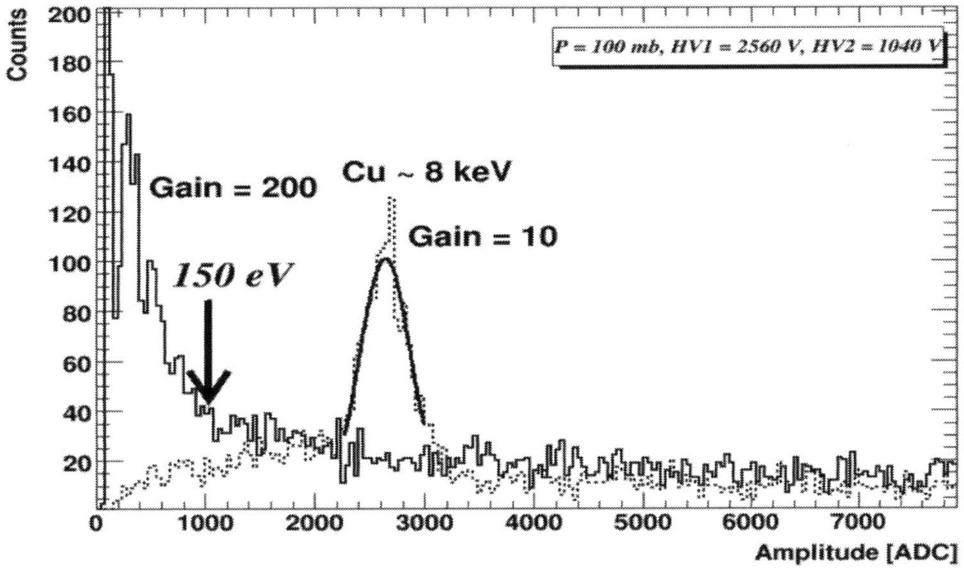

Figure 4. The energy spectrum at Gain = 10 with the Copper peak of 8 keV (dashed line) and at Gain = 200 where the single electron peak at ~ 50 eV is clearly visible (full line).

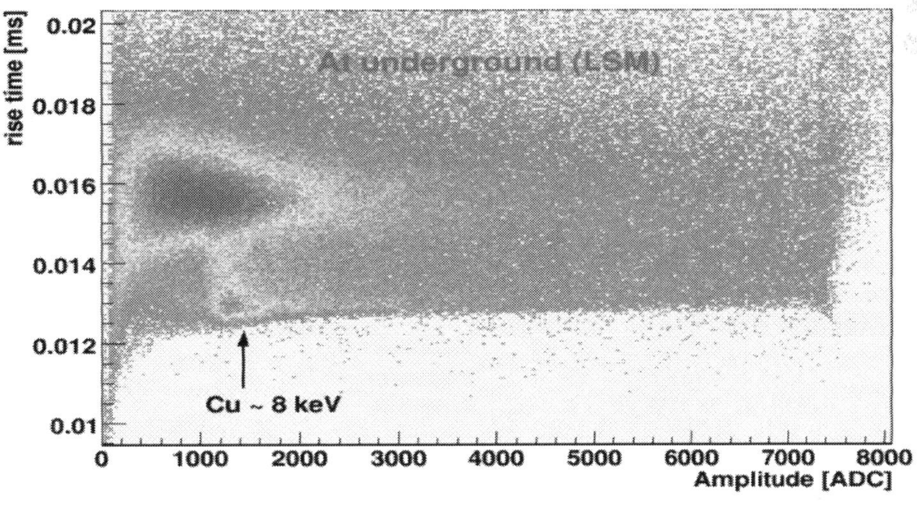

Figure 5. The rise time versus amplitude without source, taken at underground (P = 57 mb, HV1 = 1750 V, HV2 = 713 V at LSM).

3.2. *Results using a UV flash lamp*

A UV window made by MgF_2 crystal was installed in one of the sphere openings. A hydrogen relaxation flash lamp has been used to produce UV photons in the far ultra violet range. The discharge is also producing a fast signal which is observed through a capacitor and after adequate attenuation can be used as a fast trigger. For the present study, all results have been obtained without trigger. Photons are crossing the UV entrance window and hitting copper of the internal vessel producing photoelectrons by photoelectric effect. Electrons are being extracted and drifting through the radial field to the central ball where they are amplified and collected.

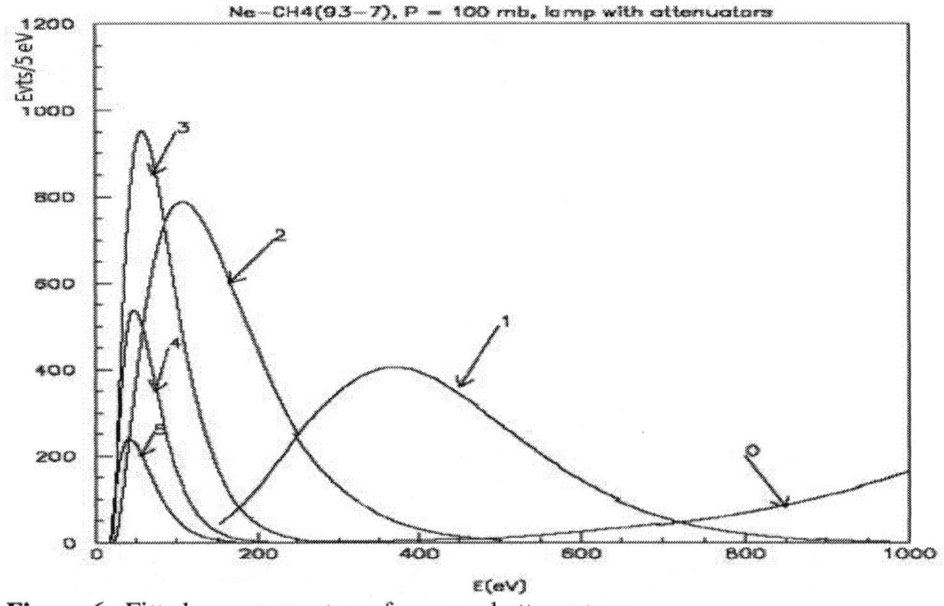

Figure 6. Fitted energy spectrum for several attenuators.

The energy calibration has been done using the 22 keV photon *X* decay of ^{109}Cd source and the 8 keV Cu fluorescence. The UV light flux of the lamp was adjusted by introducing adequate light attenuators down to the single photo-electron extraction on the Copper. The subtraction of the amplitude spectrum of the sample lamp -off from that of lamp -on suppresses the huge background due to cosmic rays. The gas used was a mixture of Ne with 7% CH_4 at a pressure of 100 mbar (the complete study can be found in [1]).

The lamp light was then gradually attenuated by 1,2, up to 5 identical attenuators. Each distribution is well fitted by a Polya function with an offset. With less than 3 attenuators, the number of events is stable, which means for those sample all the flashes are detected with a mean frequency close to 45 Hz. In fact the probability not to detect the flash is very low, 2.5% with 2 attenuators. It is compatible with the probability to get zero in the Poisson distribution with a mean value close to the number of photoelectrons per detected flash given by the fit. With more attenuators the mean frequency of detected flashes is decreasing because the probability of producing zero photoelectrons per flash is increasing exponentially. The mean number of extracted electron per flash decrease exponentially with a reduction factor close to 3 as expected, even with a large number of attenuators, pointing to a negligible inefficiency to detect the single photelectrons. The various distributions for all used attenuators are shown in Figure 5. We observe a decrease of the mean value of each distribution up to about 3 attenuators. After that the mean value is not decreasing any more assuring we are at the single electron level (the number of photoelectrons per detected flash given by the fit decreases asymptotically to one as expected) .

4. Future developments and applications

The spherical proportional counter can be applied for low-background experiments, which require a low energy threshold. An example is the detection of the coherent neutrino-nucleus scattering producing sub keV ion recoils. A spherical detector of radius 4 m and employing Xe gas at a pressure of 10 Atm will detect about 1000 events for a typical supernova explosion at 10 kpc [19]. A world wide network of several such simple, stable and low cost supernova detectors is proposed.

It is also important to study the behavior of this spherical proportional counter using well known terrestrial neutrino sources. One such source is Spallation Neutron Source (SNS) at the Oak Ridge National Laboratory [20] that provides an excellent opportunity to employ and test this new detector [21].

For underground tests another similar detector of 1.3 m in diameter has been installed in the LSM laboratory in Modane (Frejus) under 1700 m of rock (4800 meters water equivalent) providing protection from cosmic rays. Comparisons between measurements in the sub-keV energy region taken with both detectors are carried out in order to understand the background level and optimize the detector in terms of sensitivity and noise background. The development of a new spherical prototype 70 cm in diameter made of low radioactivity materials is in progress and will be installed in the LSM laboratory. By using a mixture of gas containing ^{3}He, it will be possible to measure not only the thermal neutron component [22] but also the fast neutron energy spectrum by taking profit of the 1 to few barn cross section of capture above 100 keV neutron energy. By adding a lead and Polyethylene shields, it will also be possible to measure the residual background at low energy and assess potential sensitivity to dark matter particle search and coherent neutrino-nucleus scattering.

5. Conclusions

A new detector has been developed that combines large mass and low sub-keV energy threshold with the capability of sensing even single electrons. This is a new type of radiation detector based on the radial geometry with spherical proportional amplification read-out. Its moderate cost, simplicity and robustness, make this technology a promising approach to NC based detection of reactor and astronomical neutrinos and opens a new window in dark matter searches.

REFERENCES

[1] E. Bougamond et al., arXiv: 1010.4132 [physics.ins-det], to be published at JMP (2010).
[2] K. Nakamura et al. (Particle Data Group), J. Phys. G 37: 075021 (2010).
[3] I. Giomataris and J.D. Vergados, Nucl. Instrum. Meth. A530: 330-358 (2004).

[4] J.I. Collar, Y. Giomataris, Nucl. Instrum. Meth. A471: 254-259 (2000).
[5] R. Bernabei et al., Eur. Phys. J. C67: 39-49 (2010).
[6] A. Dedes, I. Giomataris, K. Suxho, J.D. Vergados, Nucl. Phys. B826: 148-173 (2010).
[7] J. Monroe and P. Fisher, Phys. Rev. D76: 033007 (2007).
[8] D. Santos et al., arXiv: 0810.1137 [astro-ph] (2008).
[9] C. Grignon et al., JINST 4: P11003 (2009).
[10] S. Ahlen et al., Int. J. Mod. Phys. A25: 1-51 (2010).
[11] D.Z. Freedman et al., Ann. Rev, Nucl. Sci. 27: 167 (1977).
[12] A. Drukier and L. Stodolsky, Phys. Rev. D30: 2295 (1984).
[13] J.D. Vergados et al., Phys. Rev. D69: 113001 (2009).
[14] H.T. Wong, Mod. Phys. Lett. A23: 1431-1442 (2008).
[15] C.E. Aalseth et al., arXiv:1002.4703 [astro-ph.CO] (2010).
[16] S. Aune et al., AIP Conf. Proc. 785: 110-118 (2005).
[17] I. Giomataris et al., Nucl. Phys. Proc. Suppl. 150: 208-213 (2006).
[18] I. Giomataris et al., JINST 3: P090007 (2008).
[19] I. Giomataris and J.D. Vergados, Phys. Lett. B634: 23-29 (2008).
[20] K. Scholberg, Phys.Rev.D73: 033005 (2006).
[21] J.D. Vergados, F.T. Avignone III and I. Giomataris, Phys .Rev. D79: 113001 (2009).
[22] I. Giomataris et al., J. Phys.: Conf. Ser. 179: 012003 (2009).

Journal of Physics: Conference Series **309** (2011) 012024 doi:10.1088/1742-6596/309/1/012024

Low Energy Neutrino and Dark Matter Physics with sub-keV Germanium Detectors

Henry T. Wong
Institute of Physics, Academia Sinica, Taipei 11529, Taiwan
E-mail: htwong@phys.sinica.edu.tw

Abstract. The theme of the TEXONO-CDEX research program is on the studies of low energy neutrino and dark matter physics at Kuo-Sheng Reactor Neutrino Laboratory and China Jin-Ping Underground Laboratory. The current goal is to open the "sub-keV" detector window with germanium detectors. The three main scientific subjects are neutrino magnetic moments, neutrino-nucleus coherent scattering, and dark matter searches. We highlight the status, results and plans in this article.

1. Physics Motivations and Goals

Results from recent neutrino experiments provide strong evidence for neutrino oscillations due to finite neutrino masses and mixings[1]. Their physical origin and experimental consequences are not fully understood. Experimental studies on the neutrino properties and interactions can shed light on these fundamental questions and constrain theoretical models, from which unexpected surprises may arise. It is therefore highly motivated to look for and establish alternatives of neutrino sources and detection channels, especially in regions of parameter space which are experimentally unexplored.

In another arena, there are compelling evidence that about 20% of the energy density in the universe is composed of Cold Dark Matter[2] due to a not-yet-identified particle, generically categorized as Weakly Interacting Massive Particle (WIMP, denoted by χ). A direct experimental detection of WIMP is one of the biggest challenges in the frontiers of particle physics and cosmology.

The TEXONO Collaboration has contributed in formulating the physics program and in making technical advances to open a detector window in the previously unexplored "sub-keV" regime[3] with low-energy germanium detectors. There is by now increasing interest worldwide. The generic "benchmark" goals in terms of detector performance are: (1) modular target mass of order of 1 kg; (2) detector sensitivities reaching the range of 100 eV; (3) background at the range of 1 $kg^{-1}keV^{-1}day^{-1}$ (cpkkd). The neutrino physics program is being pursued at the established Kuo-Sheng Reactor Neutrino Laboratory (KSNL), while dark matter searches will be conducted at the new China Jin-Ping Underground Laboratory (CJPL)[4] where construction was recently (Fall 2010) completed. The two facilities are depicted schematically in Figures 1a&b.

1.1. Low Energy Neutrino Physics at KSNL

The various neutrino-induced interactions at KSNL are shown in Figure 2a. The TEXONO program has contributed to the studies of neutrino magnetic moments[5, 6] with a 1-kg high-

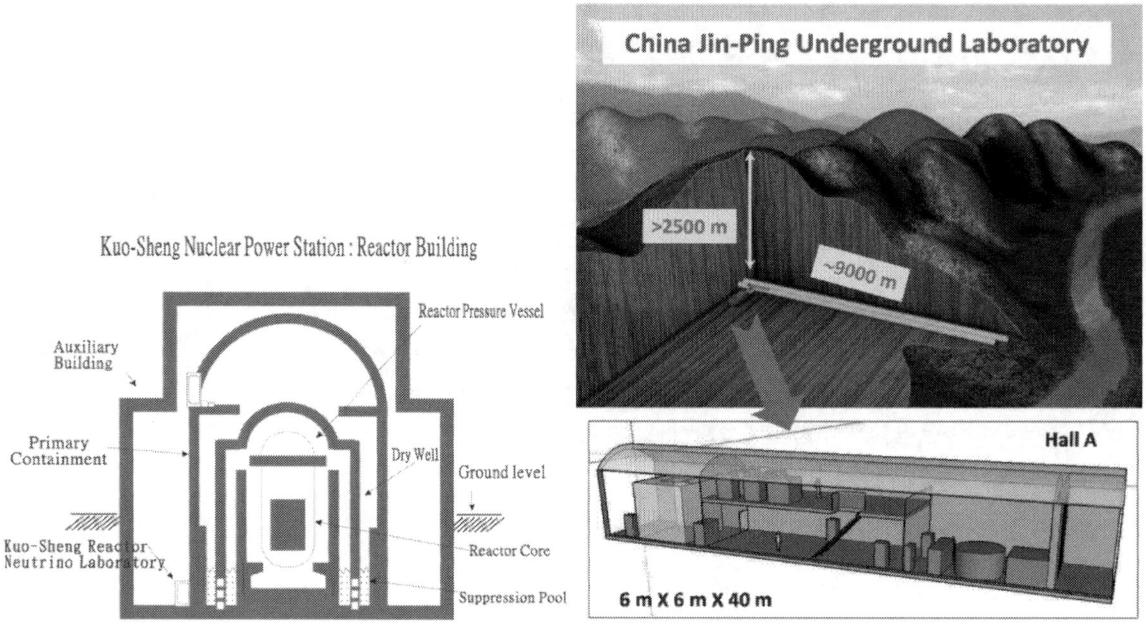

Figure 1. Schematic diagrams of the two facilities where our experiments are conducted: (a) Left: Kuo-Sheng Reactor Neutrino Laboratory (KSNL) (b) Right: China Jin-Ping Underground Laboratory (CJPL).

purity germanium detector (HPGe) and the measurement of neutrino-electron cross-section[7] with a 200-kg CsI(Tl) scintillator array. An important next goal with sub-keV germanium detectors is to study neutrino coherent scattering with the nucleus (νN)[8]: $\nu + \mathrm{N} \rightarrow \nu + \mathrm{N}$, a fundamental neutrino interaction which has never been observed. Measurement of the νN coherent scattering is a sensitive test to the Standard Model[9]. It provides a probe to the weak nuclear charge and radiative corrections due to possible non-standard neutrino interactions or additional neutral gauge bosons. The coherent interaction plays important role in astrophysical processes by being the dominant energy loss process. It is significant to the neutrino dynamics, energy transport and possible collective effects in supernovae and neutrons stars[10]. It may be a promising avenue towards a compact and relatively transportable neutrino detector, an important application of which can be the real-time monitoring of nuclear reactors and supernovae detection[11].

The maximum nuclear recoil energy at momentum transfer much larger than neutrino masses is given by: $\mathrm{T_N^{max}} = 2\mathrm{E}_\nu^2/(\mathrm{M_N} + 2\mathrm{E}_\nu)$ The differential cross section for coherent scattering versus nuclear recoil energy for Ge target (A=72.6) with typical reactor $\bar{\nu_e}$ spectra is displayed in Figure 2a, where $\mathrm{T_N^{max}} \sim 2$ keV.

In ionization detectors like Ge, the measure-able energy is only a fraction of the energy deposited in nuclear recoils due to their large dE/dx. The *quenching factor (QF)*, defined as the ratio of the measure-able to the deposit energy, is about 0.2-0.25 for Ge in the <10 keV region[12]. Accordingly, the maximum measure-able energy for νN in Ge due to reactor $\bar{\nu_e}$ is about 400-500 eV. At QF=0.25, the typical event rate at KSNL will be 11 $\mathrm{kg^{-1}day^{-1}}$ or 4000 $\mathrm{kg^{-1}yr^{-1}}$. The signal-to-background ratio will be >22 at the benchmark 1 cpkkd level.

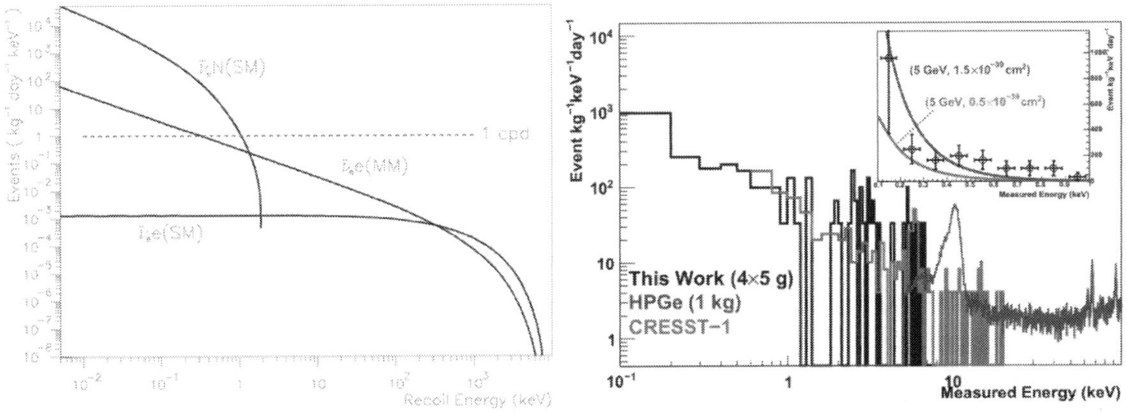

Figure 2. (a) Left: The differential cross-section of the various neutrino interaction channels at KSNL with Ge as target, where $\phi(\bar{\nu}_e)=10^{13}$ $cm^{-2}s^{-1}$ and $\mu_\nu = 10^{-10}$ μ_B. The background level of 1 cpkkd ($kg^{-1}keV^{-1}day^{-1}$) is also shown. (b) Right: The measured spectrum of ULEGe with 0.338 kg-day of data, after various background suppression procedures. Background spectra of the CRESST-I experimentand the HPGe are overlaid for comparison.

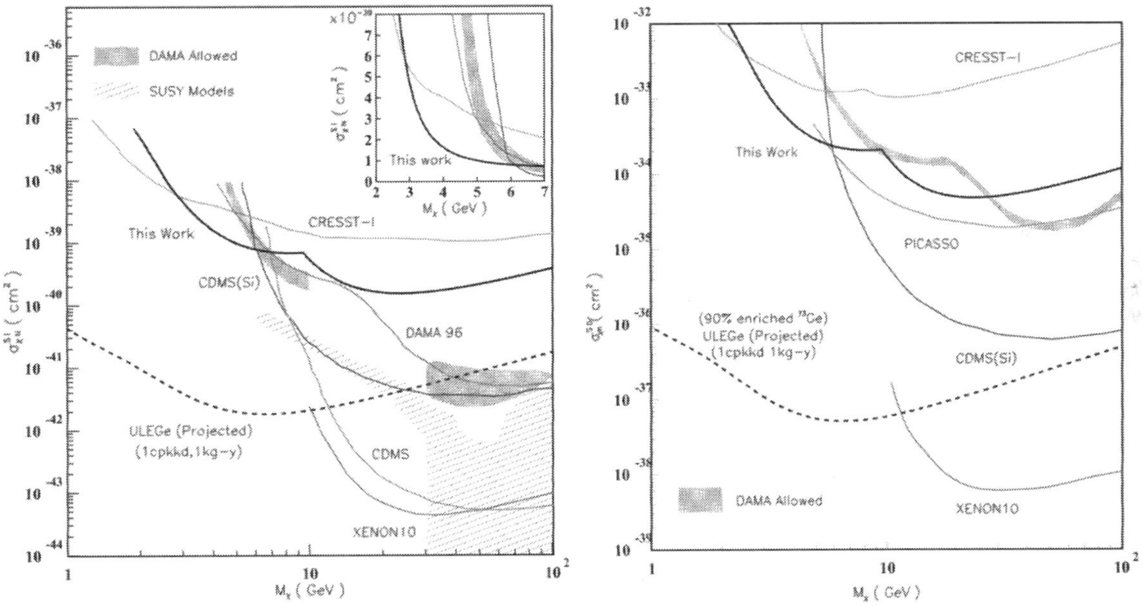

Figure 3. Exclusion plots of (a) Left: spin-independent χN and (b) Right: spin-dependent χN cross-sections versus WIMP-mass, displaying the KSNL-ULEGe limits and those defining the current boundaries The DAMA allowed regions are superimposed. The striped region is that favored by SUSY models. Projected reach of experiments at benchmark sensitivities are indicated as dotted lines. The relevant region is presented with linear scales in the inset.

1.2. Cold Dark Matter Searches at KSNL and CJPL

The WIMPs interact with matter pre-dominantly via the same coherent scattering mechanism like the neutrinos: $\chi + N \rightarrow \chi + N$. There may be both spin-independent and spin-dependent interactions between WIMP and matter.

Supersymmetric (SUSY) particlesare the leading WIMP candidates. The popular SUSY models prefer WIMP mass (m_χ) of the range of ~100 GeV. Most experimental programs optimize their design in the high-mass region and exhibit diminishing sensitivities for $m_\chi < 10$ GeV, where there is an allowed region if the annual modulation data of the DAMA experiment[13] are interpreted as WIMP signatures. There are increasing theoretical interest in this light-WIMP region[14, 15], which include models on light neutralinos, non-pointlike SUSY candidates like Q-balls, as well as WIMPless, mirror, asymmetric, and singlet fermionic dark matter. To probe the low-mass region, detector with sub-keV threshold is necessary. Such threshold presents a formidable challenge to detector technology and to background control.

A detector with 100 eV threshold will therefore open a window for Cold Dark Matter WIMP searches[2] in the unexplored mass range down to several GeV[3]. The uniqueness and advantages of having a low-threshold detector are twofold. Firstly, a new window of observation is opened for low-mass WIMPs. Secondly, the minimum velocity of WIMPs that produces χN is proportional to the recoil energy. Therefore, a lower threshold allows a larger range of WIMPs to contribute in an observable interaction and hence results in better sensitivities for all values of m_χ.

Based on data taken at KSNL with the 20-g prototype Ultra-Low-Energy Germanium detector (ULEGe), competitive limits were derived in the low WIMP mass region ($3 < m_\chi <$ 6 GeV)[16] exceeding those from the previous CRESST-I experiment[17]. The $\sigma_{\chi \mathrm{N}}^{\mathrm{SI}}$ versus m_χ and $\sigma_{\chi \mathrm{N}}^{\mathrm{SD}}$ versus m_χ exclusion plots are depicted in Figures 3a&b, respectively. The various results[18] defining the exclusion boundaries are also shown. The shaded region is a still-disputed allowed parameter space claimed by the DAMA group[13]. Projected sensitivities of a 1-kg ULEGe detector are also displayed. The CoGeNT experiment[19] first realized the Point-Contact Germanium detectors (PCGe)[20] which offer sub-keV sensitivities with detector of kg-size modular mass. Their 2010 results[19] at the Soudan Underground Laboratory, where both limit and allowed region were reported, inspire a host of theoretical investigations on possible low-mass WIMPs[15].

The underground facility CJPL[4], shown in Figure 1b, is located at Sichuan, China, with ~2500 meter of rock overburden and tunnel drive-in access. It is owned by the Ertan Hydropower Development Company, and managed by Tsinghua University, China. Construction of the first experimental hall ("Hall A") of dimension 6 m(width)X6 m(height)X40 m(length) and the first shielding structures were completed in September 2010. By the end of 2010, the necessary infrastructures, office and dormitory spaces are being installed. The first experiment with the 20 g ULEGe is being set up, followed by the 1-kg PCGe to be installed in 2011. Upgrades of detector to the 10-kg mass range at the 2012−2013 frame are planned. Potential reaches with benchmark sensitivities are depicted by dotted lines in Figures 3a&b.

2. Sub-keV Germanium Detector

Several R&D directions are intensely pursued towards improvement on the threshold and background for sub-keV germanium detectors:

(i) **Pulse Shape Analysis of Near Noise-Edge Events:**
It has been demonstrated that by studying the correlation of the Ge signals in two different shaping times[16] as depicted in Figure 4a, the threshold can be further reduced below the hardware noise edge via Pulse Shape Discrimination (PSD). The achieved thresholds at 50% signal efficiency are 220 eV and 310 eV for 20-g ULEGe and 500-g PCGe, respectively. As illustrations, the relative timing between the PCGe and anti-Compton (AC) NaI(Tl) detectors is shown in Figure 5a, for "sub-noise edge" events at 200-400 eV before and after PSD. Events in coincidence with AC at the "50−200 ns" window are due to multiple Compton scatterings, which are actual physical processes having similar pulse shapes as the neutrino and WIMP signals. The PSD selection efficiencies depicted in Figure 5b

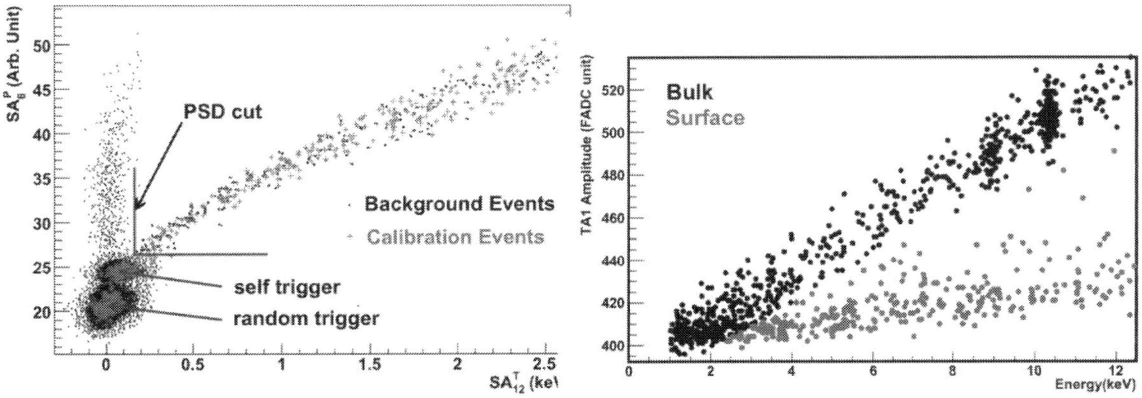

Figure 4. (a) Left: Scattered plots of the SA_6^P (shaping time 6 μs with partial integration) versus SA_{12}^T (shaping time 12 μs with partial integration) signals, for both calibration and physics events. The PSD selection is shown. (b) Right: Rise time plots, as characterized by the amplitude of timing amplifier (TA) signals, showing different behaviour between surface (faster) and bulk (slower) events.

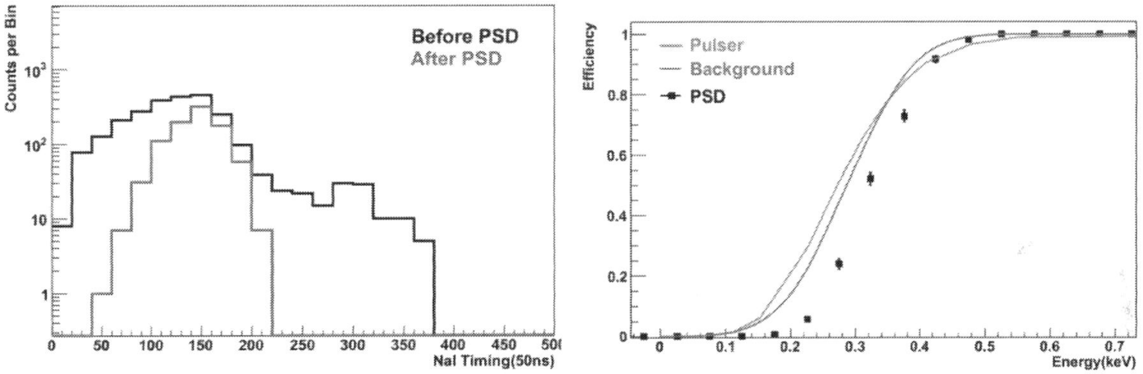

Figure 5. (a) Left: Selection of relative timing between AC-NaI(Tl) and PCGe systems, before and after PSD selection. The "50−200 ns" is the coincidence window where the signals at PCGe are mostly due to physics events correlated with the AC detector. (b) Right: The trigger efficiencies of the 500 g PCGe detector, as derived from the test pulser and *in situ* background events, respectively. PSD selection efficiencies were derived from the survival probabilities of AC-tagged events of (a).

were derived from the survival probabilities of these AC-tagged samples in the coincidence window. The trigger efficiencies were measured with two methods. The fractions of calibrated pulser events above the discriminator threshold provided the first measurement, while the studies on the amplitude distributions of *in situ* background contributed to the other. Further improvement are being made both on PSD algorithms and on efficiency measurement.

(ii) **Pulse Shape Analysis of Surface Vs Bulk Events:**
The surface and bulk events in PCGe can be separated by the rise time of the pulses as characterized by the amplitude of timing amplifier (TA) signals. It is illustrated in Figure 4b.

(iii) **Background Understanding and Suppression:**
The MeV-range background was understood to the percent level in our previous neutrino-electron measurement with CsI(Tl) scintillating crystal array[7]. However, the measured sub-keV spectrum of Figure 2b at KSNL could not be explained with standard background modeling on ambient radioactivity. Intense efforts on hardware cross-checks, further simulation and software analysis are underway. Data taking at CJPL, where the cosmic-induced background will be absent, will also elucidate the origin of the observed sub-keV events.

3. Prospects and Outlook

A detector with 1 kg mass, 100 eV threshold and 1 cpkkd background level has important applications in neutrino and dark matter physics, as well as in the monitoring of reactor operation. Crucial advances have been made in adapting the Ge detector technology towards these requirements. Relevant limits have been achieved in prototype studies at KSNL on the WIMP couplings with matter. The sub-keV events are still to be understood. Intensive research programs are being pursued along various fronts towards realization of experiments which can meet all the technical challenges. Detectors with kg-scale are being deployed at KSNL and CJPL.

References

[1] Nakamura K and Petcov S 2010 *Review of Particle Physics J. Phys.* **G 37** 164, and references therein.
[2] Drees M and Gerbier G 2010 *Review of Particle Physics J. Phys.* **G 37** 255, and references therein.
[3] Wong H T et al 2006 *J. Phys. Conf. Ser.* **39** 266; Wong et al 2008 *J. Phys. Conf. Ser.* **120** 042013.
[4] Normile D 2009 *Science* **324** 1246; Kang K J et al 2010 *J. Phys. Conf. Ser.* **203** 012028; Feder T 2010 *Physics Today* **Sept 2010** 25.
[5] Wong H T and Li H B 2005 *Mod. Phys. Lett.* **A 20** 1103, are references therein
[6] Li H B et al 2003 *Phys. Rev. Lett.* **90** 131802; Xin B et al 2005 *Phys. Rev.* **D 72** 012006; Wong H T et al 2007 *Phys. Rev.* **D 75** 012001.
[7] Deniz M et al 2010 *Phys. Rev.* **D 81** 072001; Deniz M et al 2010 *Phys. Rev.* **D 82** 033004.
[8] Freedman D S 1974 *Phys. Rev.* **D 9** 1389; Gaponov Y V and Tikhonov V N 1977 *Sov. J. Nucl. Phys.* **26** 31; Sehgal L H and Wanninger M 1986 *Phys. Lett.* **B 171** 107.
[9] Krauss L M 1991 *Phys. Lett.* **B 269** 407; Barranco J, Miranda O G and Rashba T I 2005 *JHEP* **12** 021; Scholberg K 2006 *Phys. Rev.* **D 73** 033005; Barranco J, Miranda O G and Rashba T I 2007 *Phys. Rev.* **D 76** 073008.
[10] Wilson J R 1974 *Phys. Rev. Lett.* **32** 849; D.Z. Freedman D Z, Schramm D N and Tubbs D L 1977 *Annu. Rev. Nucl. Sci.* **27** 167.
[11] Horowitz C J, Coakley K J and McKinsey D N 2003 *Phys. Rev.* **D 68** 023005; Learned J 2005 *Nucl. Phys. B (Proc. Suppl.)* **143** 152.
[12] Jones K W and Kraner H W 1971 *Phys. Rev.* **C 1** 125; Jones K W and Kraner H W 1975 *Phys. Rev.* **A 11**, 1347; Shutt T et al 1992 *Phys. Rev. Lett.* **24** 3425; Messous Y et al 1995 *Astropart. Phys.* **3** 361.
[13] Savage C, Gondolo P and Freese K 2004 *Phys. Rev.* **D 70** 123513; Gondolo P and Gelmini G 2005 *Phys. Rev.* **D 71** 123520.
[14] Bottino A et al 2005 *Phys. Rev.* **D 72** 083521;
[15] Fleming G, Kusenko A and Nussinov S 2002 *Phys. Rev. Lett.* **89** 101302; He X G et al 2007 *Mod. Phys. Lett.* **A 22** 2121; Feng J L, Kumar J and Strigari L E 2008 *Phys. Lett.* **B 670** 37; Kaplan D E, Luty M A and Zurek K M 2009 *Phy. Rev.* **D 79** 115016; Foot R 2009 *Phys. Rev.* **D 80** 091701; Kim Y G and Shin S 2009 *JHEP* **0905** 036; Fitzpatrick L, Hooper D and Zurek K M 2010, arXiv:1003.0014.
[16] Wong H T 2008 *Mod. Phys. Lett.* **A 23** 1431; Lin S T et al 2009 *Phys. Rev.* **D 76** 061101(R).
[17] Angloher G et al 2002 *Astropart. Phys.* **18** 43.
[18] Bernabei R et al 1996 *Phys. Lett.* **B 389** 757; Akerib D S 2006 *Phys. Rev. Lett.* **96** 011302; Angle J et al 2008 *Phys. Rev. Lett.* **100** 021303.
[19] Aalseth C E et al 2008 *Phys. Rev. Lett.* **101** 251301; Aalseth C E et al 2010 arXiv:1002.4703.
[20] Luke P N et al 1989 *IEEE Trans. Nucl. Sci.* **36** 926; Barbeau P A, Collar J I and Tench O 2007 *JCAP* **09** 009.

Journal of Physics: Conference Series **309** (2011) 012025 doi:10.1088/1742-6596/309/1/012025

A Parameterized Model for Supernova Electron Antineutrino Emission and Its Applications

Francesco Vissani[1], **Giulia Pagliaroli**[1], **Marialaura Costantini**[2]
[1]INFN, Laboratori Nazionali del Gran Sasso, Assergi (AQ), Italy
[2]ICRANet, Pescara (PE), Italy

Abstract. We discuss a parameterized model that describes the electron antineutrino flux emitted by core collapse supernovae. We illustrate its flexibility showing that it can be used to obtain: 1) evidence for an initial increase of the neutrino luminosity from SN1987A; 2) refined neutrino mass bounds; 3) a tool to simulate and analyze data; 4) the proof that neutrinos are an effective trigger for GW search; 5) expectations and theoretical errors for the diffuse neutrino background. Some results obtained with this model are shown in Table 1.

Table 1. Sample results based on the model of refs. [1, 3].

Analysis of SN1987A	Duration of the accretion	$\tau_a = 0.55$ s	[1]
	Neutrinosphere radius	$R_c = 16$ km	[1]
	Neutrino mass	$m_\nu < 5.8$ eV	[2]
Future supernova	Duration of the accretion	$\delta\tau_a/\tau_a$=10%	[3]
	Neutrinosphere radius	$\delta R_c/R_c = 5\%$	[3]
	Neutrino mass	$m_\nu < 1$ eV	[2]
Gravity Wave Burst	Time of the burst	$\delta t < 15$ ms	[3]
Diffuse Background	Counting rate in SK	(0.3-0.9)/yr	[4]
	Counting rate in SK+	(1.1-2.9)/yr	[4]

1. An improved parameterized model

1.1. Necessity of improvements

There is a high chance that, in the next 50 ± 20 years, neutrinos from a galactic core collapse supernova will bless Super-Kamiokande, KamLAND, Baksan, LVD, Borexino, MiniBOONE, IceCUBE and/or future detectors. Super-Kamiokande (SK) alone will see *1,000-100,000* events, at least 2 orders of magnitudes greater than seen from SN1987A. This will lead to progresses that we can be barely imagine today; what it is easy to imagine is that we will use models for the neutrino emission to extract information from the data.

Journal of Physics: Conference Series **309** (2011) 012025 doi:10.1088/1742-6596/309/1/012025

Many papers on supernova neutrinos adopt, without an adequate discussion, the same simplified model for the neutrino flux. Up to non-thermal effects, this model describes just a black body emission:

$$\Phi_{\bar{\nu}_e} \propto R_c^2 \times E^2 \, e^{-E/T} \tag{1}$$

and it takes into account the fact that the signal has a limited temporal extension by prescribing $T = T_c \; e^{-\frac{t}{4\tau_c}}$ and R_c=const. or similar time evolutions. We refer to this description as the "exponential cooling model". Often, the model is simplified even further: time evolution is ignored altogether and only the energy spectrum is considered.

The exponential cooling model has been almost universally used for the analyses of SN1987A data (with just two exceptions, see below). Our criticisms to this model are the following:

- It *does not resemble* the expectations.
- It is likely to be *inadequate* for future supernova.

These are not minor concerns. The upper panel of Fig. 1 shows the luminosity found in a numerical simulation of the emission [5]. We have a smooth evolution of the luminosity, that somewhat resembles an exponential cooling, but only after half a second. In the first half second, the luminosity is much larger. The exponential cooling model, alone, does not provide us with an adequate modeling of the flux; while this model could be adequate for some applications, it is evident that some basic information about the time distribution is missing.

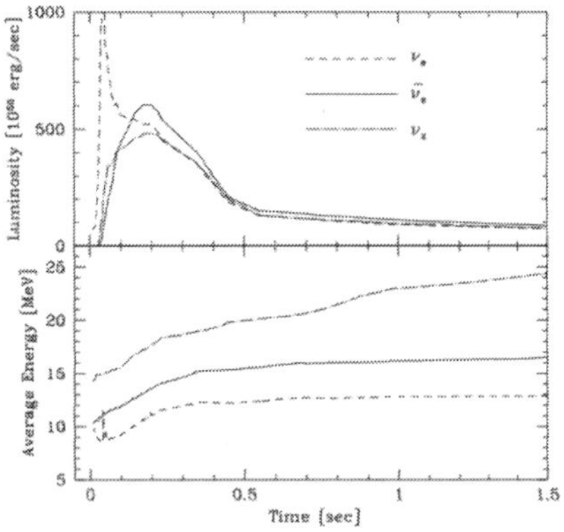

Figure 1. A typical outcome of a numerical simulation. In the upper panel, the luminosity (i.e., power) radiated in neutrinos of all types; in the lower panel, their average energy. From Totani, Sato, Dalhed, Wilson [5].

1.2. The improved model

Let us begin by outlining the physical picture of the emission. Following Loredo & Lamb [6], we aim to describe the transition from an intense initial emission to a black-body emission. The nature of the neutrino emission is assumed to change according to this scheme:

$$\text{Volume} \rightarrow \text{Surface}$$

i.e., we pass from an emission by a transparent atmosphere to the usual surface emission of a black-body. We illustrate the conditions during the initial stages of neutrino emission, as depicted in Figure 2. Just after the collapse, rapid accretion processes will produce a thermalized electromagnetic plasma composed of γ, e^-, e^+, which yields copiously $\bar{\nu}_e$ through

$$e^+ + n \rightarrow p + \bar{\nu}_e. \tag{2}$$

We assume that the $\bar{\nu}_e$ can escape freely. This means that only a certain fraction of the neutrons, the ones around the nascent neutron star, will effectively act as targets for the collisions of the

Journal of Physics: Conference Series **309** (2011) 012025 doi:10.1088/1742-6596/309/1/012025

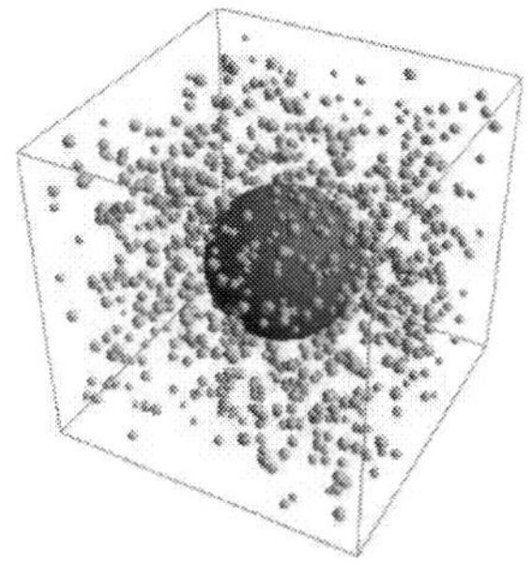

Figure 2. Sketch of the center of the collapsed star during the initial neutrino emission; in gray the nascent neutron star, in yellow the extra emission from its atmosphere.

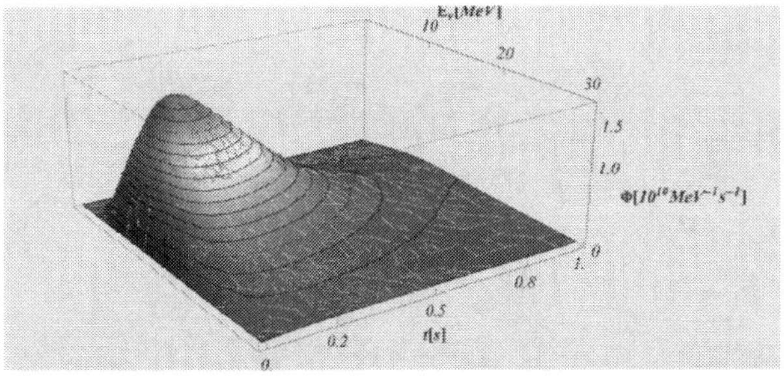

Figure 3. Our electron antineutrino flux $\Phi_{\bar{\nu}_e}(t, E_\nu)$ in the first second of emission and for a fixed choice of astrophysical parameters. The distance is 10 kpc and the intensity is shown in linear scale. The initial peak is unmistakable.

positrons. In order to check this hypothesis, we examine the expectations for the luminosity, assuming reasonable values for the number of neutrons N_n and for the temperature of positrons T_{e^+} [7]. In a transparent medium, we have from Eq. 2:

$$L_{\bar{\nu}_e} \sim N_n \cdot \sigma_{e^+n} \cdot T_{e^+}^4 \sim 5 \times 10^{52} \ \frac{\mathrm{erg}}{\mathrm{s}} \tag{3}$$

similar to the maximum luminosity resulting from Fig. 1. Later, we expect the fainter emission from the surface of the neutron star that follows from Eq. 1:

$$L_{\bar{\nu}_e} \sim R_c^2 \cdot T_{\bar{\nu}_e}^4 \sim 5 \times 10^{51} \ \frac{\mathrm{erg}}{\mathrm{s}} \tag{4}$$

This second emission corresponds to the conventional exponential cooling model discussed in the previous section. From Eqs. 3 and 4, we see that it is not difficult to improve the model by accounting for the initial peak of luminosity as indicated by the existing numerical calculations of the neutrino emission. We need two phases of emission, each one with three parameters that describe the intensity, the average energy and the time scale, namely: N_n, T_{e^+}, τ_a and $R_c, T_{\bar{\nu}_e}, \tau_c$. We join smoothly the two emission phases and include oscillations in normal hierarchy. The result is illustrated in Fig. 3 for a fixed choice of the parameters. Considering that this is a rather flexible parameterization of the flux, the agreement with Fig. 1 can be considered satisfactory.

2. Applications

2.1. A new fit of SN1987A neutrino data

A fit to the SN1987A data collected by Kamiokande-II, IMB and Baksan, based on our model, has been performed in [1]. The fit returns an antineutrino flux quite similar to the simulations. In particular, the luminosity is an order of magnitude larger in the first half of a second, as expected in the standard scenario for the emission. The total energy is estimated to be 2.2×10^{53} erg;[1] for a discussion of the importance of this quantity, see [8]. The new model describes the data 2.5 σ better than the exponential cooling model after accounting for the larger number of parameters. A sensible question is: what is the reason of the improvement? In particular, is it caused by some subtlety in the statistical analysis or does it depend on a specific feature of the data themselves?

[1] However, this result is heavily based on the hypothesis of equipartition. It will be an important experimental challenge to extract this quantity, with minimum theoretical bias, from the data of a future supernova.

The correct interpretation is the second one, as clear from the time distribution of the observed events given in Figure 4: the initial peak, comprising 30% of the detected events, indicates an increased electron antineutrino luminosity in the first instants of the emission.

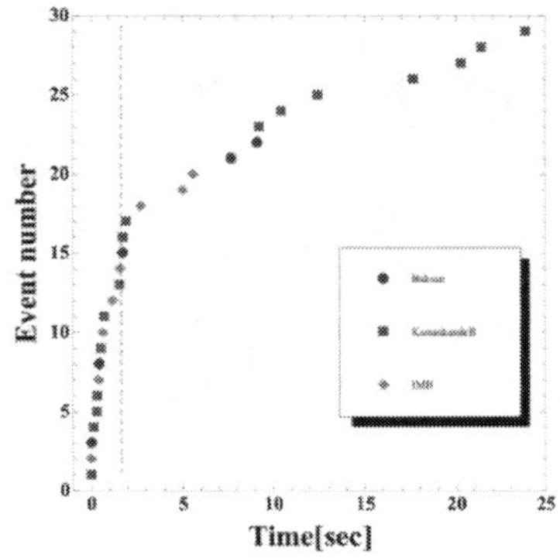

Figure 4. Cumulative distribution of SN1987A events as a function of the time from the first one [9]. The remarkable, initial peak of events is readily explained by the increased luminosity expected in the standard emission. We show only the events of Kamiokande-II, IMB and Baksan that, within errors, can be interpreted as almost simultaneous. The first event of each detector coincide: in fact, the clocks of Kamiokande-II and Baksan gave the absolute time only within large errors, thus the times among different detector acquire meaning only after the fit. We do not attempt an interpretation of LSD data.

2.2. A tool to simulate and analyze the response to a future supernova

Once we have a parameterized model [10], we can produce sets of simulated events and use them to validate a procedure of analysis or to estimate whether an experiment will permit us to understand a certain phenomenon. We show in Fig. 5 an example of a simulated dataset in a detector with 22.k kton mass of water. Its analysis suggests that, with a future supernova, a precise determination of the astrophysical parameters will be possible. The physics of the emission and, in particular, the existence of the intense initial emission of the Bethe & Wilson+Nadyozhin scenario, will be probed much more accurately. In fact, even from 20 kpc, we find $\delta\tau_a/\tau_a \approx 10\%$. The small SN1987A dataset leads to a much larger error, $\tau_a = 0.55^{+0.58}_{-0.17}$ s, though the central value agrees remarkably well with the expectations of the Bethe & Wilson+Nadyozhin scenario. For details, the main reference is [3].

2.3. Neutrino mass and supernova events

A finite mass implies that neutrinos with different energies have different velocities. If we know when the emission began, and if we see a neutrino event shortly after, we can put a strong bound on the mass, especially if its energy is low.[2] However, in order to learn something about the masses, it is necessary to know something about the emission. Actually, it is enough to know the shape of the signal, up to the unknown parameters of the emission. Thus our parameterized model offers an appropriate tool to perform such an analysis. The duration of the accretion determines the bound on the mass shown in Tab. 1. For a future supernova, the limit can be improved up to about 1 eV. See [2, 7] for details and bibliography.

2.4. A remark on supernova neutrino astronomy

The main number of events in conventional detectors (including those which observed SN1987A) are those due to "inverse beta decays" (IBD) which is, actually, the inverse of the reaction in Eq. 2:

$$\bar{\nu}_e + p \to e^+ + n \tag{5}$$

[2] Generally, we do not know the beginning of neutrino emission. Thus, in the analysis of the data, we should consider as a free parameter the amount of time between the first neutrino arrived in the detector and the first neutrino signal that was detected. However, the delay due to neutrino mass should be determined by the same data. The two effects act in opposite senses, which creates numerical difficulties in the analysis.

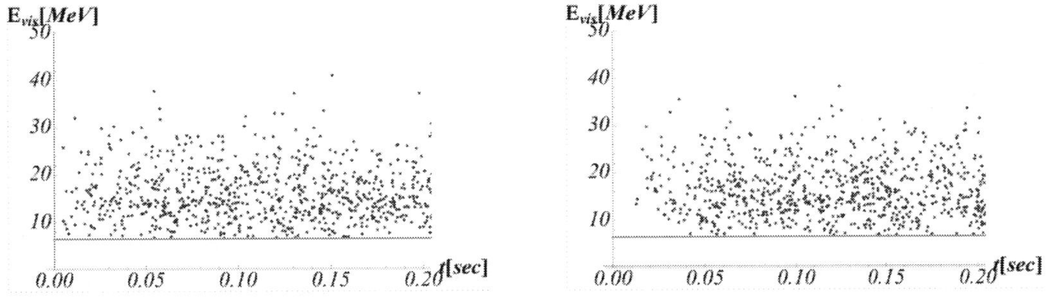

Figure 5. Early simulated antineutrino events from a supernova at 10 kpc; in the left graph, the neutrino mass is set to $m_\nu = 0$, in the right one, to 2 eV, see [2]. The horizontal line indicates the energy threshold.

This reaction has the largest cross section for detection but is only weakly directional. The "elastic scattering" (ES) reaction

$$\nu + e \to \nu + e \tag{6}$$

has opposite properties in this respect: We can identify the direction of the neutrino source, just as in ordinary astronomy, thanks to the fact that the electrons are scattered almost in the same direction of the neutrinos. Of course, the limited data set does not permit a very precise identification; though, we expect to be able to locate a supernova at the border of the Milky Way with a precision of some degrees.[3] For details and refererences, see [11] and [3].

2.5. Search for gravity wave burst

There are various moments when gravity waves (GW) could be emitted during a core collapse. One possibility is during the formation of the neutron star, when the inner core reaches the highest densities and bounces. Several simulations find that this moment is associated with a GW burst emission, even if, sometimes, this emission is found to be rather weak. The process of emission is pretty fast, lasting only some tens of milliseconds, and should almost coincide with the beginning of neutrino emission. The question that we would like to consider is the following: how effective are neutrinos to indicate the moment of the emission?[4] The time of the bounce, when a GW burst emission should occurr, can be obtained from:

$$T_{\mathsf{bounce}} = T_{\mathsf{1st}} - t_{\mathsf{mass}} - t_{\mathsf{GW}} - t_{\mathsf{fly}} - t_{\mathsf{resp}} \tag{7}$$

where the first term in the right hand side is the time of the first neutrino event and:

(i) the delay t_{mass} due to the mass is small, if the cosmological limit of about 1 eV applies;

(ii) the intrinsic delay t_{GW} is just 1.5-4.5 ms;

(iii) the error on the geometrical delay t_{fly} is below 5 ms using ES events;

(iv) the most relevant term is t_{resp}, the delay between the first neutrino and the first event.

[3] In order to improve the identification of the ES signal, one could impose the correlation of the observed directions of arrival with the galactic plane. Moreover, it is possible to adopt some criteria to increase the purity of the ES sample: e.g., kinematics imposes that the energy is shared between the final-state electron and neutrino, thus reducing the observable energy. Also, according to recent simulations, the initial ES emission has a very high average energy, which would be in agreement with the first event observed by Kamiokande-II.

[4] The traditional attitude was "the GW burst will give the t_0 of the neutrino emission"; the new attitude is motivated by the expectation that we will be able to observe neutrinos with high statistics. There is no real contradiction between these views; indeed it would be important to know "t_0" independently from the neutrinos, e.g., to investigate neutrino masses. Moreover, considering SN1987A findings with open mind, we should not neglect the possibility of major surprises, also thanks to the new types of GW detectors [13].

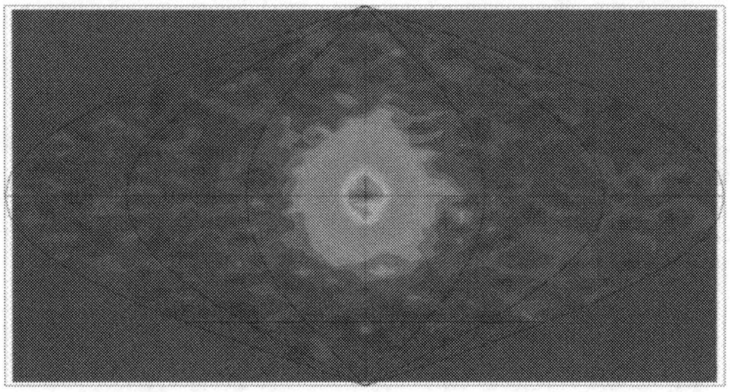

Figure 6. A famous plot obtained by Super-Kamiokande Collaboration, which proves that solar neutrinos (that are less directional than supernova neutrinos) permit us to identify the position of the Sun.

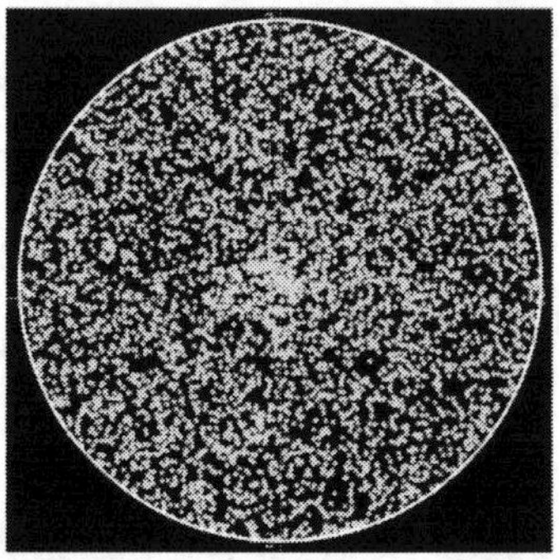

Figure 7. Directions of arrival of simulated supernova neutrino events for Super-Kamiokande. In the center, cluster of ES events [7].

The last quantity can be obtained by fitting the neutrino data. This is where the parameterized model enters [3]. Considering, as an example, Super-Kamiokande we find this quantity with an error less than 10 ms. In [12], IceCUBE is considered using a much simpler fit to the simulated data and yields a similar result. Finally, by summing the errors, we conclude that we can estimate the time of the bounce with a remarkable precision:

$$\delta T_{\mathsf{bounce}} = \sqrt{\delta t^2_{\mathsf{mass}} + \delta t^2_{\mathsf{GW}} + \delta t^2_{\mathsf{fly}} + \delta t^2_{\mathsf{resp}}} \approx 15 \text{ ms} \tag{8}$$

This accuracy in the measurement of the time is well-matched to the duration of the GW burst, ≈ 30 ms. Thus, neutrino observatories should provide gravity wave observatories, such as Virgo and LIGO, with a good trigger to search for a GW burst.

2.6. Diffuse supernova neutrino background

The diffuse supernova neutrino background (DSNB), or relic supernova neutrinos, is the result of the cosmic supernova emissions cumulated in the course of time. In principle this is visible in an energy window approximatively above $E_{th} = 20$ MeV (or 10 MeV, depending on detector capabilities). Unfortunately, DSNB gives a feeble signal;[5] but, in view of its interest in astrophysics and particle physics, it is the target of experimental search. The best limit, to date, is due to SK [14]. There are interesting prospects to proceed further in the experimental investigation, e.g., improving the tools of analysis or adding Gadolinium, which corresponds to the entry denoted by SK+ in Table 1. Assuming that the SN1987A was representative of the standard electron antineutrino emission, ref. [4] has used our parameterized model to predict the expected DSNB signal and the error-bar in the prediction (i.e., the uncertainty). The results of the analysis are given in Table 1. It is reassuring to note that a completely different procedure [16], based on a comparison of different theoretical models for the emission, yields very similar results for the error-bar.

3. Summary and discussion

We discussed a parameterized model for the analysis of supernova neutrinos [1, 3, 10], which elaborates on the earlier proposal of Loredo & Lamb [6]. We argued that our model is adequate for the analysis of SN1987A observations and it is significantly superior to those used in previous

[5] The number of DNSB events per year is roughly the same as those from a supernova as SN1987A, exploded 6 times farther from us and detected with an efficiency $\epsilon(E) = E/E_{th} - 1$ for $E > E_{th}$ [4].

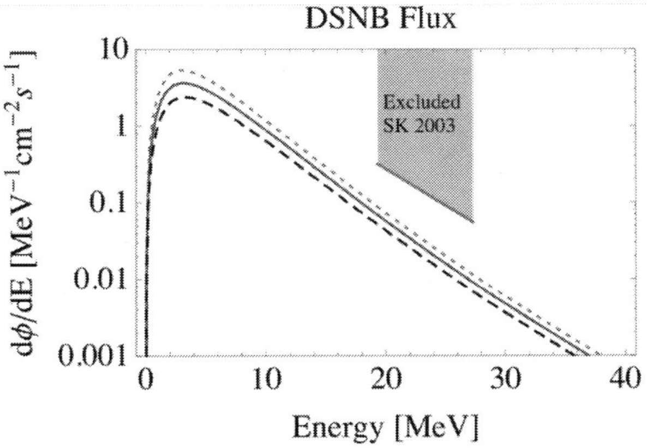

Figure 8. Expected DSNB flux derived from SN1987A [4]. The three lines correspond to various cosmological models for the rate of supernova explosion [15]. Also shown the upper limit of Super-Kamiokande 2003 [14].

analyses, as those described in [17] or those presented in [18].[6] We illustrated its usefulness, by showing several applications in particle physics and astrophysics.

We know that this model could be still improved–or even corrected, if necessary–with the contribution of astrophysicists. Such a step will be most probably necessary to interpret the result of a future observation, but it is not excluded that it could be useful also to understand even better the existing ones. (This statement is independent from the dubious aspects of SN1987A, see [9] for a review; in the dry language of statististics, we are just stressing the need to correctly formulate what is the null-hypothesis to interpret the data collected by Kamiokande-II, IMB and Baksan.)

However, the problem of designing parameterized models to describe neutrino emission is not felt as urgent among theorists interested in supernovae: astrophysicists concentrate towards understanding the explosion (see e.g. [20]), whereas, particle physicists concentrate on the oscillations (see e.g. [21, 18]). In the field of supernova neutrinos, experimentalists and theorists are working independently as happens in many other fields in physics; but it is curious that there is so little interaction among theorists with different expertises, though they are interested in understanding the same thing.

A historical account permits us to see the dangers of this attitude, see also [18]. Just after SN1987A, the average energy of the observed events–especially those of Kamiokande-II and Baksan–were generally considered below the expectations, see e.g., [17, 22] and the lower panel of Fig. 1. This opinion contributed to a diffuse feeling that SN1987A could be non-standard [18]. It also stimulated a long discussion on oscillations (see [9] for a critical review) eventually terminated when the average energies of the non-electronic antineutrinos were found be lower than previously expected [23]. Recently, also the average energies of electron antineutrinos started to diminish significantly, see [20, 18]. This implies that, after 20 years, the theory begins to match the only observation we have. In our view, this story also witnesses that we need more theoretical work on the neutrino fluxes, devoted to assess the theoretical errors in the predictions, and to identify the generic features of supernova neutrino emission.

We close recalling that many supernova neutrino detectors are operational and more will come [24]. They should be considered as telescopes, aimed at observing one of the most important chain of events in astronomy: the gravitational collapse of a stellar core, most likely followed by an explosion and by the formation of a compact star. Its observation by means of neutrinos, and possibly of gravitational waves, will permit us to progress greatly in astrophysics, nuclear

[6] The comparison with the results of Loredo & Lamb is discussed in [1] and [19]. Despite a large number of differences in the assumptions and in the model, we subscribe their main conclusion: the data of Kamiokande-II, IMB and Baksan suggest that the initial luminosity was much larger than the average one.

Figure 9. Illustration of how an excessive confidence or diffidence in the theory may paradoxically lead to the same result, namely, of stopping any further investigation of the observations.

physics and particle physics. A reliable, complete and flexible model for neutrino emission would be a precious tool to examine the expectations and to plan the next steps.

References

[1] Pagliaroli G, Vissani F, Costantini ML, Ianni A, Astropart. Phys. **31**, 163 (2009)
[2] Pagliaroli G, Rossi-Torres F, Vissani F, Astropart. Phys. **33**, 287 (2010)
[3] Pagliaroli G, Vissani F, Coccia E, Fulgione W, Phys. Rev. Lett. **103**, 031102 (2009)
[4] Vissani F, Pagliaroli G, Astronomy and Astrophysics **528**, L1 (2011)
[5] Totani T, Sato K, Dalhed HE, Wilson JR, Astrophys. J. **496**, 216 (1998)
[6] Loredo TJ, Lamb DQ, Phys. Rev. **D65**, 063002 (2002)
[7] Vissani F, Pagliaroli G, Rossi-Torres F, arXiv:1005.3682 (2010)
[8] Talk by Horowitz at this workshop
[9] Vissani F, et al., arXiv:1008.4726 (2010)
[10] A sample code can be downloaded at `http://theory.lngs.infn.it/astroparticle/sn.html`
[11] Tomas R, Semikoz D, Raffelt GG *et al.*, Phys. Rev. **D68**, 093013 (2003)
[12] Halzen F, Raffelt GG, Phys. Rev. **D80**, 087301 (2009)
[13] Talk by Cavalier at this workshop
[14] Malek M *et al.* [Super-Kamiokande Collaboration], Phys. Rev. Lett. **90**, 061101 (2003)
[15] Horiuchi S, Beacom JF and Dwek E, Phys. Rev. D **79** 083013 (2009)
[16] Ando S, Sato K, New J. Phys. **6**, 170 (2004)
[17] Bahcall JN, chapter 15 of "Neutrino Astrophysics," Cambridge Univ. Press (1989)
[18] Talk by Raffelt at this workshop
[19] Ianni A *et al.*, Phys. Rev. **D80**, 043007 (2009)
[20] Talk by Janka at this workshop
[21] Talks by Friedland, Fuller, Volpe at this workshop
[22] Schramm DN, Truran JW, Phys. Rept. **189**, 89 (1990)
[23] Keil MT, Raffelt GG, Janka H-T, Astrophys. J. 590, 971 (2003)
[24] Talks by Fulgione, Gil-Botella, Ianni, Koepke, Patzak, Scholberg, Vigorito,Virtue at this workshop

Journal of Physics: Conference Series **309** (2011) 012026 doi:10.1088/1742-6596/309/1/012026

SNEWS - The Supernova Early Warning System

C.Vigorito on behalf of the SNEWS Working Group
Dip.to di Fisica Generale, Università di Torino & INFN, via P. Giuria 1, 10125 Torino
E-mail: `vigorito@to.infn.it`

Abstract. The SuperNova Early Warning System is a cooperative project involving many experimenters coming from several, past and present, neutrino detectors. The primary goal of the collaboration is to provide the astronomical community with a prompt alert at the time of next galactic supernova. Requiring the online coincidence among neutrino detectors, world-wide distributed and with different detection techniques, SNEWS will generate a reliable, prompt and automated alert by eliminating the potential local false triggers.
In this paper I will give a concise overview on the project motivations, network implementation, working procedures and current status.

1. Introduction

Gravitational stellar collapses are astrophysical events of great interest. Because of the complexity of the problem, the modeling of the physical processes is still in evolution, but it is in general accepted that the role of neutrinos is critical to allow the supernova to form out of a collapse [1]. Moreover the confirmed detection (e.g. in [2], [3] and [4]) of the neutrino signal from the SN 1987A in the Large Magellanic Cloud marked the beginning of the neutrino astrophysics era. In spite of some unresolved controversies [5] the observation opened the way for a new method of investigation: the **neutrino astronomy**.
Even in the lack of a complete theory of the core collapse supernova explosion the correlated neutrino emission is believed to be well established and should be detected to maximize the scientific feed back with different active detectors when next event will occur within the Milky Way boundaries. More than 99% of the binding energy of the resulting neutron star is released in the form of $\nu/\bar{\nu}$ of all flavors at energy of tens of MeV, while the remaining 1% in the form of optically visible radiation and kinetic energy of the expanding remnant. The time scale of the neutrino emission is estimated few tens of seconds, promptly after the collapse, $\sim$ hours before the photon signal can emerge from the stellar envelope giving rise to the Type II supernova event.
A core collapse supernova in our Galaxy will bring therefore a wealth of scientific information. The early observation of the supernova light curve turn-on will bring informations about the progenitor and its environment, giving the astrophysicists the opportunity to learn more about the nature of the core collapse (e.g. the study of the explosion mechanism, black hole formation, accretion disks, explosive nucleo-synthesis) and, to particle physicists, valuable feed backs on neutrino properties themselves (e.g. absolute mass, mixing, magnetic moment).

The aim of SNEWS network is therefore to provide first the early warning of a supernova's occurrence and second to improve global sensitivity to the neutrino burst via inter-experiment collaboration and downtime coordination.

2. The network strategy

From the experimental point of view any neutrino detector can provide its own supernova trigger but, usually, the warning generated by dedicated analysis gets hours , or even days, to be carefully checked. From experience a standalone of a positive observation announcement is a complicated and long job involving open discussions within experiment collaboration before any public release can be performed.
The idea of a blind central coincidence computer receiving signals from several experiments is the basis of the SNEWS network and, beyond the early warning advantages, includes many parallel benefits as: increased sensitivity to weak signals in the case of distant SuperNova (SN) event, suppression of non-Poissonian background fluctuations which are unlikely correlated in different laboratories and detectors, automated alert without any human interference and finally the possibility to determine, even with a poor resolution, the incoming direction.

To make the best use of a neutrino burst supernova alert we follow the astronomical community requirements summarized by the following three P -**Prompt, Pointing, Positive**- scheme.

- *Prompt* Time is mandatory for a supernova early warning and thus the alert must be as prompt as possible to catch the early stages of the shock breakout. We estimate a potential alert dissemination of minutes for the automated alert, depending on the expected delay in the alarm formation at each single detector. A timescale of 5 minutes or even less have been demonstrated during the commissioning phase of the network and is periodically checked out by using test signals.

- *Pointing* Clearly, the more accurately we can point to a core collapse event using neutrino information, the more likely it will be that early light turn-on will be observed by astronomers. Even for the common case when no directional information is available (e.g. for only scintillator detectors online) it is still useful for astronomers to know that a gravitational collapse event has occurred. However, any pointing information is extremely valuable. There are two ways of pointing with neutrinos: (1) individual detectors can make use of asymmetric reactions for which the products remember the direction of the incoming neutrino, tipically the ν_e scatterig; (2) the timing of the neutrino burst via triangulation of signals recorded in different detectors around the globe.

- *Positive* It is clear that there must be no false supernova alerts to the world community. Requiring an inter-experiment coincidence the false alert rate can be decreased to nearly zero. We have chosen the nominal acceptable average false alert rate to be one per century, accordingly to the expected SN Type II event within the Milky Way boundaries.

When the emitted neutrino burst from a SN somewhere in the Milky Way will arrive at Earth the supernova monitor of each experiment forms the trigger and send it through a secure TCP/IP connection to the SNEWS server. If the server sees a $\geq$ 2-fold coincidence within 10 seconds, with at least 2 involved detectors at different laboratories, an automated supernova alert will be sent out by e-mail. This procedure eliminates the human time scale delay which affects procedure of verification by shifter and expert people at each single client detector. Sources of noise (i.e. electronic spikes, human errors, DAQ faults) that can mimic a SN real neutrino event are unlikely in coincidence at the same time at different laboratories, so far naturally

Journal of Physics: Conference Series **309** (2011) 012026 doi:10.1088/1742-6596/309/1/012026

suppressed. The trigger logic is summarized in figure 1.

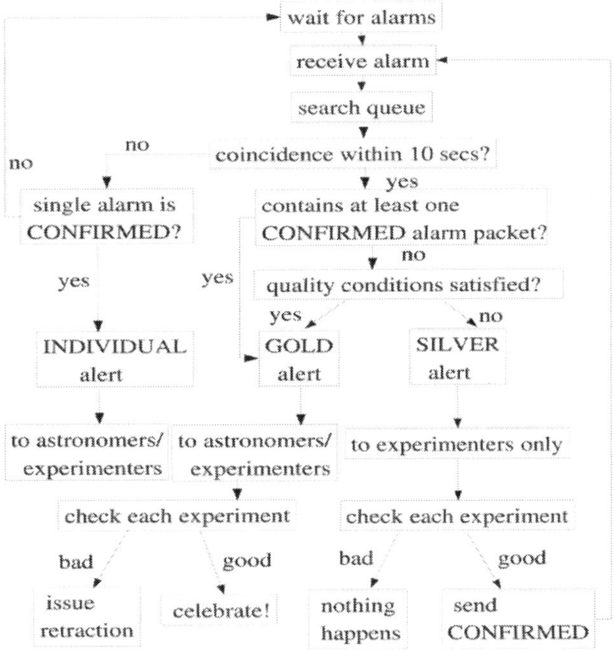

Figure 1. Flowchart of the decision procedure to determine the type of the global alert dispatch whatever the type *Gold*, *Silver* or *Individual*. Details on the alert type definition and logic are fully discussed in [6].

Once the alert is generated a dispatch containing the essential informations (e.g. the involved detectors, the absolute time and, if available, the error box of the direction in the sky) is sent out via e-mail in order to join as many people as possible.
That is why SNEWS maintains a mailing list of interested parties including professionals and amateurs to be alerted in the case of a positive signal. Any one interested can be freely sign-in at our central web site www.snews.bnl.gov. Amateurs astronomers are an integral part of SNEWS. They have wide area viewing capability, enthusiasm and expertise. In the absence of precise neutrino-driven pointing information, amateurs may be the first (as it was at time of SN1987A) to locate the supernova optical counterpart in the sky. Also the *Sky and Telescope* magazine manages an Astro-Alert mailing list to which the SNEWS dispatch will be automatically forwarded.

Details on the SNEWS network principles, adopted technical solutions and test results can be found in [6].

3. The SNEWS Status

Currently operating and contributing experiments to the alert network are Super-K [7], LVD [8], IceCube [9] and Borexino [10]. Two servers process client datagrams, i.e. the experiment alarms, the main one located at Brookhaven National Laboratory (BNL) and a second one at central data center of Istituto Nazionale di Fisica Nucleare (INFN) in Bologna, Italy. Only one server at time is enabled to dispatch the alert in case of a positive coincidence while both receive and process data from client detectors. For robustness of the system there are duplicate servers in both BNL and INFN hosting sites: they can be switched on if necessary.
The time evolution of SNEWS (i.e. participating members and planned activities) is shown in

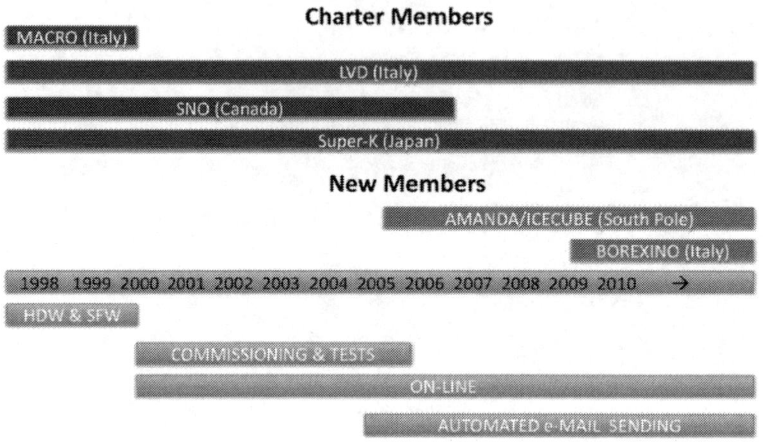

Figure 2. Status of the 12 years long story of the SNEWS project: past and present members are shown, being in automated alert mode online since July 2005.

figure 2. Since July 1st 2005 the network is online with the automated alert mode enabled. We expect future experiments to join as they come online.

4. Conclusions

At present many running experiments are sensitive to the gravitational stellar collapses in the Galaxy looking to the neutrino signal and they make possible the early warning. Requiring the online coincidence among them a clear and automated real time alert will be provided by SNEWS. The network after a long commissioning and test phase comes to the effective automated mode in July 2005. Capabilities have been deeply investigated.
Up to day no signal has been recorded and we are still waiting for next supernova event in the Galaxy.

Acknowledgments

We would like to thank the members of the experimental collaborations for the help given to the SNEWS project development as well the responsible people at BNL and INFN Bologna laboratories for providing secure hosting sites and continuos maintenance of the coincidence servers.

References

[1] Bethe, H. A. & Wilson, J. R., 1985, Astrophys. J.,295, 14
[2] Hirata, K., et al., 1987, Phys. Rev. Lett., 58, 1490 (1987)
[3] Bionta, R. M., et al. (IMB collaboration). 1987, Phys. Rev. Lett., 58, 1494
[4] Alekseev, E. N., et al., 1987, J. Exp. Theor. Phys. Lett., 45, 589
[5] Aglietta M. et al., 1987, Europhys. Letters, 3, 1315
[6] Antonioli, P. et al. 2004, New J. Phys. 6, 114
[7] Ikeda M. et al., 2007, The Astrophysical Journal 669, 519
[8] Aglietta M. et al. Nuovo Cimento A 105,1793 (1992)
[9] Ahrens J. et al., Astropart. Phys., 16, 345 (2002)
[10] Alimonti G. et al., 2009, Nucl. Instr. Meth. Phys. Reserach A, 600 568

Journal of Physics: Conference Series **309** (2011) 012027 doi:10.1088/1742-6596/309/1/012027

Supernova Neutrino Detection With Liquid Scintillators

Aldo Ianni

I.N.F.N. Gran Sasso Laboratory, S.S. 17bis, 67100 , Assergi, Italy

E-mail: aldo.ianni@lngs.infn.it

Abstract. Core collapse supernovae are a remarkable source of neutrinos. These neutrinos can also be detected by means of massive liquid scintillators located underground. Observations of supernova neutrinos can shed light on the explosion mechanism and on neutrino properties. In this paper we review the detection channels for neutrinos in liquid scintillators. We consider present and future experiments for supernova neutrino searches.

1. Underground massive liquid scintillator detectors

The technology of massive (> 100 tons) organic liquid scintillators in operation underground has been developed over the last 30 years through a number of detectors: the Baksan observatory [1] in Russia; LVD [2] and Borexino [3] at the Gran Sasso laboratory in Italy; KamLAND [4] at Kamioka, Japan; and, MiniBooNE [5] at Fermilab. This latter in particular is locate in a shallow site. The physics search of these detectors can be divided as follows.

- Solar neutrinos to search for neutrino properties and understand the physics of the sun.
- Electron antineutrinos from reactors to search for neutrino properties.
- Detection of neutrinos from a far away accelerator.
- Geo-neutrinos to understand the origin of the heat radiated by the earth and the reservoirs of U and Th within the mantle of the earth.
- Rare processes:
  - Neutrinos from core collapse supernovae.
  - Non-standard neutrino interactions
  - Diffuse neutrino background
  - Muon-induced neutrons and radioactive isotopes
  - Neutrinoless double beta decay.

So in general supernova neutrinos is not the main research goal of these massive projects due to the fact that a galactic supernova is a rare event. These detectors make use of a large volume of organic solvent, C_nH_{2n}, with $\rho \sim 0.8$ g/cm^3. The detector's design can be segmented or unsegmented. LVD is an example of a segmented detector which consists of stainless steel tanks ($1\times1\times1.5$ m^3) each equipped with 3 photomultipliers; the total mass is about 1 kton with 9.34×10^{31} protons and 4.23×10^{31} C nuclei. Borexino is an example of an unsegmented detector with 300 tons of scintillator contained inside a thin nylon spherical vessel; the scintillator is viewed by about 2000 photomultipliers. In this case, in order to reduce the external background,

Journal of Physics: Conference Series **309** (2011) 012027 doi:10.1088/1742-6596/309/1/012027

the scintillator is shielded by a buffer volume and an outer water Cherenkov veto. The detection threshold of each apparatus depends on the main physics goal: in LVD it is 4 MeV; in KamLAND 1 MeV; in Borexino 200 keV.

A galactic core collapse supernova is a rare event. Therefore, a supernova observatory must be in operation for more than a decade. The design in this case is important to guarantee a constant high duty cycle. The Baksan observatory has been in operation for 25 years and LVD for 15 years. A number of new projects are under consideration for the future: SNO+ [6] at SNOlab in Canada and LENA [7] proposed for an underground laboratory in Finland. In Tab. 1 we summarize some of the main characteristics of the present and future massive liquid scintillator detectors. In particular, among the future projects SNO+ is expected to be operational after 2012. The next observation of supernova neutrinos hopefully should give the opportunity to measure the binding energy and the temperature (average energy) of the different kind of neutrinos (ν_e, $\bar{\nu}_e$ and $\nu_{x=\mu,\tau}$).

Table 1. Summary of the main characteristics of massive liquid scintillator detectors for supernova neutrino searches.

	Type	Target mass [tons]	Threshold [MeV]	Rate above threshold [s^{-1}]	$\Delta E/E$ @10MeV [%]
Baksan Obs.	segmented	330			
		inner core 130	8	0.02	15
LVD	segmented	1000	4	0.03(>10 MeV)	14
		9.34×10^{31} p			
		4.23×10^{31} C			
Borexino	unsegmen.	300	0.25	0.02(>1 MeV)	1.4
		1.81×10^{31} p			
		1.35×10^{31} C			
KamLAND	unsegmen.	1000	1	0.02	2
				>1 MeV	
MiniBooNE	unsegmen.	800	5	0.03	2.4
				>5 MeV	
SNO+	unsegmen.	780	0.25	-	-
			design		
LENA	unsegmen.	50000	0.25	-	-
		design	design	-	-

2. Interaction channels for supernova neutrino detection with liquid scintillators

In this Section we briefly review the detection channels for supernova neutrinos in liquid scintillators. Before discussing the different reactions we report about the reference supernova chosen in this paper. In our case the binding energy is equal to $E_b = 3 \times 10^{53}$ ergs with $\langle E_{\nu_e} \rangle =$ 12 MeV, $\langle E_{\bar{\nu}_e} \rangle = 15$ MeV and $\langle E_{\nu_x} \rangle = 20$ MeV. In Fig. 1 we show the time-integrated energy spectra for the reference supernova. For more details on supernova neutrino spectra we invite the reader to refer to more specifc papers in this proceedings.

Journal of Physics: Conference Series **309** (2011) 012027 doi:10.1088/1742-6596/309/1/012027

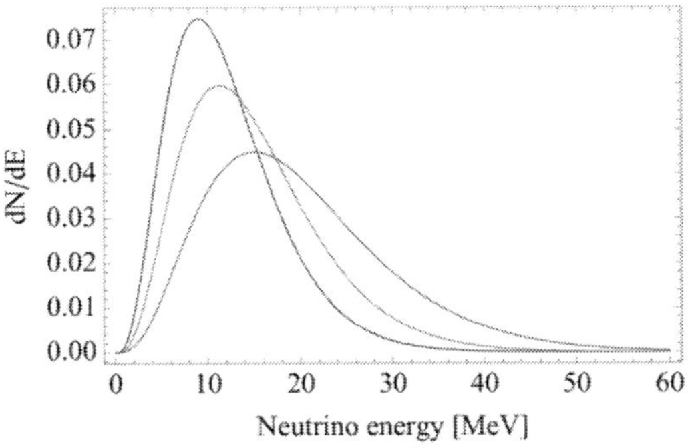

Figure 1. Reference supernova neutrino time integrated energy spectra used in the paper. From left to right: ν_e, $\bar{\nu}_e$ and ν_x.

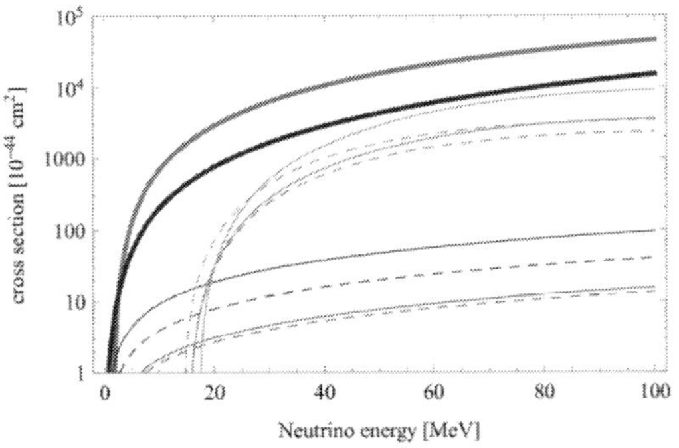

Figure 2. Cross sections of the main detection reactions for supernova neutrinos in liquid scintillators. See text for details.

2.1. Inverse-beta decay

In this case the reaction is: $\bar{\nu}_e + p \rightarrow e^+ + n$ with a threshold of 1.806 MeV. The neutron is captured on hydrogen in about 250 μs producing a 2.22 MeV gamma-ray. The prompt (positron) and delayed (2.22 MeV gamma-ray) signals increase the tagging power of this reaction. The cross section as a function of energy is shown in Fig. 2 (first red thick solid line from left). About 340 events without oscillations are expected for 10^{32} target protons for a supernova at 10 kpc. A target mass of 10^{32} protons corresponds to about 1 kton detector. In Fig. 3 we show the predicted spectra for three scenarios: no oscillations, normal hierarchy and inverted hierarchy. As it can be seen neutrino oscillations affect the neutrino spectra and fluence. However, the possibility to detect such an effect depends at some extent on the supernova model taken as reference and on the statistics of the detected neutrino sample.

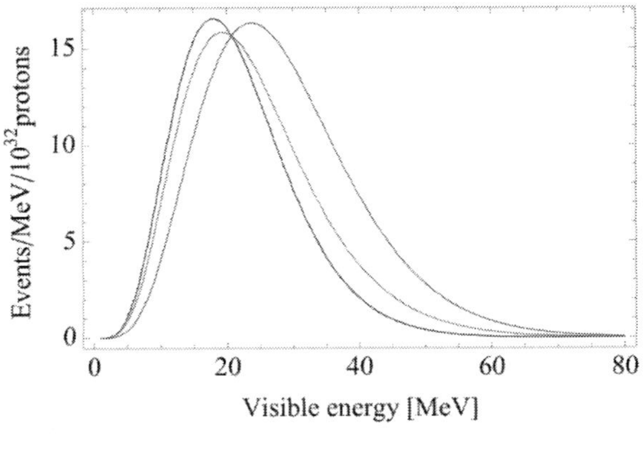

Figure 3. Neutrino spectra predicted for the inverse-beta decay detection channel for three scenarios. From left to right: no oscillations, normal hierarchy and inverted hierarchy (small θ_{13})

2.2. Charged current on ^{12}C

Liquid scintillators are rich in ^{12}C. The cross section for neutrino interactions on carbon are large ($\sim 10^{-42}$ cm^2, solid and dashed lines in green below the thick lines in Fig. 2). This offers the possibility to make use of neutrino on carbon interaction channels to search for supernova neutrinos. In particular, we have two charged-current reactions: $\nu_e +^{12} C \rightarrow e^- +^{12} N$ with threshold equal to 17.34 MeV and $\bar{\nu}_e +^{12} \rightarrow e^+ +^{12} B$ with threshold equal to 14.39 MeV. Both reactions can be tagged by means of a prompt-delayed sequence due to the ^{12}N and ^{12}B β decays with 11 ms and 20.2 ms half-life, respectively. The expected number of events for 10^{32} target nuclei and a 10 kpc supernova are less than 30 in each case at most. The prompt-delayed tagging could be used to distinguish ν_e against $\bar{\nu}_e$. However, as it is shown in Fig. 4, this goal is very difficult to achieve without a very large sample of events. Therefore, considering the mass of present detectors this possibility could be exploited by future super-massive experiments such as LENA.

2.3. Neutral current on ^{12}C

The neutral current on ^{12}C, $\nu_x +^{12}C \rightarrow \nu_x +^{12}C^*$, produces an unambiguous supernova neutrino signal by means of a gamma-ray at 15.11 MeV. The cross section is large but smaller than the charged current reactions (magenta solid and dashed lines below the charged current ones in Fig. 2). However, the degeneracy between the average neutrino energy and the supernova luminosity does not allow to measure these parameters independently. We predict about 70 events for 10^{32} target nuclei and a 10 kpc supernova.

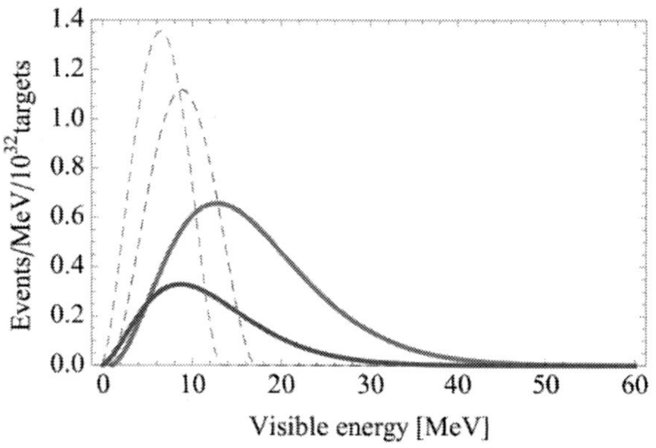

Figure 4. Spectrum of prompt (solid lines) and delayed (dashed lines) signal for charged current interations on ^{12}C.

2.4. Neutrino-electron elastic scattering

The neutrino-elastic scattering have charged current and neutral current channels. These interactions have smaller cross section with respect to all others: lower curves in Fig. 2. There is no specific tagging. Therefore, this interaction channel gives only a minor contribution to the present detectors in supernova neutrino search.

2.5. Neutrino-proton elastic scattering

The neutrino-proton elastic scattering for supernova neutrinos in liquid scintillators was first proposed in 2002 [8]. This is a special detection channel for high purity liquid scintillators. The visible energy due to the scintillation produced by recoiled protons shifts the observed spectrum toward lower energies. In Fig. 5 we show the visible energy against the real proton energy measured for KamLAND and Borexino. As it can be seen the effect of quenching is large and this needs a sub-MeV detection threshold. In Fig. 6 we show the expected spectrum using the function in Fig. 5. This figure shows that above 0.3 MeV only ν_x's give a contribution. Because of this effect the degeneracy between the binding energy and the average energy of ν_x's is broken. This is clearly shown in Fig. 7 where we have determined the ratio for a given E_x and T_x to the standard case (0.5×10^{52} ergs and 20 MeV). About 100 events are expected above 0.25 MeV for 10^{32} target protons and a 10 kpc supernova.

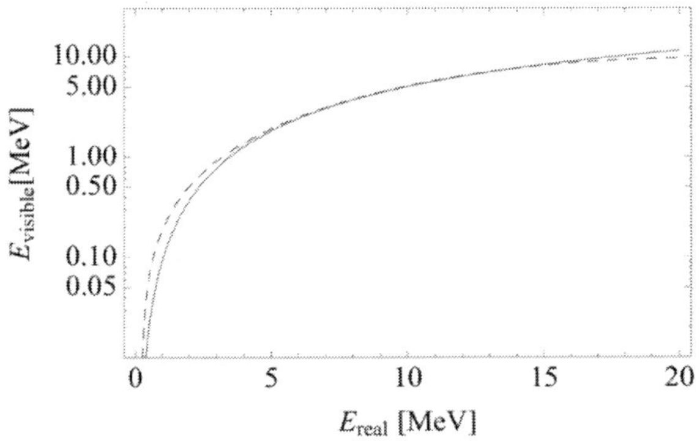

Figure 5. Visible energy against real energy due to quenching for proton scintillation prompt energy. Solid line: KamLAND. Dashed line: Borexino (preliminary).

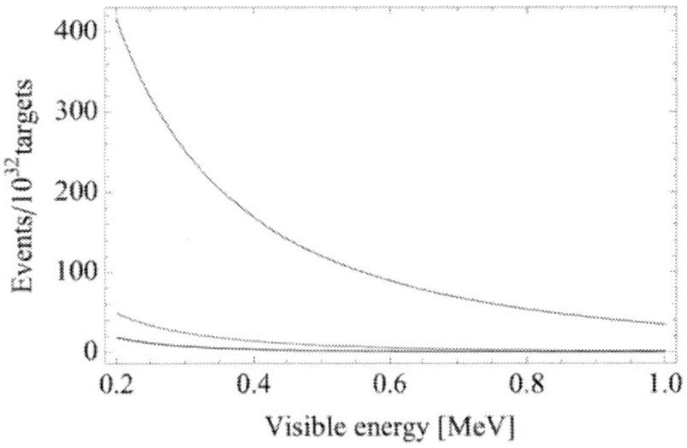

Figure 6. Spectrum of prompt signal due to neutrino-proton elastic scattering. From right to left: ν_x, $\bar{\nu}_e$ and ν_e.

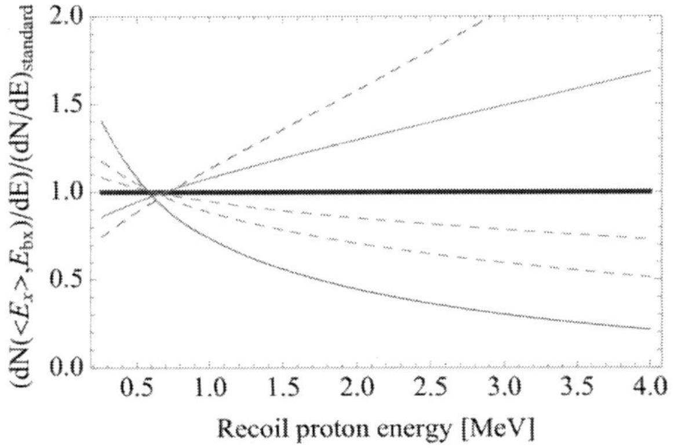

Figure 7. Ratio of spectra from neutrino-proton elastic scattering for given binding energy E_x and temperature T_x to the standard case described in the paper.

3. Regeneration effect through the earth

One feature of neutrino oscillations which is of particular value in this context is the regeneration through the earth. This effect produces a distortion of the spectrum which is a clear signal of neutrino oscillations. In Fig. 8 we show an example of the regeneration process. The possibility to detect supernova neutrinos with more than one experiment from different locations offers the opportunity to compare different spectra and enhance the oscillation process for the case where neutrinos cross the earth. The magnitude of this effect depends on the supernova location with respect to the earth, on the fluence, on the detector performances (such as energy resolution and efficiency) and on the oscillation scenario. At present, considering the target masses of the detector in operation and the uncertainty on the supernova parameters (binding energy and neutrino temperatures), it is not clear how well this regeneration process can be observed.

4. Conclusions

We have summarized the main features of the interaction channels for supernova neutrinos in massive liquid scintillators. Fig. 9 reports the predicted rate with and without oscillations (normal and inverted hierarchy) against the detector-supernova distance. A galactic supernova will have a distance between 1 and 20 kpc. Fig. 9 shows that present liquid scintillator detectors

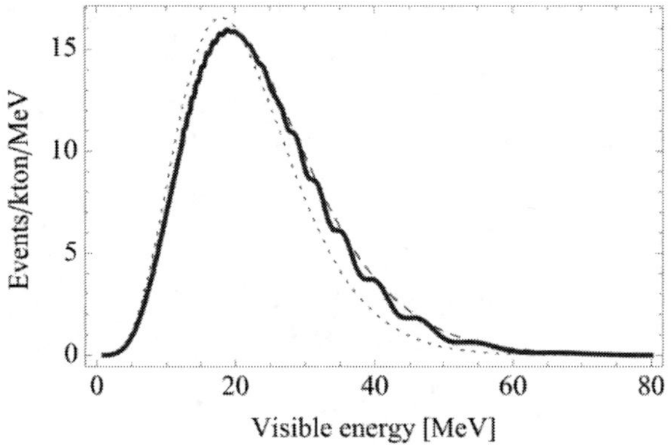

Figure 8. Regeneration effect due to neutrino travelling through the earth. Dotted line: no oscillations. Dashed line: normal hierarchy. Solid thick line: neutrino travelling through the mantle of the earth.

will observe about 100-400 events for a 10 kpc supernova. The golden channel for supernova neutrinos in liquid scintillators is the inverse-beta decay. The neutrino-proton elastic scattering offers a great opportunity for high purity liquid scintillators as discussed in the text.

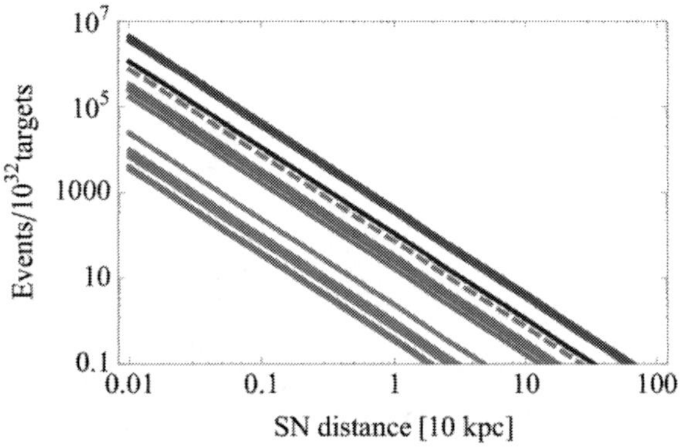

Figure 9. Rate expected for supernova neutrinos in massive underground detectors against the source-earth distance. From top to botton: inverse-beta decay; neutrino-proton elastic scattering; neutral current on ^{12}C; charged current on ^{12}C; neutrino-electron elastic scattering.

Acknowledgments
The author wishes to acknowledge the organizing committee for the warm atmosphere and wide discussions.

References

[1] E.N. Alexeyev and L.N. Alexeyeva, astro-ph/0212499v1.
[2] LVD collaboration, Astropart.Phys.27:254-270,2007. LVD collaboration, Nucl.Phys.Proc.Suppl. 110 (2002) 410-413.
[3] Borexino collaboration, NIM A 600, 568, 2010.
[4] KamLAND collaboration, Phys.Rev.Lett.90:021802,2003. P. Vogel, Twenty years after SN1987A, Hawaii, Feb. 25, 2007.
[5] MiniBooNE collaboration, Phys.Rev.D81:032001,2010.
[6] M. Chen, The SNO+ experiment, ICHEP08.
[7] M. Wurm et al., arXiv:1004.3474.
[8] J. Beacom at al., Phys. Rev. D 66 (2002).

TPC2010 IOP Publishing
Journal of Physics: Conference Series **309** (2011) 012028 doi:10.1088/1742-6596/309/1/012028

Supernova Neutrino Detection in Water Cherenkov Detectors

Kate Scholberg
Department of Physics, Duke University, Durham, NC, 27708
E-mail: schol@phy.duke.edu

Abstract. This talk will describe sensitivity of past, current and future water Cherenkov detectors to a burst of supernova neutrinos.

1. Features of Water Cherenkov Detectors

Water Cherenkov detectors, in the form of large volumes of ultrapure water surrounded by photomultiplier tubes (PMTs), are sensitive to the charged particles produced by interactions of core collapse supernova neutrinos in the few to few tens of MeV range. Charged particles moving faster than the speed of light in a medium produce Cherenkov photons if $\beta > 1/n$, where n is the refractive index of the medium. Water is a convenient and cheap detector material, suitable for neutrino detection because very large volumes can be deployed cheaply, even though light yields are typically much lower than for scintillator. With index of refraction $n \sim 1.34$, the (total energy) Cherenkov threshold for electrons is 0.8 MeV; for muons the threshold is 160 MeV and for protons it is 1400 MeV. Energy loss is proportional to the number of photons detected, and one may reconstruct the charged particle's interaction vertex and direction via the Cherenkov ring pattern (which has angle of 42° for relativistic particles). However it is important to remember that heavy particles may be invisible in Cherenkov detectors, and signals from low energy electrons, positrons and gammas (which are detected via Compton-scattered electrons) may be lost.[1]

2. Past, Current and Future Water Cherenkov Detectors

Table 1 lists past, current and future supernova-neutrino-sensitive water Cherenkov detectors.

Famous past water Cherenkov detectors are the IMB and Kamiokande-II detectors, which both observed the burst from SN1987A, with 19 events collected between them [1, 2]. SNO, with 1.7 kton of light water and 1 kton of heavy water, ran from 1999 to 2006, although no supernova burst was observed during that time [3]. The only current example of a large water Cherenkov detector is Super-Kamiokande (Super-K), which has been running since 1996 [4]; Super-K's data-taking phases are shown in Tab. 1. Super-K is currently running with an energy threshold of about 4-5 MeV. A number of very large next-generation detectors are currently under

[1] Note that long string ice- or water-based detectors like IceCube and Antares also detect Cherenkov photons from supernova bursts. Although these detectors have very good timing capabilities, they do not reconstruct interactions event by event. IceCube's supernova capabilities were described in a separate presentation at this symposium; therefore long string water Cherenkov detectors will not be addressed here.

Table 1. Summary of past, current and future water Cherenkov detectors with supernova neutrino sensitivity.

Detector	Location	Fiducial Mass (kton)	PMTs (diameter, cm)	Effective coverage	pe/MeV	Live dates
IMB-1	US	3.3	2048 (12.5)	1%	0.25	1982-1985
IMB-2	US	3.3	2048 (20)	4.5%	1.1	1987-1990
Kamiokande I	Japan	0.88	1000/948 (50)	20%	3.4	1983-1985
Kamiokande II	Japan	1.04	1000/948 (50)	20%	3.4	1986-1990+
Super-K I	Japan	22.5	11146 (50)	39%	6	1996-2001
Super-K II	Japan	22.5	5182 (50)	19%	3	2002-2005
Super-K III+	Japan	22.5	11129 (50)	39%	6	2006-
SNO	Canada	1/1.7 (D_2O/H_2O)	9438 (20)	54%	9	1999-2006
LBNE WC	US	100-300	TBD	10-40%	2-6	202x?
MEMPHYS	Europe	440	TBD	30-40%	~4-6	202x?
Hyper-K	Japan	540	TBD	~20%	5-6	202x?

development: these include the Long Baseline Neutrino Experiment (LBNE) water Cherenkov detector in the US (100-300 kton) [5], Hyper-K in Japan (540 kton) [6] and MEMPHYS in Europe (450 kton) [7]. For these, photosensors tend to be a driving cost; improvements to light collection and quantum efficiency, currently a focus of worldwide effort, are of great interest.

3. Interactions in Water at Supernova Neutrino Energies

The total number of events expected scales as $1/R^2$, where R is the distance to the supernova (a distance of 10 kpc, which is a bit beyond the center of the Milky Way, is usually taken as a standard). A few hundred events per kton within a few tens of seconds are expected at this distance; at underground sites under consideration for next-generation detectors, one expects a clean signal out to Andromeda. See Fig. 1.

The cross-sections for relevant interactions in water are shown in Fig. 2. Some of these cross-sections– in particular, charged current inverse beta decay $\bar{\nu}_e + p \rightarrow e^+ + n$ (IBD) and elastic scattering (ES) of neutrinos on electrons $\nu_{e,x} + e^- \rightarrow \nu_{e,x} + e^-$, which proceeds via both neutral current (NC) and charged current (CC) channels, are known to few percent or better level. In contrast, other interactions on ^{16}O nuclei have relatively large uncertainties, and cross-sections have never been measured in the few tens-of-MeV energy range.

IBD is overwhelmingly dominant in the supernova neutrino energy regime: water Cherenkov detectors are primarily sensitive to the $\bar{\nu}_e$ component of the flux. The primary observable is the Cherenkov radiation of the IBD positron. In principle one may also exploit the delayed coincidence between the positron signal and a signal from capture of a neutron to tag the interaction. For neutron capture on free protons, the time interval for neutron thermalization and capture is about 200 μs, and a 2.2 MeV gamma is released from the capture. A gamma of such low energy is difficult to detect in a large water Cherenkov detector (and the gammas from positron annihilation are below Cherenkov threshold and hence their Compton scatters are invisible).

A potential enhancement to improve detectability of the neutrons for IBD tagging is addition of a Gd-containing solute to the water [8]. (Gd addition has been used successfully in scintillation detectors). Gd has a huge cross-section for neutron capture (49,000 barns for natural Gd, in contrast to a 0.3 barn neutron capture cross-section for free protons); after swallowing a neutron,

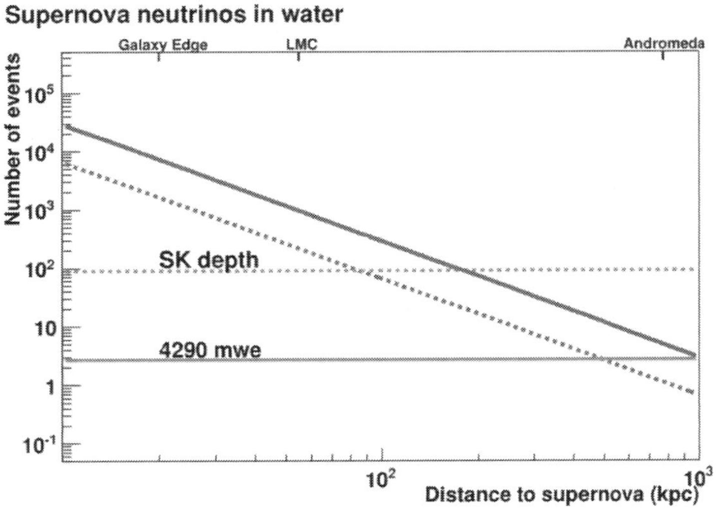

Figure 1. Approximate number of events detected as a function of distance to the supernova for Super-K (dashed line) and a 100 kton water detector (solid line). The horizontal green lines indicate cosmic muon rates at SK depth (~2300 meters water equivalent (mwe)) and the Homestake mine depth (4290 mwe). (Note that cosmic muons can be effectively vetoed through several orders of magnitude.)

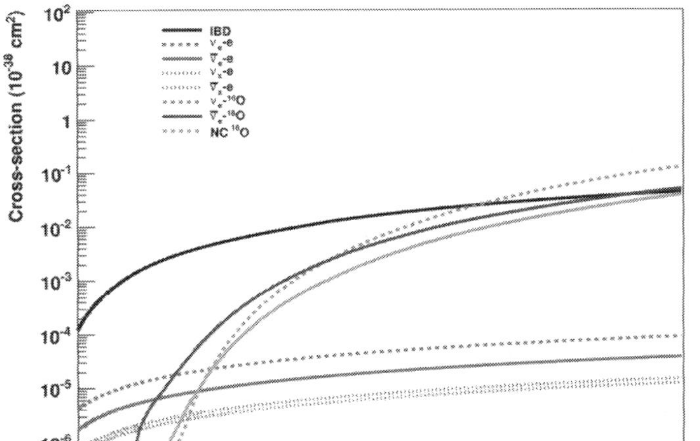

Figure 2. Cross-sections for relevant supernova neutrino interactions in water.

a Gd nucleus will release a cascade of gamma rays totalling about 8 MeV. Detection efficiency has been measured using a test setup in Super-K to be about 67% [9]. However note that because only about 4.5 MeV of energy per neutron capture is observed (because the gammas often Compton scatter electrons below Cherenkov threshold), good photocoverage is necessary to achieve this efficiency. Possibly improved analysis techniques, such as the use of isotropy information (as has been used for SNO in salt mode [10]), could be developed to improve neutron tagging efficiency. The Gd technique is under study for Super-K using a test tank [11], in order to answer questions of compatibility and purity of materials, feasibility of addition and removal of Gd, and neutron background.

Elastic scattering, although a relatively small component of the signal, is of significance because of its directionality. IBD positrons are emitted nearly isotropically, so are of little use for pointing. In contrast, electrons are kicked in the direction of the incoming neutrino, and the Cherenkov light cone allows determination of the charged particle direction. The pointing quality scales approximately as $\sim 25°/\sqrt{N}$; degradation by isotropic background results in about 5 degrees pointing accuracy for a few hundred kton detector at 10 kpc [12, 13]. The pointing quality can be improved somewhat with the addition of Gd to reduce the isotropic background. The angular information can be also used to select a flavor-enhanced sample (see Section 5). Furthermore, if the direction to the supernova is known, neutrino energy can be more precisely reconstructed for ES events.

There are also non-negligible contributions from charged current interactions on oxygen, $\nu_e + ^{16}\mathrm{O} \rightarrow e^- + ^{16}\mathrm{F}$, $\bar{\nu}_e + ^{16}\mathrm{O} \rightarrow e^+ + ^{16}\mathrm{N}$, as well as neutral current excitations [14]. These reactions have diverse final states, including ejected nucleons and deexcitation gammas. These interactions are also asymmetric, and this asymmetry could potentially be of use in disentangling flavor components. However the cross-sections and final states for oxygen interactions in the supernova neutrino energy regime are relatively poorly understood both theoretically and experimentally.

4. Interaction Rate Calculation

I report here preliminary results from a study of event rates in a future large water Cherenkov detector, to be published in [15].

The predicted event rate from a supernova burst may be calculated by folding expected neutrino differential spectra with cross-sections for the relevant channels, and with detector response. We have performed the event rate computation by using estimated detector responses, making use of the GLoBES software [16]. We employ only the front-end rate engine part of GLoBES, and not the oscillation sensitivity part. GLoBES takes as input fluxes, cross-sections, "smearing matrices" and post-smearing efficiencies. The smearing matrices incorporate both interaction product spectra and detector response, determined using parameterized response functions or detector simulation output, or both. We are able to modify the smearing matrices to evaluate different detector configurations, such as for varied PMT coverages; in particular, we looked at 100 kton detectors with 30% and 15% coverages of high-quantum-efficiency PMTs (roughly corresponding to Super-K I and Super-K II capabilities). The software package used for the study will be made publicly available; the results should be generally applicable to any large water detector (as well as other detector types).

In this study, only the lepton in the final state for the CC interactions is considered, taking into account the energy threshold of the interaction. For the NC interaction with ^{16}O, $\nu_x + ^{16}\mathrm{O} \rightarrow \nu_x + ^{16}\mathrm{O}^*$, we are using a simplified model of the resulting deexcitation gammas by assuming relative final energy levels according to reference [14]. Because this reference does not provide differential final state information, we approximate the distribution of these levels to be independent of neutrino energy. The resulting gamma cascade was simulated using relative probabilities of the transitions for a given excited state; the resulting gamma spectrum was then run through the LBNE water Cherenkov detector simulation. We found rather poor efficiency for detecting these gammas, in contrast to the results in reference [17], due to the fact that gammas frequently scatter electrons below Compton threshold.

Figure 3 shows the resulting differential energy spectra for the different channels. The plot on the left shows the interaction rates as a function of neutrino energy. The plot on the right shows the distribution of observed event energies in the detector. Table 2 shows the breakdown of detected event channels, for two different specific supernova neutrino flux models, the "Livermore" [18] and "GKVM" models [19]. We note that different flux models can give substantially different event rates. In particular, because of the thresholds of the ^{16}O interactions, the rates of the CC interactions on oxygen are quite sensitive to the ν_e and $\bar{\nu}_e$ spectra.

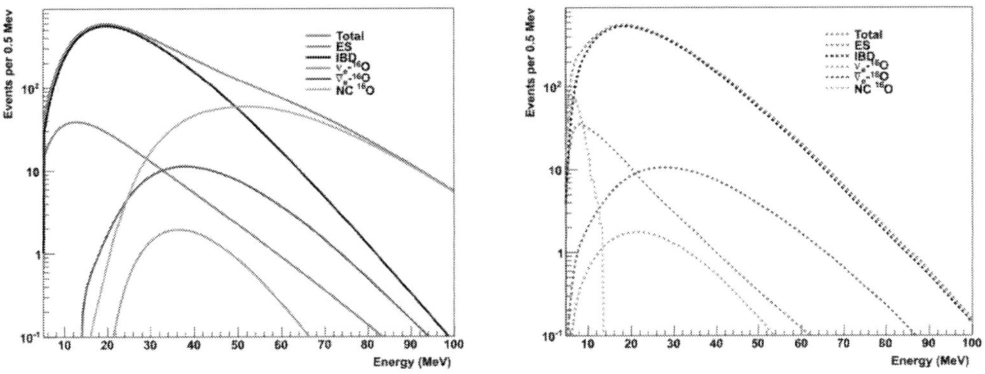

Figure 3. Event rates in water, for the Livermore model and a detector configuration with 30% high-quantum-efficiency PMT coverage (events per 0.5 MeV).

Channel	Events, "Livermore" model	Events, "GKVM" model
$\bar{\nu}_e + p \rightarrow e^+ + n$	27116	16211
$\nu_x + e^- \rightarrow \nu_x + e^-$	868	534
$\nu_e +^{16} \mathrm{O} \rightarrow e^- +^{16} \mathrm{F}$	88	379
$\bar{\nu}_e +^{16} \mathrm{O} \rightarrow e^+ +^{16} \mathrm{N}$	700	490
$\nu_x +^{16} \mathrm{O} \rightarrow \nu_x +^{16} \mathrm{O}^*$	513	124
Total	29285	17738

Table 2. Event rates for different models in 100 kton of water, for the 30% PMT coverage detector configuration.

Figure 4 shows the difference in observed event rates between the 15% and 30% PMT coverage configurations. For the 15% configuration, one loses about 10% of self-triggered events below ~10 MeV. The loss includes most of the NC excitation events. (We note that clever triggering may mitigate this loss.)

5. Disentangling Flavors in Water

The addition of Gd to a water detector will not substantially change event rates, but will enhance ability to determine the flavor composition of an observed signal by allowing tagging of IBD events (although note that interactions on ^{16}O may produce ejected neutrons as well). To get some general idea of the value of neutron tagging of $\bar{\nu}_e$ we performed a simple study: we looked at flavor composition for tagged and untagged events. We assume that 67% of the true

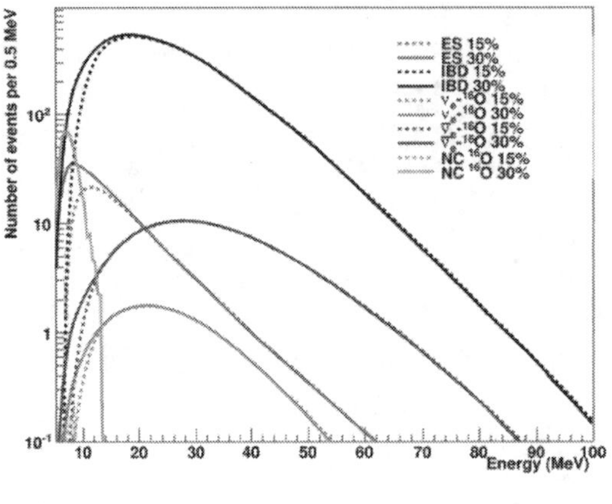

Figure 4. Comparison of event rates for 15% and 30% PMT coverage configurations in 100 kton of water.

IBD events will be tagged; we also assume that no events without a neutron will be falsely tagged as having a neutron (the false tagging rate should be $\sim 10^{-4}$ according to reference [9]). We also take into account CC and NC reactions of neutrinos on ^{16}O, for which some final states have neutrons; to estimate this contribution we use tables II, III and IV from reference [14]. Figure 5 shows the contributions of the different interaction channels for tagged and untagged events, for the GKVM flux. The neutron-tagged event rate is a nearly-pure IBD sample. The untagged event rate has contributions from elastic scattering (ES), and from CC and NC interactions on ^{16}O, but is dominated by untagged IBD.

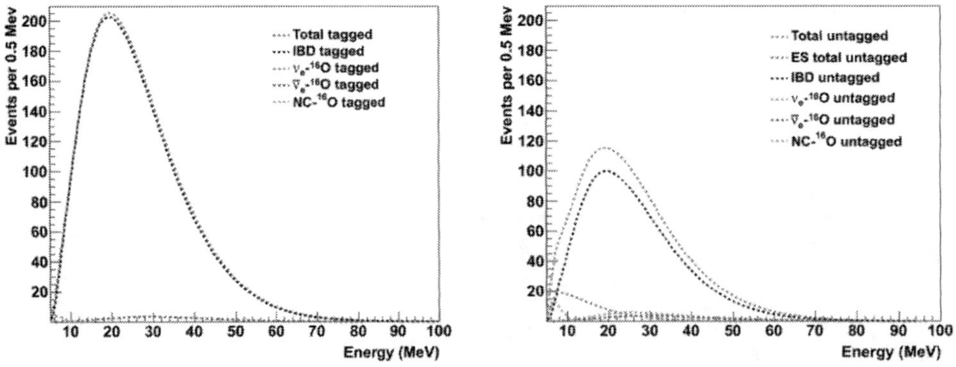

Figure 5. Total events in 100 kton of water showing contribution from the different interaction channels, for neutron-tagged (left) and untagged (right) events.

Figure 6 shows the contributions of the different neutrino flavors for tagged and untagged events. The tagged sample is nearly pure $\bar{\nu}_e$. The untagged sample has contributions from other flavors, and large contamination from untagged IBD $\bar{\nu}_e$.

Another potential method for selecting a flavor-enhanced sample in a water Cherenkov detector is to use the directionality of the neutrino-electron scattering signal: the ES sample is

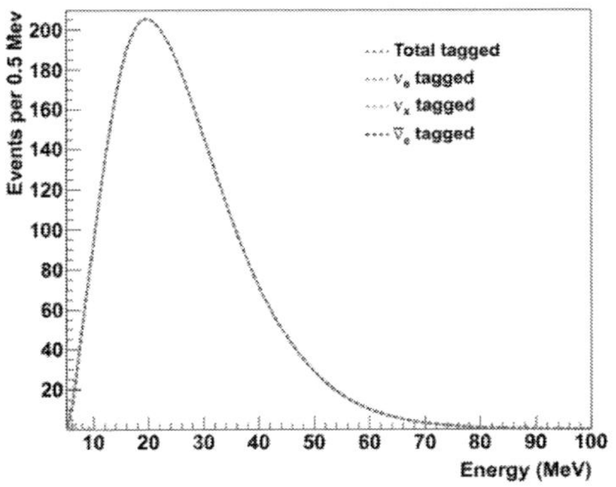

Figure 6. Total events in 100 kton of water showing contribution from the different flavors, for neutron-tagged (left) and untagged (right).

enriched in ν_e and ν_x relative to $\bar{\nu}_e$ We estimate the quality of a flavor-enriched ES sample by assuming that 66% of ES events will have $\cos\theta > 0.9$, where θ is the reconstructed angle of the scattered event [12]. Such a cut will reduce isotropic background by 95%. Figure 7 shows the interaction and flavor compositions of the "ES-enriched" sample selected by an angular cut, for the GKVM model. The non-ES background can be determined by counting events outside of the angular cut window.[2] Figure 8 shows the background-subtracted ES-enriched signal, with statistical error bars, for the GKVM flux. From Fig. 8 we can see that the ES component can be identified with high statistical significance. However the sample contains contributions from many flavors, and the relative amounts of the different contributions depend on the supernova flux.

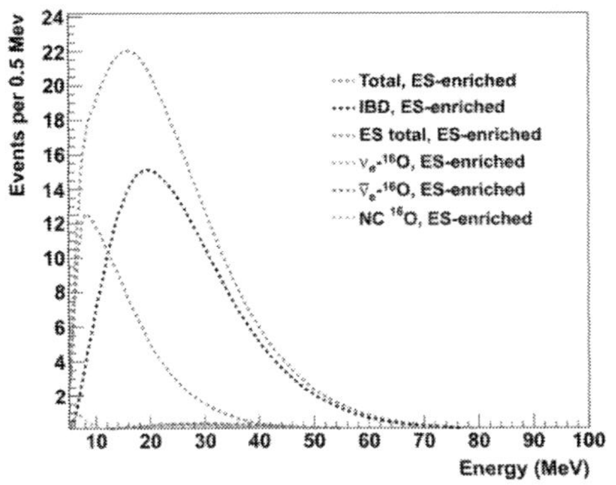

Figure 7. For the GKVM model, interaction (left) and flavor (right) composition of the ES-enriched sample in 100 kton of water.

One can also combine neutron-tagging and ES information, and imagine that there were an independent measurement of the ν_e flux from a liquid argon detector (or some other ν_e-sensitive

[2] We assume for the purpose of determining statistical uncertainty on the ES background subtraction that non-ES events have isotropic background, although that is not completely true– IBD events have a weak asymmetry, and interactions on ^{16}O have a backwards asymmetry.

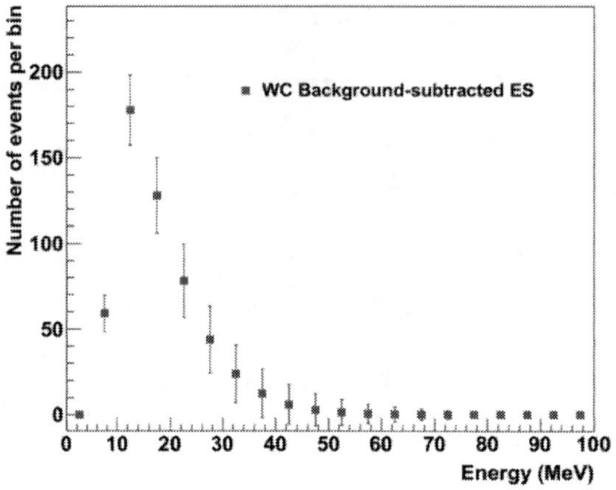

Figure 8. Background-subtracted ES signal in 100 kton of water in 5 MeV bins, for the GKVM flux. Error bars are statistical.

detector). In this case, one could in principle determine the ν_x flux, of considerable interest in itself. Figure 9 summarizes the total ES, $\nu_{e,xES}$, and ν_{xES} scattering rates for the GKVM flux, assuming both neutron-tagging and angular selection.

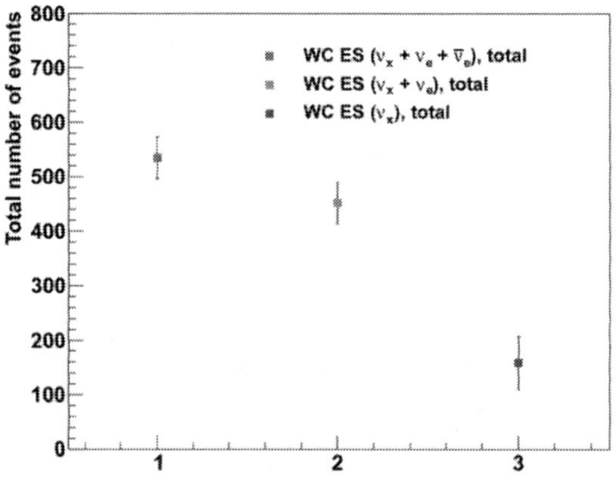

Figure 9. Inferred flavor components of a water Cherenkov ES signal, assuming neutron-tagging, angular selection, and a ν_e measurement from LAr, for the GKVM flux. Error bars include statistical and systematic contributions.

6. Summary

In summary, supernova neutrino detection with water Cherenkov detectors has a glorious past (IBM/Kamiokande-II detection of 1987A), a glorious present (Super-K) and, potentially, a glorious future: next-generation multi-hundred-kton scale detectors, possibly with neutron tagging using dissolved Gd, will provide an enormous signal for a nearby supernova. One expects a few hundred events/kton; most of these will be $\bar{\nu}_e$ from IBD, but other interactions contribute. ES is especially interesting because of its clean directionality. The supernova neutrino physics capabilities of a water Cherenkov detector will be enhanced by combination with other experiments' different flavor sensitivities.

Acknowledgments

The author wishes to thank the Long Baseline Neutrino Experiment Physics Working Group (and particularly the Supernova Burst Topical Group members Alex Beck, Farzan Beroz, Rachel Carr, Huaiyu Duan, Alex Friedland, Jim Kneller, Zander Moss, Diane Reitzner and David Webber) for contributions to this work.

References

[1] R. M. Bionta *et al.*, Phys. Rev. Lett. **58**, 1494 (1987).
[2] K. Hirata *et al.* [KAMIOKANDE-II Collaboration], Phys. Rev. Lett. **58**, 1490 (1987).
[3] C. J. Virtue [SNO Collaboration], Nucl. Phys. Proc. Suppl. **100**, 326-331 (2001). [astro-ph/0103324].
[4] M. Ikeda *et al.* [Super-Kamiokande Collaboration], Astrophys. J. **669**, 519-524 (2007). [arXiv:0706.2283 [astro-ph]].
[5] http://lbne.fnal.gov/index.shtml
[6] K. Nakamura, Int. J. Mod. Phys. **A18**, 4053-4063 (2003).
[7] A. de Bellefon, J. Bouchez, J. Busto *et al.*, [hep-ex/0607026].
[8] J. F. Beacom and M. R. Vagins, Phys. Rev. Lett. **93**, 171101 (2004) [arXiv:hep-ph/0309300].
[9] H. Watanabe *et al.* [Super-Kamiokande Collaboration], arXiv:0811.0735 [hep-ex].
[10] B. Aharmim *et al.* [SNO Collaboration], Phys. Rev. C **72**, 055502 (2005) [arXiv:nucl-ex/0502021].
[11] A. Kibayashi, f. t. S. -K. Collaboration, [arXiv:0909.5528 [astro-ph.IM]].
[12] J. F. Beacom and P. Vogel, Phys. Rev. D **60**, 033007 (1999) [arXiv:astro-ph/9811350].
[13] R. Tomas, D. Semikoz, G. G. Raffelt, M. Kachelriess and A. S. Dighe, Phys. Rev. D **68**, 093013 (2003) [arXiv:hep-ph/0307050].
[14] E. Kolbe, K. Langanke and P. Vogel, Phys. Rev. D **66**, 013007 (2002).
[15] LBNE Physics Working Group Study document, in preparation.
[16] http://www.mpi-hd.mpg.de/personalhomes/globes/
[17] K. Langanke, P. Vogel, E. Kolbe, Phys. Rev. Lett. **76**, 2629-2632 (1996). [nucl-th/9511032].
[18] T. Totani, K. Sato, H. E. Dalhed and J. R. Wilson, Astrophys. J. **496**, 216 (1998) [arXiv:astro-ph/9710203].
[19] J. Gava, J. Kneller, C. Volpe and G. C. McLaughlin, Phys. Rev. Lett. **103**, 071101 (2009) [arXiv:0902.0317 [hep-ph]].
[20] H. Duan and A. Friedland, arXiv:1006.2359 [hep-ph].

Journal of Physics: Conference Series **309** (2011) 012029 doi:10.1088/1742-6596/309/1/012029

Supernova Neutrino Detection with IceCube

Lutz Köpke for the IceCube Collaboration
University of Mainz, D-55099 Mainz, Germany
E-mail: lutz.koepke@uni-mainz.de

Abstract. IceCube was completed in December 2010. It forms a lattice of 5160 photomultiplier tubes that monitor a volume of ~ 1 km^3 in the deep Antarctic ice for particle induced photons. The telescope was designed to detect neutrinos with energies greater than 100 GeV. Owing to subfreezing ice temperatures, the photomultiplier dark noise rates are particularly low. Hence IceCube can also detect large numbers of MeV neutrinos by observing a collective rise in all photomultiplier rates on top of the dark noise. With 2 ms timing resolution, IceCube can track subtle features in the temporal development of the supernova neutrino burst. For a supernova at the galactic center, its sensitivity matches that of a background-free megaton-scale supernova search experiment. The sensitivity decreases to 20 standard deviations at the galactic edge (30 kpc) and 6 standard deviations at the Large Magellanic Cloud (50 kpc). IceCube is sending triggers from potential supernovae to the Supernova Early Warning System. The sensitivity to neutrino properties such as the neutrino hierarchy is discussed and simulations of tantalizing signatures, such as the formation of a quark star or a black hole as well as the characteristics of shock waves are presented. All results are preliminary.

1. Introduction

It was recognized early by [1] and [2] that neutrino telescopes offer the possibility to monitor our Galaxy for supernovae. IceCube is uniquely suited for this measurement due to its location and 1 km^3 size. The noise rates in IceCube's photomultiplier tubes average around 540 Hz since they are surrounded by inert and cold ice with depth dependent temperatures ranging from −43 °C to −20 °C. At depths between (1450 – 2450) m they are partly shielded from cosmic rays. Cherenkov light induced by neutrino interactions will increase the count rate of all light sensors above their average value. Although this increase in individual light sensor is not statistically significant, the effect will be clearly seen once the rise is considered collectively over many sensors. The 5160 photomultipliers are sufficiently far apart such that the probability to detect light from a single interaction in more than one DOM is small. With absorption lengths exceeding 100 m, photons travel long distances in the ice such that each DOM effectively monitors several hundred cubic-meters of ice. The inverse beta process $\bar{\nu}_e + p \rightarrow e^+ + n$ dominates supernova neutrino interactions with $O(10\,\mathrm{MeV})$ energy in ice or water, leading to positron tracks of about $0.6\,\mathrm{cm}\cdot E_\nu/\,\mathrm{MeV}$ length. Considering the approximate E_ν^2 dependence of the cross section, the light yield per neutrino roughly scales with E_ν^3. The detection principle was demonstrated with the AMANDA experiment, IceCube's predecessor [3]. Since 2009, IceCube has been sending real-time datagrams to the Supernova Early Warning System (SNEWS) [6] when detecting supernova candidate events.

2. Detector

The Digital Optical Module (DOM) is the fundamental element in the IceCube architecture. Housed in a 13" (33 cm) borosilicate glass pressure sphere, it contains a Hamamatsu 10" hemispherical

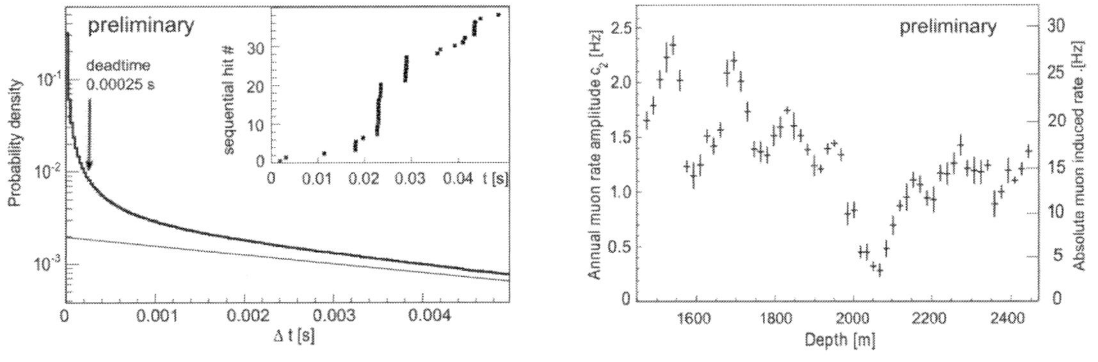

Figure 1. Left: Probability density distribution of time differences between pulses for noise (bold line) and the expectation for a Poissonian process fitted in the range 15 ms $< \Delta T <$ 50 ms (thin line). The excess is due to bursts of correlated hits, as indicated by the 50 ms long snapshot of hit times shown in the insert. Right: Parameter c_2 (see section 4) and estimated muon induced rate as function of depth. The variation with depth is mostly due to the optical properties of the ice and muons ranging out.

photomultiplier tube [4] as well as several electronics boards containing a processor, memory, flash file system and realtime operating system that allows each DOM to operate as a complete and autonomous data acquisition system [5]. It stores the digitized data internally and transmits the information to a surface data acquisition system on request. The supernova detection relies on continuous measurements of photomultiplier rates. The rate information is stored and buffered on each DOM in a 4-bit counter in 1.6384 ms time bins (2^{16} cycles of the 40 MHz clock). For the real-time processing, the information is synchronized with the help of a GPS clock and regrouped in 2 ms bins. The South Pole is out of reach for most communication satellites and high bandwidth connectivity is available only for about 6 hours per day. Therefore, a dedicated Iridium-satellite [7] connection is used by the SNDAQ host system to transmit urgent alerts. In that case, a short datagram is sent to the northern hemisphere. The receiving system parses the message and forwards information on the supernova candidate event to the international SNEWS group. The time delay between photons hitting the optical module and the arrival of the datagram at SNEWS stands at about 6 min, providing close to real-time monitoring and triggering. Due to satellite bandwidth constraints, the data are re-binned in 0.5 s intervals and then subjected to a statistical online analysis; the fine time information in 2 ms intervals is transmitted for a period starting 30 s before and ending 60 s after a trigger flagging a candidate supernova explosion.

Several effects contribute to the prevailing low noise rate of 540 Hz: a Poissonian noise contribution from radioactivity, atmospheric muons and remaining thermal noise, as well as correlated noise from Cherenkov radiation and scintillation originating in the glass of the photomultiplier and the pressure vessel. It is suspected that residual β and α decays from trace elements in the uranium/thorium chain powers cerium-based scintillation, causing a series of pulses. The observed time difference between noise hits deviates from an exponential distribution as expected for a Poissonian process (see Fig. 1). With typical times between correlated noise pulses of $O(100\,\mu s)$, the signal-to-noise ratio of the measurement can be improved by adding an artificial dead time that is configurable by a field programmable gate array in the DOM. The optimal setting for the dead time with respect to the signal over noise ratio for supernovae was found to be $\tau \approx 250\,\mu s$.

3. Real-time analysis method

Counting N_i pulses during a given time interval Δt, rates $r_i = N_i/\Delta t$ for DOM i, are derived. The index i ranges from 1 to the total number of operational optical modules N_{DOM}. With sufficiently large Δt's, the distributions of the r_i's can be described by lognormal distributions (see Fig. 2) that, for simplicity, are approximated by Gaussian distributions with rate expectation values $\langle r_i \rangle$ and corresponding standard deviation expectation values $\langle \sigma_i \rangle$. These expectation values are computed from moving 300 s time intervals before and after the investigated time interval. At the beginning and the end of a SNDAQ-run, asymmetric intervals are used. The time windows exclude 30 s before and after the investigated bin in order to reduce the impact of a wide signal on the mean rates. The most likely *collective rate deviation* $\Delta\mu$ of all DOM noise rates r_i from their individual $\langle r_i \rangle$'s, assuming the null hypothesis of no signal, is obtained by maximizing the likelihood

$$\mathcal{L}(\Delta\mu) = \prod_{i=1}^{N_{\mathrm{DOM}}} \frac{1}{\sqrt{2\pi}\,\langle \sigma_i \rangle} \exp\left(-\frac{(r_i - (\langle r_i \rangle + \epsilon_i\,\Delta\mu))^2}{2\langle \sigma_i \rangle^2}\right) . \quad (1)$$

Here ϵ_i denotes a correction for module and depth dependent detection probabilities. An analytic minimization of $-\ln \mathcal{L}$ leads to

$$\Delta\mu = \sigma_{\Delta\mu}^2 \sum_{i=1}^{N_{\mathrm{DOM}}} \frac{\epsilon_i\,(r_i - \langle r_i \rangle)}{\langle \sigma_i \rangle^2} \quad \text{with} \quad \sigma_{\Delta\mu}^2 = \left(\sum_{i=1}^{N_{\mathrm{DOM}}} \frac{\epsilon_i^{\,2}}{\langle \sigma_i \rangle^2} \right)^{-1} . \quad (2)$$

Assuming uncorrelated background noise and a large number of contributing DOMs, the significance $\xi = \Delta\mu/\sigma_{\Delta\mu}$ should approximately follow a Gaussian distribution with unit width centered at zero. The likelihood that a deviation is caused by an isotropic and homogeneous illumination of the ice can be calculated from the χ^2-probability $-2\ln(\mathcal{L}) = \chi^2_{\Delta\mu} = \sum_{i=1}^{N_{\mathrm{DOM}}} (r_i - (\langle r_i \rangle + \epsilon_i\,\Delta\mu)/\langle \sigma_i \rangle)^2$.

To cover model uncertainties, analyses with time bases of 0.5 s, 4 s and 10 s are run in parallel. The collective rate deviation $\Delta\mu$ and its uncertainty $\sigma_{\Delta\mu}$ in the time bases of 4 s and 10 s are calculated using sliding windows in 0.5 s steps and extracting the maximal significance. This procedure ensures that the signal detection efficiency is not reduced by binning effects.

4. Detector performance

The DOM rates $r(t)$ are characterized by an exponential rate decrease over long time periods and a slight seasonal modulation that is represented quite well by the formula $r(t) = r_0 + c_1 e^{-t/\tau} + c_2 \sin(2\pi(t/\mathrm{year}))$. Interpreting the seasonal modulation as being due to stratospheric temperature variations, the averaged muonic contribution to single DOM rates is ≈ 16 Hz, with a strong depth dependence due to variations in absorption length and muons ranging out (see right plot of Fig. 1). The slightly skewed rate distribution of a single DOM is better described by a lognormal distribution than by a Gaussian (see left plot of Fig. 2 with 250 μs dead time applied). Thanks to the tight quality control, the average noise rates between DOMs vary only by 10%.

The expected signal sensitivity in IceCube is somewhat reduced due to two types of correlations between pulses that introduce supra-Poissonian fluctuations. The first correlation concerns single photomultiplier tubes and arises from a burst of photons produced by radioactive decays in the pressure vessel discussed above. The artificial deadtime suppresses - but does not completely remove - the effect. This leads to a broadening of a factor of $\approx$ 1.28 in the standard deviation of single photomultiplier tubes w.r.t. to the Poissonian expectation. The second correlation arises from the cosmic-ray muon background discussed above: a single cosmic ray shower can produce a bundle of muons which is seen by hundreds of optical modules. As evident from the right plot of Fig. 2, the measured significance distribution ξ is broader than expected and can be fairly well fitted by a Gaussian with width $\sigma = 1.27$. The broadening increases with the size of the detector and has reached $\sigma = 1.43$ with 79 operating strings. Offline, roughly half of the hits associated with triggered cosmic ray muon can be removed, lowering the broadening to $\sigma = 1.06$. The standard deviation of the rate sum of all DOMs suffers from both effects and turns out to be between 1.3 – 1.7 times larger than the Poissonian expectation for 2 ms and 500 ms bins, respectively. Subtracting hits associated with cosmic ray muons lowers the standard deviation to $(1.24 - 1.32)\sqrt{\sum_i r_i}$, slightly dependent on the binning. It may be possible to further reduce this broadening in the future using a better identification of bursts of correlated hits.

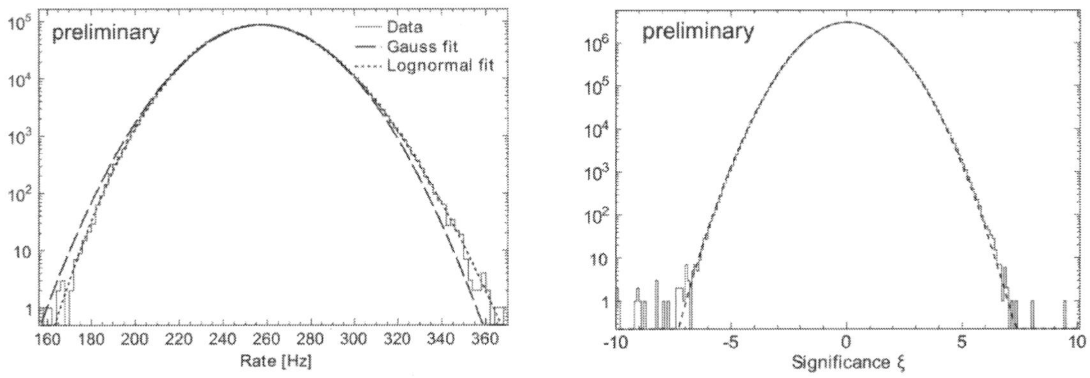

Figure 2. Left: Rate distribution of a typical standard efficiency DOM taken over 29 consecutive days. Each measurement corresponds to 0.5 s integration time. The application of an artificial deadtime of 250 μs lowers the mean rate of this DOM to $\approx$ 260 Hz. Gaussian and lognormal fits are shown. Right: Significance distribution in 0.5 s binning for a detector uptime of 556 days with 22 and 40 strings deployed. The two outliers at $\xi = 8$ and 9.5 occurred during test runs employing artificial light. The dashed line shows a Gaussian fit with $\sigma = 1.27$.

5. Production, interaction and detection of supernova neutrinos

We use the Lawrence-Livermore [10] and Garching models [11] as benchmarks, but will also refer to more specific models that were selected to demonstrate IceCube's physics performance. The spherically symmetric Lawrence-Livermore simulation was performed from the onset of the collapse to 18 s after the core bounce, encompassing the complete accretion phase and a large part of the cooling phase. It is modeled after SN 1987A and assumes a 20 $M_\odot$ progenitor. The total emitted energy is 2.9×10^{53} erg, of which 16 % is carried by $\bar{\nu}_e$ with 15.3 MeV energy on average. The newer spherically symmetric Garching simulations include more detailed information on neutrino energy spectra and use a sophisticated neutrino transport mechanism. They cover 0.80 s following the collapse of an O-Ne-Mg 8 – 10 $M_\odot$ progenitor star, that is destabilized due to rapid electron capture on neon and magnesium. This class of stars may represent up to 30 % of all core collapse supernovae.

Neutrino oscillations in the supernova environment play an important role, modify the detected rate and thus provide sensitivity on neutrino properties. Neutrinos streaming out of the core will encounter matter densities ranging from 10^{13} kg/m^3 to zero and through two MSW-resonance layers. Both mix the initial fluxes of ν_e, $\bar{\nu}_e$ and all other flavors ν_x depending on the survival probabilities. We consider two limiting cases as benchmarks to discuss the effect of the assumed neutrino hierarchy on the spectra observed with IceCube. Scenario A describes the normal neutrino hierarchy case and Scenario B represents the inverted hierarchy case with a static density profile of the supernova, both paired with a relatively large mixing angle $\theta_{13} > 0.9°$. The oscillation scenario B for an inverted neutrino mass hierarchy shows the largest signal for the Lawrence-Livermore and Garching models because energetic $\bar{\nu}_x$ will oscillate into $\bar{\nu}_e$, harden their spectrum and thus increase the detection probability. The scenario without any oscillation is given as a reference and leads typically to the weakest signal.

In the simulation we include ν_e and $\bar{\nu}_e$ interactions on protons, electrons and $^{16/17/18}$O as well as positron annihilation and neutron capture. Reactions producing electrons or positrons in the final state radiate $N_\gamma = 325.4$ cm$^{-1} \cdot x$ (x in cm) Cherenkov photons in the (300 - 600) nm wavelength range along their flight path x, as long as their kinetic energies exceed the Cherenkov threshold of 0.272 MeV. The mean travel path for $O(10\,\text{MeV})$ from $\bar{\nu}_e$, including secondary leptons with energies above the Cherenkov threshold as well as positron annihilation, was determined to be $\bar{x} = (0.577 \pm 0.005\,(\text{stat.}) \pm$

0.029 (syst.)) cm $\cdot E_{e^+}$ / MeV.

Neutrinos are detected in IceCube by registering predominantly single Cherenkov photons radiated by the neutrino interaction products. The optical scattering and absorption in glacial ice at the South Pole has been studied extensively [15]. The mean number of photons recorded by an optical module averaged over energy is given by $N_\gamma^{\rm detect} = \epsilon_{\rm deadtime} \cdot n_\nu^{\rm interact} \cdot \overline{N_\gamma} \cdot V_\gamma^{\rm eff}$, where $n_\nu^{\rm interact}$ is the neutrino density, $\epsilon_{\rm deadtime} \approx 0.87/(1 + r_{SN} \cdot \tau)$ is the loss in signal rate r_{SN} due to the artificial deadtime of $\tau = 250\,\mu$s, $\overline{N_\gamma}$ is the energy averaged number of Cherenkov photons, and $V_\gamma^{\rm eff}$ is the so-called single photon effective volume, that varies strongly with the photon absorption. As a first approximation, it can be estimated by the product of average Cherenkov spectrum and DOM sensitivity weighted absorption length ($\approx$ 100 m), DOM geometric cross section (0.0856 m^2), Cherenkov spectrum weighted optical module sensitivity ($\approx$ 0.071), average angular sensitivity including cable shadowing effects ($\approx$ 0.32), and the fraction of single photon hits passing the electronic DOM threshold ($\approx$ 0.85).

We use two alternative procedures to calculate the number of detected signal hits from the number of neutrinos crossing the detector: the first approach relies on predetermined tables [16], created to track photons across the Antarctic ice and separates simulations of particle interactions, Cherenkov photon creation, propagation and detection. We obtain $V_\gamma^{\rm eff} = 0.160 \pm 0.004$ (stat.) $\pm$ 0.020 (syst.) m^3. For positrons with a cross section weighted average energy of $\overline{E_{e^+}}$=20 MeV one obtains $\overline{N_\gamma} \cdot V_\gamma^{\rm eff} \approx (590 \pm 80)$ m^3. This volume corresponds to an envisioned sphere of $\approx$ 5.2 m radius centered at the optical module position, with full sensitivity inside and zero outside. To give an example, a study of the initial 380 ms of the burst in the Lawrence Livermore model (see Table 1) at distances of 10 kpc (5 kpc) would require a 0.45 (1.6) Mton background free detector to statistically compete with IceCube. The second GEANT GCALOR-based simulation combines all the steps in one program, which allows one e.g. to determine a 20 % dependence of the detector sensitivity on the incoming neutrino direction for neutrino interactions on electrons. Fig. 3 shows the clustering of detected inverse beta neutrino interactions at the position of the detector strings to visualize the effective volumes.

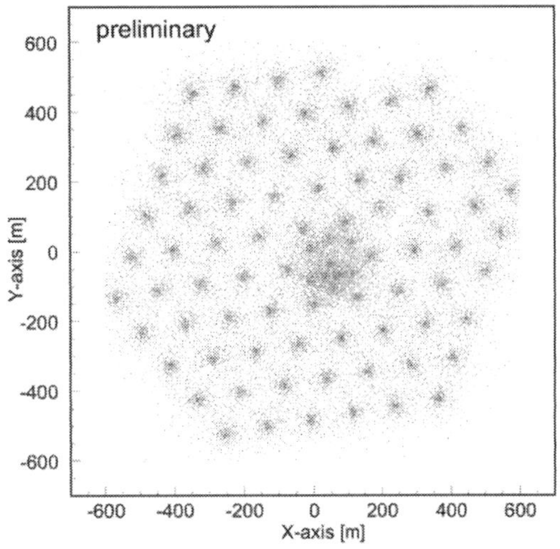

Figure 3. Detected neutrino inverse beta decay interaction vertices projected onto horizontal plane (GEANT simulation with 10 million neutrino interactions).

6. Performance Simulations

All simulations are performed for the final IceCube array with 4800 standard and 360 high efficiency DOMs. We assume that 2 % of the DOMs are excluded from the analysis, either because they are not working or they give unstable rates. A likelihood ratio method was used to determine the range within which models can be distinguished. From sets of several thousand test experiments, we determine limits at the 90 % confidence level, while requiring that the tested scenario is detected in at least 50 % of the cases. Note that the ranges obtained should be interpreted as optimal as we assume that the model shapes are perfectly known and only the overall flux is left to vary; we also disregard the possibility that multiple effects, such as matter induced neutrino oscillations and neutrino self-interactions, could co-exist and thus may be hard to disentangle.

IceCube is particularly well suited to study fine details of the neutrino flux as function of time. While the expected ν_e deleptonization signal will be hard to disentangle with IceCube, forward and reverse shocks as well as stationary accretions shock instabilities (SASI) can be observed. Fig. 4 bottom left shows a simulation based on a model [8] that predicts a sudden spike in the $\bar{\nu}_e$ flux lasting for a few ms while the neutron star turns to a quark star. The likelihood ratio test gives a deviation larger than 5 σ from the hypothesis of no quark star formation for distances up to 30 kpc. The height and shape of the peak depend on the neutrino hierarchy. Scenarios A and B can be distinguished at 90 % C.L. up to distances of 30 kpc.

Fig. 4 bottom right shows a simulation based on the prediction of [9] for the formation of a black hole following a collapse of a 40 solar mass progenitor star. Neutrinos reach energies up to 27 MeV (ν_e and $\bar{\nu}_e$) and 40 MeV (ν_μ and ν_τ), carry a correspondingly large detection probability and thus produce very clear evidence for the formation of the black hole after 1.3 s. The corresponding drop in ν rate can be identified at higher than 90 % C.L. for all stars in our Galaxy and the Magellanic Clouds. The simulation

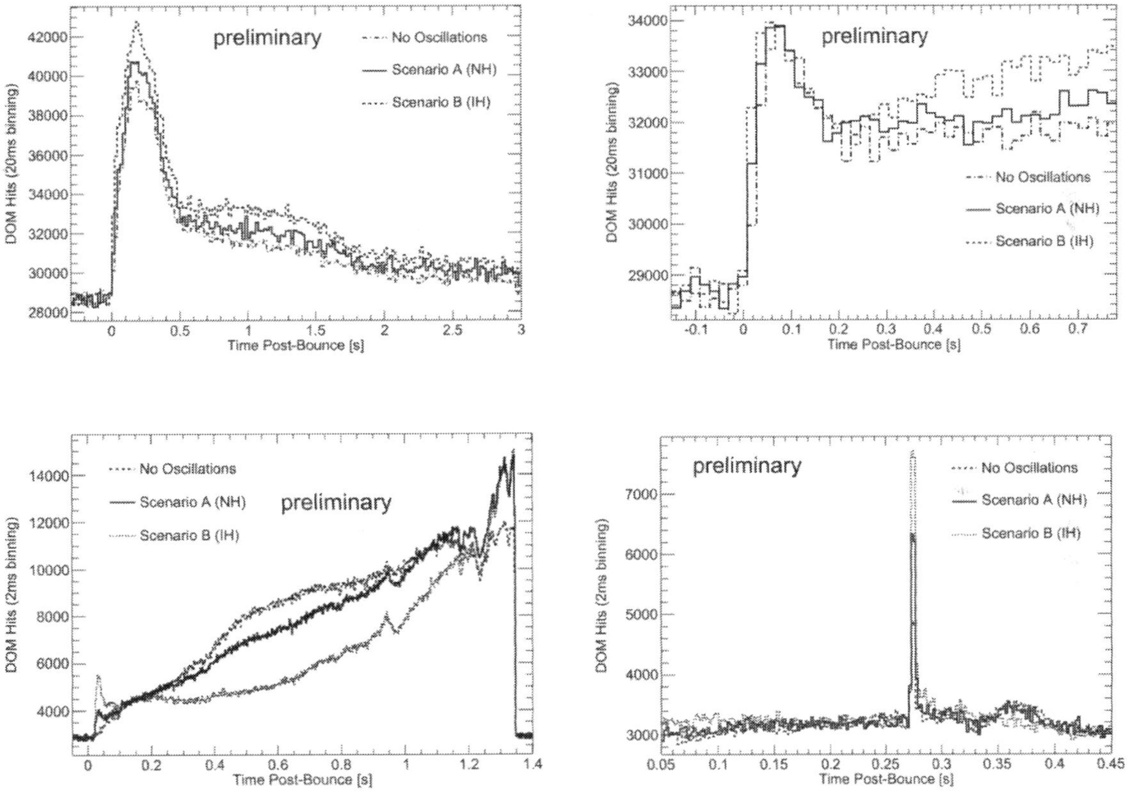

Figure 4. Top: Expected rate distribution at 10 kpc supernova distance for oscillation scenarios A (normal hierarchy) and B (inverted hierarchy). Fluxes and energies in the left plot are taken from the Lawrence-Livermore model [10] and in the right plot from the Garching model [11]. The case of no oscillation is given as a reference. Bottom: Expected neutrino signal from the gravitational collapse of a non rotating massive star of 40 solar masses into a black hole following [9]. Comparison of the neutrino light curve with quark-hadron phase transition for a progenitor star with 10 solar masses. The observation of the sharp $\bar{\nu}_e$ induced burst 257 ms $< t <$ 261 ms after the onset of neutrino emission would constitute direct evidence of quark matter. The 1 σ-bands corresponding to measured detector noise (hatched area) have a width of about ± 215 counts for a 20 ms binning and ± 70 counts for a 2 ms binning.

of an expected signal from a supernova within the Milky Way has to take into account the number of likely progenitor stars in the Galaxy as a function of the distance from Earth. The expected significances of supernova signals according to the Lawrence-Livermore model for three oscillation scenarios are shown in the left plot of Fig. 5. The number of standard deviation with which normal and inverted ν hierarchies (Scenarios A and B) can be distinguished are plotted in Fig. 5 (right plot) as function of the supernova distance for selected models. The values represent the optimal cases when model shapes (but not necessarily the absolute fluxes) are perfectly known. Table 1 lists the number of neutrino induced

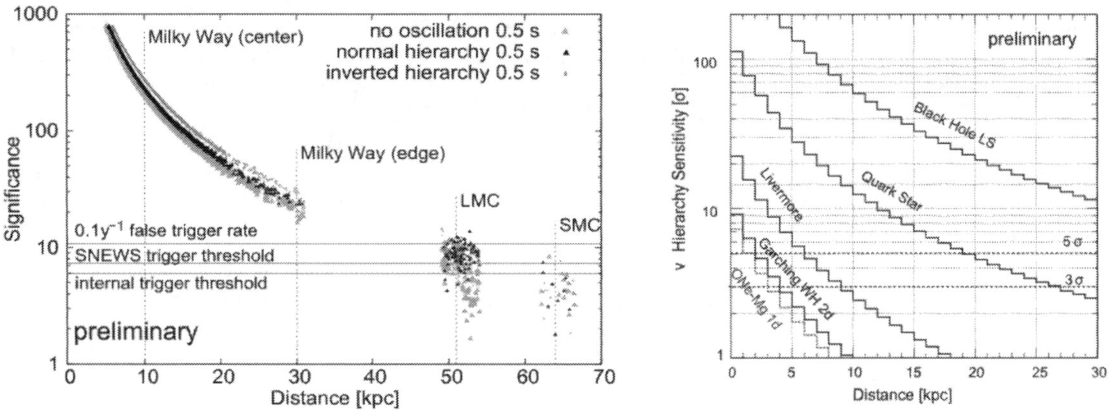

Figure 5. Left: Significance versus distance assuming the Lawrence-Livermore model. The significances are increased by neutrino oscillations in the star by typically 15 % in case of a normal hierarchy (Scenario A) and 40 % in case of an inverted hierarchy (Scenario B). The Magellanic Clouds as well as center and edge of the Milky Way are marked. The density of the data points reflect the star distribution. Right: Number of standard deviation with which scenarios A (normal hierarchy) and B (inverted hierarchy) can be distinguished in at least 50 % of all cases as function of supernova distance for some of the models listed in Table 1. A likelihood ratio method was used assuming that the model shapes are perfectly known.

photon hits that would be recorded by IceCube on top of the nominal DOM noise level for various supernova models.

Table 1. Number of recorded DOM hits in IceCube ($\approx \#\nu$'s) for various models of the supernova collapse and progenitor masses assuming a distance of 10 kpc, approximately corresponding to the center of our Galaxy. A normal neutrino hierarchy is assumed.

Model	Reference	Progenitor mass ($M_\odot$)	$\#\nu$'s $t < 380$ ms	$\#\nu$'s all times
"Livermore"	[10]	20	0.174×10^6	0.79×10^6
"Garching LS-EOS 1d"	[11]	8 – 10	0.069×10^6	-
"Garching WH-EOS 1d"	[11]	8 – 10	0.078×10^6	-
"Garching SASI 2d"	[12]	15	0.106×10^6	-
"1987A at 10 kpc"	[13]	15 – 20		$(0.57 \pm 0.18) \times 10^6$
"O-Ne-Mg 1d"	[14]	8.8	0.054×10^6	0.17×10^6
"Quark Star (full opacities)"	[8]	10	0.067×10^6	-
"Black Hole LS-EOS"	[9]	40	0.395×10^6	1.03×10^6
"Black Hole SH-EOS"	[9]	40	0.335×10^6	3.40×10^6

7. Summmary

IceCube was completed in December 2010 and monitors $\approx$ 1 km^3 of deep Antarctic ice for particle induced photons with 5160 photomultiplier tubes. Since 2009 it supersedes AMANDA in the SNEWS network. With a 250 μs artificial dead time setting, the average DOM noise rate is 286 Hz. The data taking is very reliable and covers the whole calendar year, including periods when new strings were deployed. The uptime has continuously improved toward a goal of > 98 % and reached 96.7 % in 2009. IceCube's sensitivity corresponds to a megaton scale detector for galactic supernovae, triggering on supernovae with about 200, 20, and 6 standard deviations at the galactic center (10 kpc), the galactic edge (30 kpc), and the Large Magellanic Cloud (50 kpc). IceCube cannot determine the type, energy, and direction of individual neutrinos and the signal is extracted statistically from rates that include a noise pedestal. On the other hand, IceCube is currently the world's best detector for establishing subtle features in the temporal development of the neutrino flux. The statistical uncertainties at 10 kpc distance in 20 ms bins around the signal maximum are about 1.5 % and 3 % for the Lawrence Livermore and Garching models, respectively.

Depending on the model, in particular the progenitor star mass, the assumed neutrino hierarchy and neutrino mixing, the total number of recorded neutrino induced photons from a burst 10 kpc away ranges between $\approx 0.17 \times 10^6$ (8.8 $M_\odot$ O-Ne-Mg core), $\approx 0.8 \times 10^6$ (20 $M_\odot$ iron core) to $\approx 3.4 \times 10^6$ for a 40 $M_\odot$ progenitor turning into a black hole. For a supernova in the center of our Galaxy, IceCube's high statistics would allow for a clear distinction between the accretion and cooling phases, an estimation of the progenitor mass from the shape of the neutrino light curve, and for the observation of short term modulation due to turbulent phenomena or forward and reverse shocks during the cooling phase. IceCube will be able to distinguish inverted and normal hierarchies for the Garching, Lawrence-Livermore and black hole models for a large fraction of supernova bursts in our Galaxy provided that the model shapes are known and $\theta_{13} > 0.9°$. The slope of the rising neutrino flux following the collapse can be used to distinguish both hierarchies in a less model dependent way for distances up to 6 kpc at 90 % C.L. As in the case of the inverted hierarchy, coherent neutrino oscillation will enhance the detectable flux considerably. A strikingly sharp spike in the $\overline{\nu}_e$ flux, detectable by IceCube for all stars within the Milky Way, would provide a clear proof of the transition for neutron to a quark star as would be the sudden drop of the neutrino flux in case of a black hole formation.

Further optimizations may be applied to the data acquisition and analysis in the future, e.g. by incorporating a more sophisticated method to remove correlated noise, by excluding the bin-by-bin contribution of measured cosmic ray muon hits to the rate measurement, by storing time stamps of all hits in case of a significant alarm to e.g. improve on the timing resolution, and by employing temporal templates in likelihood or cross-correlation studies.

References

[1] Pryor, C., Roos, C. & Webster, M. 1988, Astrophys. J., 329, 335
[2] Halzen, F., Jacobsen, J. E. & Zas, E. 1996, Phys.Rev.D, 53, 7359
[3] J. Ahrens, J., et al. 2002, Astropart. Phys., 16, 345
[4] Abbasi, R., et al. 2010, Nucl. Instrum. Meth. A618, 139
[5] Abbasi, R., et al. 2009, Nucl. Instrum. Meth. A601, 294
[6] Antonioli, P., et al. 2004, New J. Phys., 6, 114
[7] Pratt, S. R., Raines, R. A., Fossa Jr., C. E. & Temple, M. A. 1999, IEEE Commun. Surveys Tuts., 2, 2
[8] Dasgupta, B., et al. 2010, Phys. Rev. Lett. D, 81, 103005
[9] Sumiyoshi, K., Yamada S., & Suzuki, H. 2007, Astrop.J, 667, 382
[10] Totani, T., Sato, K., Dalhed, H. E., Wilson, J. R. 1998, Astrop. Phys., 496, 216
[11] Kitaura, F. S., Janka, H.-Th. & Hillebrandt, W. 2006, A&A, 450 1, 345
[12] Marek, A., Janka, H.-T. & Müller, E. 2009, A&A, 496 2, 475
[13] Pagliaroli, G., Vissani, F., Constantini, M. L. & Ianni, A. 2009, Astropart. Phys., 31, 163
[14] Hüdepohl, L., Müller, B., Janka, H.-T., Marek, A., & Raffelt, G.G. 2010, Phys.Rev.Lett., 104, 251101
[15] Ackermann, M., et al. 1996, J. Geophys. Res., 111, D13203
[16] Lundberg, J., et al. 2007, Nucl.Instrum.Meth.A, 581, 619

Journal of Physics: Conference Series **309** (2011) 012030 doi:10.1088/1742-6596/309/1/012030

Supernova neutrino detection in LAr TPCs

Inés Gil-Botella

CIEMAT, Basic Research Department, Avenida Complutense, 22, 28040 Madrid, Spain

E-mail: ines.gil@ciemat.es

Abstract. The neutrino burst from a core collapse supernova can provide information about the explosion mechanism and the mechanisms of proto neutron star cooling but also about the intrinsic properties of the neutrino such as flavor oscillations. One important question is to understand to which extent can the supernova and the neutrino physics be decoupled in the observation of a single supernova. The possibility to probe the neutrino mixing angle θ_{13} and the type of mass hierarchy from the detection of supernova neutrinos with liquid argon detectors is summarized in this paper. Moreover, a quantitative study about the possibility to constrain the supernova parameters is presented. A very massive liquid argon detector ($\approx$ 100 kton) is needed to perform accurate measurements of these parameters. In addition, these detectors could also provide information on the ν_e component of the diffuse supernova neutrino background.

1. Introduction

Core collapse supernovae are a huge source of all flavor neutrinos. In 1987 astrophysics entered a new era with the detection of the neutrinos from the supernova SN1987A [1], which exploded in the Large Magellanic Cloud at a distance of $\approx$ 50 kpc. The burst of light was visible to the naked eye. About three hours before the light breakout of the supernova (SN), an increase of neutrinos was detected by three water Cerenkov detectors (Kamiokande, IMB and Baksan). This observation confirmed important parts of the neutrino supernova theory such as total energy, mean temperature and time duration. However, limited quantitative information on the neutrino spectrum was obtained due to the small statistics recorded.

The flavor composition, energy spectrum and time structure of the neutrino burst from a supernova can give information about the explosion mechanism and the mechanisms of proto neutron star cooling. In addition, the intrinsic properties of the neutrino such as flavor oscillations can also be studied.

Although new data from neutrino detection have contributed to the understanding of the neutrino properties, still the neutrino mixing angle θ_{13}, the CP violation phase and the nature of the mass hierarchy, *i.e.*, normal (n.h.) or inverted (i.h.), remain unknown. On the other hand, detailed supernova simulations (see e.g. [2]) indicate that the average neutrino energy is flavor dependent and hierarchical ($E_{\nu_e} < E_{\bar{\nu}_e} < E_{\nu_{\mu,\tau},\bar{\nu}_{\mu,\tau}}$). However, great uncertainties still exist in the spectral and temporal evolution of the neutrino fluxes [3, 4]. The effects of these uncertainties can change dramatically the characteristics of the events observed in a given experiment. Indeed, this dependence can be used to determine the supernova and mixing parameters.

Several studies in the literature concern water, heavy-water or liquid scintillator targets (see e.g. [5]). This paper summarizes the capabilities of liquid argon Time Projection Chamber (LAr TPC) detectors to investigate supernova neutrinos thanks to the simultaneous observation of neutrino elastic scattering on electrons and both nuclear charged current and neutral current interactions on argon. While almost all the current detectors are primarily sensitive to $\bar{\nu}_e$ events, LAr TPCs will provide invaluable information from the ν_e detection. The complete study on supernova neutrino detection in LAr TPCs can be found in references [6].

Fundamental physics subjects as neutrino mass hierarchy, CP violation, matter stability, astroparticle physics and in particular, supernova neutrino detection, call for a new generation of very large mass detectors able to collect high statistics and perform detailed event analysis. Among them, LAr detectors of the order of 100 kton will be ideal devices for these purposes offering the widest physics output (accelerator & non-accelerator) [7].

2. Neutrino emission from supernovae

During the gravitational collapse of a supernova, electron neutrinos first and then the rest of flavor neutrinos are produced. It is believed that the 99% of the total binding energy of the star, $E_B \approx 3 \times 10^{53}$ ergs, is emitted in the form of neutrinos.

Figure 1 shows the prediction for the time evolution of the neutrino fluxes for the different flavors in absence of neutrino oscillations from [8]. The time origin is defined such that the supernova core bounce time corresponds to t = 200 ms. The early ν_e pulse expected in the first 40 ms of the collapse is called *shock breakout* or *neutronization burst* and the rest beyond the first 40 ms is considered as the (*cooling stage*) and it lasts up to ~10 s.

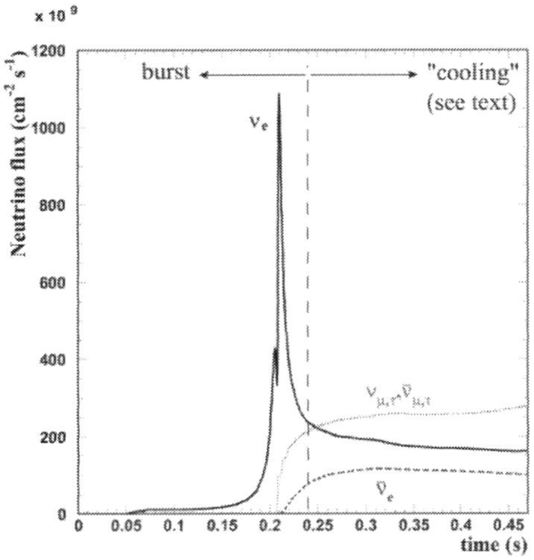

Figure 1. Prediction of [8] of the neutrino flux at Earth in absence of neutrino flavor oscillations from a supernova at a distance of 10 kpc.

The neutrinos in the cooling stage are in equilibrium with their surrounding matter density and their energy spectra can be described by a function close to a Fermi-Dirac distribution. The flux of an emitted neutrino ν_α can then be written as [4]:

$$\phi_\alpha(E_\alpha, L_\alpha, D, T_\alpha, \eta_\alpha) = \frac{L_\alpha}{4\pi D^2 F_3(\eta_\alpha) T_\alpha^4} \frac{E_\alpha^2}{e^{E_\alpha/T_\alpha - \eta_\alpha} + 1} \tag{1}$$

where L_α is the luminosity of the flavor ν_α ($E_B = \sum L_\alpha$), D is the distance to the supernova, E_α is the energy of the ν_α neutrino, T_α is the neutrino temperature inside the neutrinosphere and η_α is the "pinching" factor. We have considered $\eta_\alpha = 0$, $\langle E_{\nu_\alpha} \rangle \approx 3.15\ T_\alpha$ and normalization factor $F_3(0) \approx 5.68$.

The original ν_μ, ν_τ, $\bar{\nu}_\mu$ and $\bar{\nu}_\tau$ fluxes are approximately equal and therefore we treat them as ν_x. An energy hierarchy between the different neutrino flavors is generally believed to hold and imply $\langle E_{\nu_e}\rangle < \langle E_{\bar{\nu}_e}\rangle < \langle E_{\nu_x}\rangle$. However, the specific neutrino spectra remain a matter of detailed calculations. In particular, recent simulations seem to indicate that the energy differences between flavors could be very small and possible collective neutrino flavor conversions could arise for either mass hierarchy depending on the primary fluxes [9].

Neutrino oscillations and matter effects in the supernova will change significantly the neutrino fluxes and therefore, the number of events expected in the detectors. If the neutrino energy spectra are different, then θ_{13} and the mass hierarchy can be probed. For *small mixing angle (S)* ($\sin^2\theta_{13} < 2\times 10^{-6}$), there are no effects on θ_{13} and we can not distinguish among mass hierarchies. Only an upper bound on $\sin^2\theta_{13}$ can be set. For *intermediate* θ_{13} ($2\times 10^{-6} < \sin^2\theta_{13} < 3\times 10^{-4}$) maximal sensitivity to the angle is achieved and measurements of the angle are possible in this region. For *large mixing angle (L)* ($\sin^2\theta_{13} > 3\times 10^{-4}$) maximal conversions occur. The mass hierarchy can be probed but only a lower bound on θ_{13} can be established.

3. Supernova neutrino detection in terrestrial detectors

Most of the current and near-future supernova neutrino experiments [10] are water Cerenkov or liquid scintillator detectors and therefore, primarily sensitive to the $\bar{\nu}_e$ component of the signal, via inverse beta decay $\bar{\nu}_e + p \rightarrow n + e^+$. For supernova burst detection, not only statistics but also diversity of flavor sensitivity is needed: neutral current sensitivity, which gives access to the ν_μ and ν_τ components of the flux, and ν_e sensitivity are particularly valuable.

Only two near-future experiments will be mainly sensitive to ν_e's. The HALO detector [11] is under construction at SNOlab and it uses 80 tons of lead blocks instrumented with the unused SNO NCD counters to record neutrons and electromagnetic signals. However, this technique has some limitations since no energy or pointing information can be obtained and only rates are provided. The ICARUS detector at Gran Sasso [12] is a 600 ton LAr TPC with excellent ν_e sensitivity via ^{40}Ar CC interactions, for which de-excitation gammas will be visible.

All current supernova neutrino experiments participate to the Supernova Early Warning System (SNEWS) [13], the network of SN neutrino observatories whose main goal is to provide the astronomical community with a prompt alert for the next galactic core collapse supernova explosion.

Very promising for the future are a number of planned mega-detectors exploring essentially three technologies: megaton scale water Cerenkov detectors, like LBNE in DUSEL [14], Hyper-K in Japan [15] and Memphys in Europe [16], 100 kton scale LAr TPC detectors like GLACIER in Europe [17] or LAr LBNE in DUSEL [18] and 50 kton scale liquid scintillator detectors like LENA in Europe [19] or Hanohano in Hawai [20]. Some such detectors can hope to collect individual neutrino events every few years from beyond the Local Group of galaxies (few Mpc), assuming that background can be reduced sufficiently.

The LAGUNA [21] project in Europe is studying the performances of these three technologies for detecting supernova neutrinos. The three proposed large volume detector neutrino observatories can guarantee continuous exposure for several decades so that a high statistics supernova neutrino signal could be eventually observed. The expected number of events for GLACIER, LENA and MEMPHYS are reported in [22], including the neutronization burst rates and diffuse supernova neutrino background. LAr TPCs provide fundamental information from the ν_e detection with a very low energy threshold.

4. Supernova neutrino detection in Liquid Argon TPCs

In a liquid argon TPC supernova neutrinos ($E_\nu < 100$ MeV) can be detected through four different channels [6]:

(i) Charged current (CC) interactions on argon

$$\nu_e \, ^{40}Ar \to e^- + ^{40}K^* \tag{2}$$

$$\bar{\nu}_e \, ^{40}Ar \to e^+ + ^{40}Cl^* \tag{3}$$

The neutrino energy thresholds of these reactions are 1.5 and 7.48 MeV, respectively.

(ii) Neutral current (NC) interactions on argon

$$\overset{(-)}{\nu} \, ^{40}Ar \to \overset{(-)}{\nu} + ^{40}Ar^* \tag{4}$$

The energy threshold of this reaction is 1.46 MeV.

(iii) Elastic scattering (ELAS) on electrons

$$\overset{(-)}{\nu} \, e^- \to \overset{(-)}{\nu} \, e^- \tag{5}$$

Figure 2 shows the cross sections of all the processes as a function of the neutrino energy. It is possible to separate the different channels by measuring the associated photons from the K, Cl or Ar de-excitation, or by the absence of them in the case of elastic scattering.

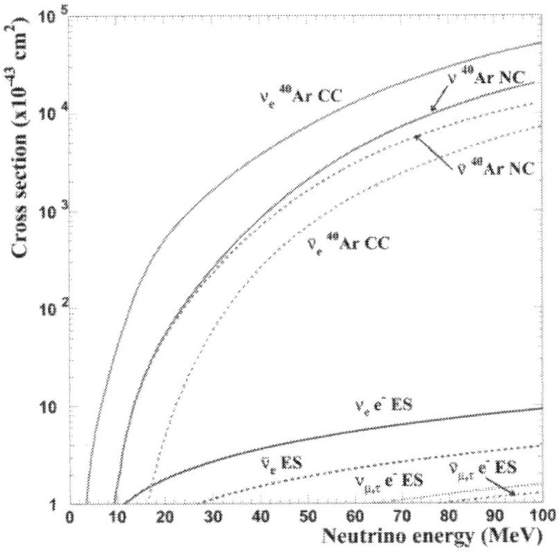

Figure 2. Neutrino cross sections relevant to the supernova detection with a liquid argon TPC.

4.1. Neutronization burst in LAr TPCs

The prompt ν_e burst is in principle observable and represents a diagnostic of the fundamental collapse supernova behavior. LAr TPCs provide unique information about the early breakout pulse due to their excellent sensitivity to ν_e neutrinos mainly through the CC process. The analysis of the time structure of the supernova signal during the first few tens of milliseconds after the core bounce can provide a clean indication if the full ν_e burst is present or absent and therefore allows distinguishing between different mixing scenarios.

Figure 3 shows the effect of the reduction of the ν_e peak due to oscillations. The suppression is maximal in the case of for normal hierarchy and large θ_{13} mixing angle (n.h.-L) due to the total conversion of ν_e into $\nu_{\mu,\tau}$. The energy spectrum moves slightly to higher neutrino energy values. Large LAr TPC detectors of the order of 100 kton will be needed to provide enough statistics to study the ν_e neutrino burst.

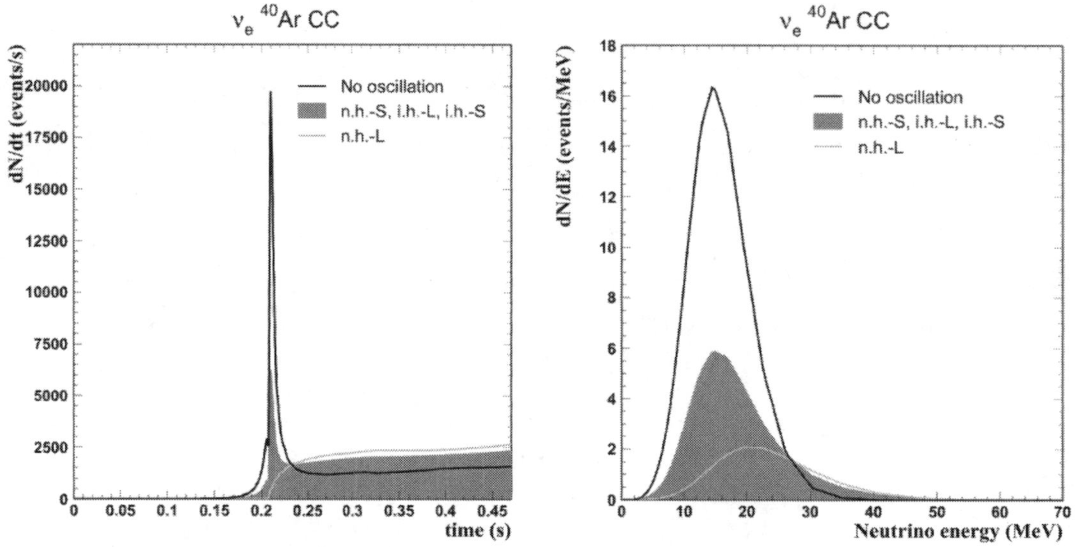

Figure 3. Time evolution of the ν_e CC event rate (left) and the corresponding time integrated event spectra (right) for different oscillation scenarios.

4.2. Cooling phase

The main neutrino emission from core collapse supernovae occurs during the cooling stage. Table 1 shows the expected number of neutrino events from a supernova at 10 kpc in a 100 kton detector. The four detection channels are considered independently and oscillation and non-oscillation cases are computed for normal and inverted hierarchies.

Table 1. Expected neutrino events in a 100 kton detector from a supernova at a distance of 10 kpc.

$\langle E_{\nu_e}\rangle = 11$ MeV, $\langle E_{\bar{\nu}_e}\rangle = 16$ MeV, $\langle E_{\nu_x}\rangle = \langle E_{\bar{\nu}_x}\rangle = 25$ MeV and luminosity equipartition					
	No	Oscillation (n.h.)		Oscillation (i.h.)	
Reaction	oscillation	Large θ_{13}	Small θ_{13}	Large θ_{13}	Small θ_{13}
ELAS	1330	1330	1330	1330	1330
ν_eCC	6240	31320	23820	23820	23820
$\bar{\nu}_e$CC	540	1110	1110	2420	1110
NC	30440	30440	30440	30440	30440
Total	38550	64200	56700	58010	56700

A total of 38550 neutrino events are expected in case of no oscillation assuming $\langle E_{\nu_e}\rangle = 11$ MeV, $\langle E_{\bar{\nu}_e}\rangle = 16$ MeV, $\langle E_{\nu_x}\rangle = \langle E_{\bar{\nu}_x}\rangle = 25$ MeV and luminosity equipartition among flavors (scenario I). CC events are very sensitive to the change in the energy spectra due to oscillations. A clear increase in the number of events is expected, being a factor 4-5 for ν_eCC events and a factor 2-4 for $\bar{\nu}_e$CC interactions. The main variations with the θ_{13} angle are expected for the ν_eCC channel in the case of normal hierarchy and for the $\bar{\nu}_e$CC channel for inverted hierarchy. NC events are not affected by oscillations and they constitute an excellent probe for the supernova properties.

Figure 4 shows the oscillation effects in the neutrino energy spectra of the expected ν_eCC and $\bar{\nu}_e$CC events in a 3 kton detector. Due to the total conversion $\nu_{\mu,\tau} \to \nu_e$ for the n.h.-L case, the $\nu_e$energy spectra is harder and this leads to a huge increase of the expected events due to the quadratic dependence of the CC cross section with energy. The same effect can be seen for $\bar{\nu}_e$ events and i.h.-L case.

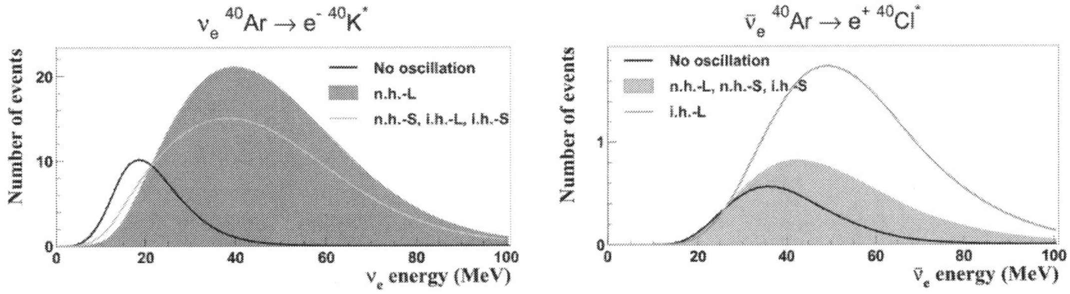

Figure 4. Expected number of ν_eCC (left) and $\bar{\nu}_e$CC (right) events in a 3 kton LAr detector from a supernova at 10 kpc.

Other models for supernova neutrino fluxes considering MSW flavor conversions have been studied in [6]. Collective oscillations giving rise to spectral swaps and possible Earth matter effects are also considered in recent papers [9]. These effects can produce distortions in the observable supernova neutrino signal and are crucially dependent on the neutrino mass hierarchy and the relative sizes of the primary neutrino spectra.

5. Decoupling supernova and neutrino oscillation physics with LAr TPC detectors

The understanding of the supernova explosion is still plagued by uncertainties which have an impact on the precision with which one can predict time, energy and flavor dependent neutrino fluxes. On the other hand, the neutrino mixing properties are not fully known and in fact large uncertainty exists on the prediction of the actual effect of neutrino oscillations in the event of a supernova explosion. The main question is to understand to which extent the supernova and the neutrino physics can be decoupled in the observation of a single supernova.

In the case of LAr TPCs, the combination of the four independent detection channels can over-constrain the features of the supernova fluxes and the mixing parameters. The capability of LAr TPC detectors to simultaneously measure the supernova and the oscillation parameters is quantitatively analyzed in next sections. A global fit to the event energy distribution for elastic and charged current events and to the rate for the neutral current events is performed.

5.1. Study of the neutrino oscillation parameters

First we investigate the sensitivity to the mixing angle θ_{13} and the type of mass hierarchy without any assumption on the supernova parameters. We exploit the information from the four neutrino detection channels on argon and we minimize a χ^2 function letting free the five supernova parameters.

In order to understand the uncertainties on the knowledge of the supernova physics, we consider two different sets of reference values for the supernova parameters, corresponding to one hierarchical energy scenario (I) and a non-hierarchical scenario (II). They differ in the assumptions on the average energies of the neutrino flavors. The luminosities of the neutrino flavors are assumed to be equal.

We first analyze the true small and large mixing angle scenarios and study the possibility of distinguishing the mass hierarchy and putting a bound on the θ_{13} value.

Figure 5 shows the variation of the minimized χ^2 value with the $\sin^2\theta_{13}$ parameter and the mass hierarchy for a 100 kton detector. In the top right corner is indicated in bold the "true" scenario considered. The left part of a given plot corresponds to the results of the fit under the assumption of normal hierarchy and the right part for an inverted hierarchy. Solid lines are computed considering the four supernova neutrino detection channels. Dashed lines do not take into account the NC processes. The shaded region illustrates the excluded region by the results of reactor experiments. The one- and two-sigma levels are shown by horizontal lines.

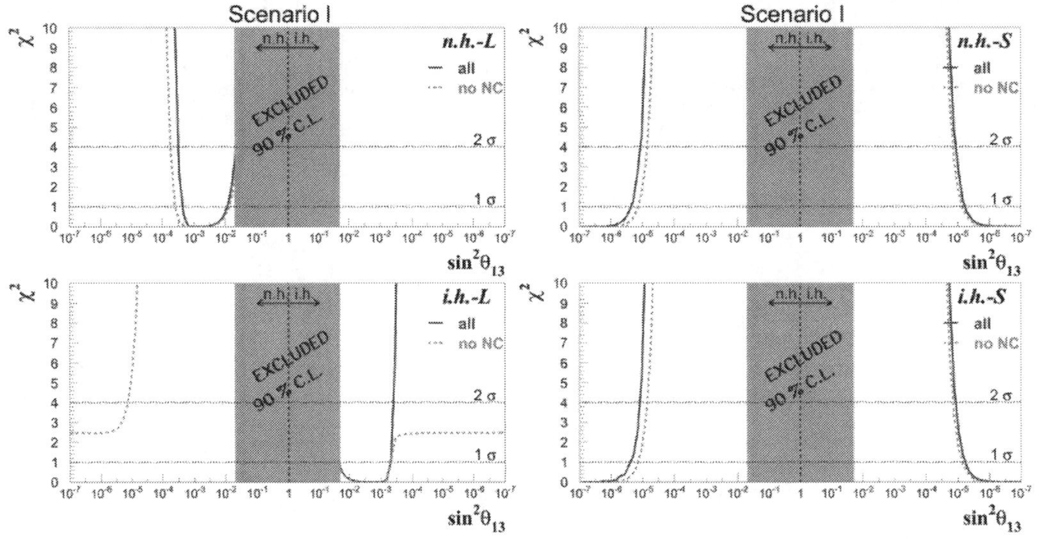

Figure 5. χ^2 value of the fit as a function of $\sin^2\theta_{13}$ for a 100 kton detector. Curves on the left (right) part of every plot correspond to the results of the fit under the assumption of a normal (inverted) hierarchy. Four limiting oscillation scenarios (n.h.-L, n.h.-S, i.h.-L, i.h.-S) are considered as "true" and the reference values of the supernova parameters correspond to scenario I. Solid curves are computed using the information of the four supernova neutrino detection processes and the dashed line does not take into account NC events.

We see that normal and inverted hierarchies are indistinguishable for small "true" θ_{13} mixing angle. The results of the fit are similar and only an upper bound on $\sin^2\theta_{13}$ can be set. If the "true" mixing angle is large, we are able to distinguish among hierarchies. Assuming the "true" hierarchy is normal and the value of θ_{13} is large (n.h.-L), we could put a lower limit on the θ_{13} angle. The same is possible for the case of inverted hierarchy. The wrong hierarchies are excluded. We illustrate the importance of the NC events. If we do not include them, we loose sensitivity in the large mixing angle cases and i.h.-L could be misidentified as n.h.-S.

The sensitivity to the oscillation parameters is almost similar for hierarchical (scenario I) and non-hierarchical scenarios (scenario II: $\langle E_{\nu_e}\rangle = 13$ MeV, $\langle E_{\bar{\nu}_e}\rangle = 16$ MeV, $\langle E_{\nu_x}\rangle = \langle E_{\bar{\nu}_x}\rangle = 17.6$ MeV) considering a 100 kton detector. Table 2 shows the results on the determination of the θ_{13} mixing angle for the different oscillation cases.

If the θ_{13} mixing angle is in the intermediate range ($2 \times 10^{-6} < \sin^2\theta_{13} < 3 \times 10^{-4}$), maximal

Table 2. Estimated limit on the θ_{13} mixing angle at 1σ for different "true" oscillation cases with a 100 kton detector. We compare the results considering as reference values for the supernova neutrino parameters scenarios I (hierarchical) and II (non-hierarchical).

Determination of $\sin^2\theta_{13}$ and mass hierarchy (SN parameters free)		
True osc. case	SN scen. I	SN scen. II
n.h.-L	$> 4.5\times10^{-4}$ n.h.	$> 1.8\times10^{-4}$ n.h.
	excluded i.h.	excluded i.h.
n.h.-S	$< 4.3\times10^{-6}$ n.h.	$< 1.2\times10^{-5}$ n.h.
	$< 5.9\times10^{-6}$ i.h.	$< 4.3\times10^{-6}$ i.h.
i.h.-L	excluded n.h.	excluded n.h.
	$> 5.0\times10^{-4}$ i.h.	$> 2.3\times10^{-4}$ i.h.
i.h.-S	$< 5.9\times10^{-6}$ n.h.	$< 1.2\times10^{-5}$ n.h.
	$< 4.0\times10^{-6}$ i.h.	$< 4.1\times10^{-5}$ i.h.

sensitivity to the angle is achieved and measurements of the value are possible. Considering a 100 kton detector, statistically accurate measurements can be performed at 1σ and 2σ levels for scenario I (assuming true $\sin^2\theta_{13} = 10^{-4}$):

$$9.1(8.1)\times10^{-5} < \sin^2\theta_{13} < 1.1(1.3)\times10^{-4} \quad 1\sigma(2\sigma) \quad (\textit{assumed n.h.}) \tag{6}$$

$$8.9(7.4)\times10^{-5} < \sin^2\theta_{13} < 1.1(1.3)\times10^{-4} \quad 1\sigma(2\sigma) \quad (\textit{assumed i.h.}) \tag{7}$$

The wrong hierarchies are excluded in both cases. For scenario II, a measurement of the angle is also possible at the 1σ level and the mass hierarchy is well determined.

5.2. Study of the supernova parameters

First we study the information about the astrophysical parameters that can be extracted from the detection of supernova neutrinos considering that the oscillation parameters $\sin^2\theta_{13}$ and the mass hierarchy have been determined from reactor and long-baseline experiments.

Assuming that the angle has been measured with a precision of 10% ($\sin^2\theta_{13} = 10^{-3} \pm 10^{-4}$) and the mass hierarchy is also known, we perform a χ^2 minimization letting the supernova parameters free in order to see how well we can constrain them. The expected accuracies at 90% C.L. are summarized in table 3. With the statistics given by a 100 kton detector, all the variables can be measured in a certain range, independently on the supernova scenario. Only the L_e/L_x parameter in the case of i.h. is less well determined for scenario II ($\sim 32\%$) than for scenario I ($\sim 9\%$).

Another possibility is that future neutrino experiments will not be sensitive enough to measure $\sin^2\theta_{13}$ and will place an upper limit on its value. Assuming the limit $\sin^2\theta_{13} < 10^{-4}$, figure 6 shows the allowed regions at 68%, 90% and 99% C.L. on different supernova parameters planes for a 100 kton detector in the case of i.h. Thanks to the high statistics available with a very massive detector, we can fit very precisely the data and the results are essentially equal for both mass hierarchies. For non-hierarchical scenarios, the uncertainties are bigger, specially for the ratio between the electron and non-electron neutrino luminosities.

Finally we investigate the information that can be extracted from supernova neutrinos if a supernova explosion occurred nowadays when the value of the oscillation parameters is unknown. Letting all parameters $\{E_B, \langle E_{\nu_e}\rangle, \langle E_{\bar{\nu}_e}\rangle, \langle E_{\nu_x}\rangle, L_e/L_x, \sin^2\theta_{13}, \mathrm{sign}[\Delta m^2_{32}]\}$ freely vary, we study how well we can determine them in the case of a 100 kton detector. The results show

Table 3. Expected accuracies at 90% C.L. in the determination of the supernova parameters using the neutrinos measured with a 100 kton detector. We have assumed that the mass hierarchy is known and the θ_{13} mixing angle has been measured by terrestrial experiments, being equal to $\sin^2\theta_{13} = 10^{-3} \pm 10^{-4}$. Supernova scenarios I and II are tested.

With constraint $\sin^2\theta_{13} = 10^{-3} \pm 10^{-4}$ from terrestrial experiment						
True hierarchy	SN scen.	$\frac{\Delta\langle E_{\nu_e}\rangle}{\langle E_{\nu_e}\rangle}(\%)$	$\frac{\Delta\langle E_{\bar{\nu}_e}\rangle}{\langle E_{\bar{\nu}_e}\rangle}(\%)$	$\frac{\Delta\langle E_{\nu_x}\rangle}{\langle E_{\nu_x}\rangle}(\%)$	$\frac{\Delta E_B}{E_B}(\%)$	$\frac{\Delta(L_e/L_x)}{(L_e/L_x)}(\%)$
n.h.	I	~ 14	~ 4	< 1	~ 2	~ 11
i.h.	I	~ 5	~ 9	~ 1	~ 2	~ 9
n.h.	II	~ 9	~ 3	< 1	~ 4	~ 12
i.h.	II	~ 6	~ 6	~ 1	~ 2	~ 32

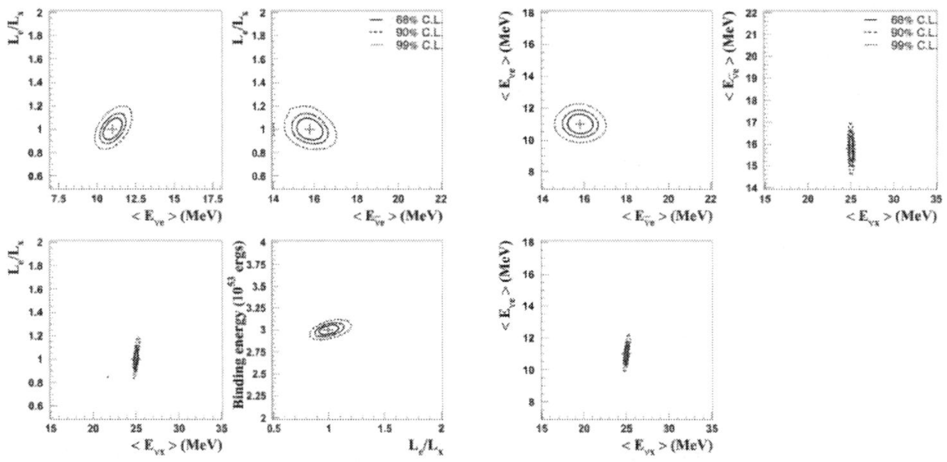

Figure 6. 68%, 90% and 99% C.L. allowed regions for the supernova parameters with a 100 kton detector, assuming that an upper limit on the value of the θ_{13} mixing angle has been set ($\sin^2\theta_{13} < 10^{-4}$) and the mass hierarchy is inverted ($\Delta m^2_{31} < 0$). Crosses indicate the value of the parameters for the best fits.

that with large statistics it is possible to decouple the supernova and oscillations physics and determine the value of the parameters with high precision (table 4).

The expected accuracies are similar to the ones obtained constraining the mixing angle and the mass hierarchy, except for the $\langle E_{\nu_e}\rangle$parameter and n.h.-L. In this scenario, the knowledge of the θ_{13} value improves the determination of the $\langle E_{\nu_e}\rangle$energy from 17% to 14% for scenario I and from 21% to 9% for scenario II.

6. Diffuse supernova neutrino background in LAr TPCs

The diffuse supernova neutrino background (DSNB) is the flux of neutrinos and antineutrinos emitted by all core-collapse supernovae occurred so far in the Universe. It will appear isotropic and time independent in feasible observations.

The DSNB has not been detected yet. The Super-Kamiokande experiment established an

Table 4. Expected accuracies at 90% C.L. in the determination of the supernova parameters using the neutrinos measured with a 100 kton detector. No conditions on the θ_{13} angle and mass hierarchy have been considered.

True osc. case	SN scen.	$\frac{\Delta\langle E_{\nu_e}\rangle}{\langle E_{\nu_e}\rangle}(\%)$	$\frac{\Delta\langle E_{\bar{\nu}_e}\rangle}{\langle E_{\bar{\nu}_e}\rangle}(\%)$	$\frac{\Delta\langle E_{\nu_x}\rangle}{\langle E_{\nu_x}\rangle}(\%)$	$\frac{\Delta E_B}{E_B}(\%)$	$\frac{\Delta(L_e/L_x)}{(L_e/L_x)}(\%)$
n.h.-L	I	~ 17	~ 4	< 1	~ 2	~ 11
i.h.-L	I	~ 5	~ 9	< 1	~ 2	~ 9
n.h.-S	I	~ 6	~ 4	< 1	~ 1	~ 11
i.h.-S	I	~ 5	~ 4	< 1	~ 2	~ 9
n.h.-L	II	~ 21	~ 3	< 1	~ 4	~ 14
i.h.-L	II	~ 6	~ 9	~ 1	~ 2	~ 34
n.h.-S	II	~ 11	~ 5	< 1	~ 2	~ 35
i.h.-S	II	~ 8	~ 4	~ 1	~ 2	~ 37

upper limit on the $\bar{\nu}_e$ flux of $\Phi(\bar{\nu}_e) < 1.2$ cm^{-2}s^{-1} for neutrino energies higher than 19.3 MeV [23], close to the predictions. If Super-Kamiokande is modified with dissolved gadolinium to reduce detector backgrounds and increase the energy range for analysis, then the DSNB could be detected at a rate of a few events per year.

LAr TPCs would be able to mainly detect the ν_e component of the DSNB signal providing complementary information with respect to Super-Kamiokande. The main background sources for these events in the relevant neutrino energy range of 10-50 MeV are solar and low energy atmospheric neutrinos. Depending on the theoretical predictions for the DSNB flux, a 100 kton LAr detector running for five years would get more than 4σ measurement of the DSNB flux [24].

7. Conclusions

Liquid argon TPC detectors have excellent capabilities to study and decouple supernova and neutrino oscillation physics by the observation of four independent channels: elastic scattering on atomic electrons, charged current interactions on argon (independently sensitive to ν_e and $\bar{\nu}_e$ neutrinos) and neutral current interactions on argon.

We have studied the sensitivity to the θ_{13} and mass hierarchy parameters and found that for large mixing angle, the mass hierarchy can be identified and lower limits on the θ_{13} value are set. For small mixing angle, it is not possible to distinguish the mass hierarchy but upper limits on θ_{13} can be obtained. If the mixing angle is in an intermediate range, measurements of its value are possible. The statistics provided by the detection of a supernova at 10 kpc with a 100 kton detector allow to quantitatively determine the neutrino oscillation parameters.

Finally we have studied the ability to measure the supernova parameters with good precision using large LAr detectors. Thanks to the high sensitivity to ν_e neutrinos, these detectors can play a crucial role providing fundamental information about the shock breakout mechanism. With a 100 kton detector, a single supernova explosion would allow determination of the supernova cooling phase quite precisely. In addition, the diffuse supernova neutrino background signal could be observed in five years at the 4σ level.

In conclusion, the information from the ν_e spectrum observed at a LAr TPC will probe features of the supernova neutrino signal that are not accesible to a $\bar{\nu}_e$ detector. The complementary physics potential of large future neutrino detectors will offer unprecedented opportunities to understand the physics of neutrinos and their conversions during a stellar collapse.

Journal of Physics: Conference Series **309** (2011) 012030 doi:10.1088/1742-6596/309/1/012030

References

[1] Hirata K *et al.* 1987 *Phys. Rev. Lett.* **58** 1490; Bionta R *et al.* 1987 *Phys. Rev. Lett.* **58** 1494; Alekseev E N *et al.* 1987 *JETP Lett.* **45** 589
[2] Takahashi K *et al.* 2003 *Astrop. Phys.* **20** 189-193
[3] Raffelt G G *et al.* 2003 *Preprint* astro-ph/0303226; Takahashi K *et al.* 2003 *Phys. Rev.* D **68** 113009
[4] Lunardini C and Smirnov A Y 2003 *J. Cosmol. Astropart. Phys.* JCAP06(2003)009
[5] Totani T *et al.* 1998 *Astrophys. J.* **496** 216; Minakata H *et al.* 2002 *Phys. Lett.* B **542** 239; Dutta G *et al.* 2001 *Phys. Rev.* D **64** 073011; Bandyopadhyay A *et al.* 2003 *Preprint* hep-ph/0312315
[6] Gil-Botella I and Rubbia A 2004 *J. Cosmol. Astropart. Phys.* JCAP08(2004)001; Gil-Botella I and Rubbia A 2003 *J. Cosmol. Astropart. Phys.* JCAP10(2003)009; Bueno A , Gil-Botella I and Rubbia A 2003 *Preprint* hep-ph/0307222
[7] Rubbia A 2003 *Preprint* hep-ph/0402110
[8] Thompson T A, Burrows A and Pinto P A 2003 *Astrophys. J.* **592** 434
[9] Choubey S *et al.* 2010 *Preprint* arXiv:1008.0308 [hep-ph]
[10] Scholberg K 2010 *J. Phys.: Conf. Series* **203** 012079
[11] Duba C A *et al.* 2008 *J. Phys.: Conf. Series* **136** 042077
[12] Aprili P *et al.* 2001 LNGS-EXP 13/89 add. 2/01, CERN-SPSC-2002-027 (SPSC-P-323)
[13] Antonioli P *et al.* 2004 *New. J. Phys.* **6** 114
[14] Maricic J 2010 *J. Phys.: Conf. Series* **203** 012109
[15] Nakamura K 2003 *Int. J. Mod. Phys.* A **18** 4053
[16] Bellefon A *et al.* 2006 *Preprint* hep-ex/0607026
[17] Rubbia A 2009 *J. Phys.: Conf. Series* **171** 012020
[18] Baller B 2010 *These proceedings*
[19] Marrodan Undagoitia T *et al.* 2006 *J. Phys.: Conf. Series* **39** 278-280
[20] Maricic J 2010 *J. Phys.: Conf. Series* **203** 012137
[21] Rubbia A 2010 *Acta Phys. Polon.* B **41** 1727-1732
[22] Autiero D *et al.* 2007 *J. Cosmol. Astropart. Phys.* JCAP0711(2007)011
[23] Malek M *et al.* 2003 *Phys. Rev. Lett.* **90** 061101
[24] Cocco A G *et al.* 2004 *J. Cosmol. Astropart. Phys.* JCAP12(2004)002

Journal of Physics: Conference Series **309** (2011) 012031 doi:10.1088/1742-6596/309/1/012031

Neutral Current Coherent Cross Sections-Implications on Gaseous Spherical TPC's for detecting SN and Earth neutrinos

J. D. Vergados

Theoretical Physics Division, University of Ioannina, Ioannina, Gr 451 10, Greece

E-mail: vergados@uoi.gr

Abstract. The detection of galactic supernova (SN) neutrinos represents one of the future frontiers of low-energy neutrino physics and astrophysics. The collapse of a neutron star liberates a gravitational binding energy of about 3×10^{53} erg, 99% of which is transferred to neutrinos and antineutrions of all the flavors and only 1% to the kinetic energy of the explosion. In other words, a core-collapse supernova represents one of the most powerful sources of neutrinos and antineutrinos of all flavors in the Universe. The neutron coherence of neutral currents (NC) allows quite large cross sections in the case of neutron rich targets, which can be exploited in detecting earth and sky neutrinos by measuring nuclear recoils. These (NC) cross sections are not dependent on flavor conversions and, thus, their measurement will provide useful information about the neutrino source. In particular they will yield information about the primary neutrino fluxes, i.e. before flavor conversions in neutrino sphere. The advantages of large gaseous low threshold and high resolution time projection counters (TPC) detectors will be discussed. These are especially promising since they are expected to be relatively cheap and easy to maintain. The information thus obtained can also be useful to other flavor sensitive detectors, e.g. the large liquid scintillation detectors like LENA. All together such detectors will provide invaluable information on the astrophysics of core-collapse explosion and on the neutrino mixing parameters. In particular, neutrino flavor transitions in the SN envelope might be sensitive to the value of θ_{13} and to the unknown neutrino mass hierarchy. Untill a real SN explosion is detected, one can use available earth neutrino sources with similar energy spectra to test the behavior of these detectors. Among them, the ORNL Neutron Spallation source (SNS) and boosted radioactive neutrino beams are good candidates.

1. Introduction

The detection of galactic supernova (SN) neutrinos represents one of the future frontiers of low-energy neutrino physics and astrophysics.[1] In this paper we are going to discuss the relevant physics for the design and construction of a gaseous spherical TPC for dedicated supernova detection, exploiting the coherent neutrino-nucleus elastic scattering due to the neutral current interaction. This detector can draw on the progress made in recent years in connection with measuring nuclear recoils in dark matter searches. It has low threshold and high resolution, it is relatively cheap and easy to maintain. Before doing this, however, we will briefly discuss

[1] see G. Raffelt, "Physics opportunities with supernova neutrinos", these proceedings.

the essential physics of neutrinos emitted in supernova explosions [Tomas et al.(2004)Tomas, Kachellriess, G.G. Raffelt, Janka, and Schreck] (for a review, see, e.g. the recent report [Janka et al.(2007)Janka, Langanke, Marek, Martinez-Pinedo, and Mueller])[2].

2. The primary supernova neutrino flux.

We will assume that the neutrino spectrum can be described be a Fermi Dirac Distribution with a given temperature T and a chemical potential $\mu = aT$. The constants T and a will be treated as free parameters. Thus

$$f_{sp}(E_\nu, T, a) = \mathcal{N}\frac{1}{1+\exp\left(E_\nu/T - a\right)} \tag{1}$$

where $\mathcal{N}$ is a normalization constant. The temperature T is taken to be 3.5, 5 and 8 MeV for electron neutrinos (ν_e), electron antineutrinos ($\tilde{\nu}_e$) and all other flavors (ν_x) respectively. The parameter a will be taken to be $0 \le a \le 5$. The average neutrino energies obtained from this

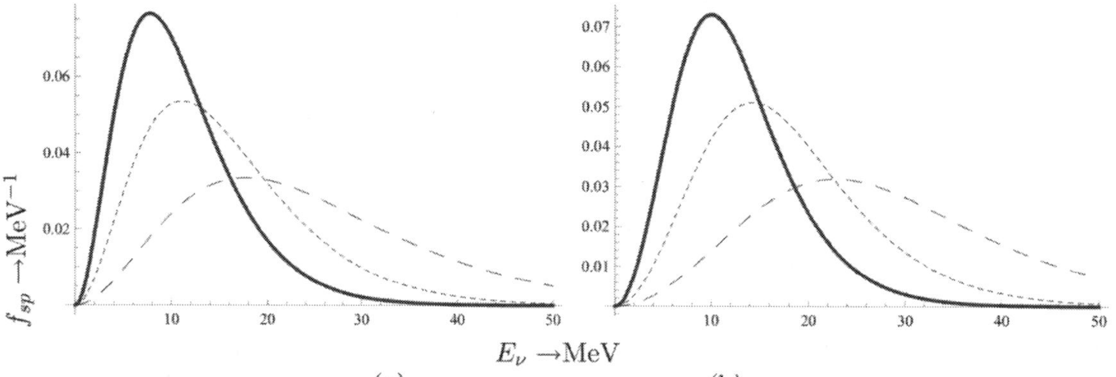

Figure 1: The normalized to unity SN spectrum for $a = 0$ (left panel) and $a = 3$ (right panel). The continuous, dotted and dashed curves correspond to T=3.5 (ν_e), 5 ($\tilde{\nu}_e$) and 8 (ν_x) respectively.

distribution are shown in Table 1. The number of emitted neutrinos [Tomas et al.(2004)Tomas, Kachellriess, G.G. Raffelt, Janka, and Schreck] can be obtained from the total emitted energy $U_\nu = 3 \times 10^{53}$ erg

$$N_\nu = \frac{U_\nu}{\prec E_\nu \succ}. \tag{2}$$

The obtained results are shown in Table 2 The (time averaged) neutrino flux $\Phi_\nu = N_\nu/(4\pi D^2)$ at a distance $D = 10$ kpc$=3.1 \times 10^{22}$cm is given in Table 3.

3. The differential and total cross section

The differential cross section for a given neutrino energy E_ν can be cast in the form[Giomataris and Vergados(2006)]:

$$\left(\frac{d\sigma}{dT_A}\right)_w (T_A, E_\nu) = \frac{G_F^2 A m_N}{2\pi}\,(N^2/4) F_{coh}(T_A, E_\nu), \tag{3}$$

[2] see also G. Fuller, Neutrinos from Supernovae, these proceedings.

Table 1: The average supernova neutrino energies as a function of the parameters a and T.

a	$\prec E_\nu \succ$ (MeV)		
	ν_e T=3.5MeV	$\tilde{\nu}_e$ T=5MeV	$\sum_x \nu_x$ T=8MeV
0	11.0298	15.7569	25.211
0.75	11.4504	16.3578	26.1724
1.50	12.0787	17.2553	27.6085
2.00	12.6194	18.0277	28.8443
3.00	13.9733	19.9619	31.9391
4.00	15.6313	22.3305	35.7288
5.00	17.5179	25.0255	40.0408

Table 2: The number of primary neutrinos emitted in a typical supernova explosion as a function of the parameters a and T in units of 10^{58}.

a	$N_\nu/10^{58}$		
	ν_e T=3.5 MeV	$\tilde{\nu}_e$ T=5 MeV	$\sum_x \nu_x$ T=8 MeV
0	0.282969	0.198079	0.495196
0.75	0.272575	0.190802	0.477006
1.50	0.258397	0.180878	0.452194
2.00	0.247326	0.173128	0.43282
3.00	0.223361	0.156353	0.390882
4.00	0.199669	0.139768	0.349421
5.00	0.178167	0.124717	0.311792

Table 3: The (time integrated) neutrino flux, in units of 10^{12}cm^{-2}, at a distance 10 kpc from the source.

a	$\Phi_\nu/10^{12}\text{cm}^{-2}$		
	ν_e T=3.5 MeV	$\tilde{\nu}_e$ T=5 MeV	$\sum_x \nu_x$ T=8 Mev
0	0.234318	0.164023	0.410057
0.75	0.225711	0.157997	0.394994
1.50	0.213971	0.149779	0.374448
2.00	0.204803	0.143362	0.358405
3.00	0.184958	0.129471	0.323677
4.00	0.16534	0.115738	0.289345
5.00	0.147534	0.103274	0.258185

with

$$F_{coh}(T_A, E_\nu) = F^2(q^2)\left(1 + (1 - \frac{T_A}{E_\nu})^2 - \frac{Am_N T_A}{E_\nu^2}\right) \quad (4)$$

Journal of Physics: Conference Series **309** (2011) 012031 doi:10.1088/1742-6596/309/1/012031

where N is the neutron number and $F(q^2) = F(T_A^2 + 2Am_N T_A)$ is the nuclear form factor. The effect of the nuclear form factor depends on the target (see Fig. 2).

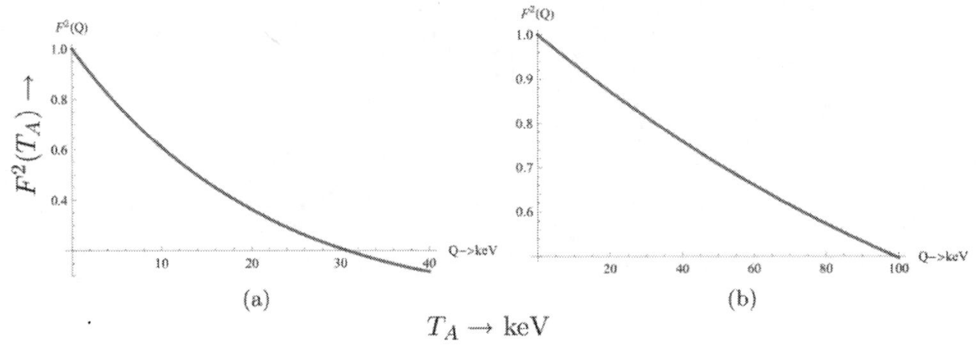

Figure 2: The square of the nuclear form factor, $F^2(T_A)$, as a function of the recoil energy for A=131 (a) and A=40 (b). Note that the maximum recoil energy is different for each target.

Since the SN source is not "'monochromatic"' the above equation can be written as:

$$\frac{d\sigma}{dT_A} = \int_0^\infty \left(\frac{d\sigma}{dT_A}\right)_w (T_A, E_\nu) f_{sp}(E_\nu, T, a) dE_\nu \tag{5}$$

Integrating the total cross section of Fig. 5 from $T_A = E_{th}$ to infinity we obtain the total cross section. The threshold energy E_{th} depends on the detector. Furthermore for a real detector the nuclear recoil events are quenched, especially at low energies. The quenching factor for a given detector is the ratio of the signal height for a recoil track divided by that of an electron signal with the same energy. We should not forget that the signal heights depend on the velocity and how the signals are extracted experimentally. The actual quenching factors must be determined experimentally for each target. In the case of NaI the quenching factor is 0.05, while for Ge and Si it is 0.2-0.3. For our purposes it is adequate, to multiply the energy scale by an recoil energy dependent quenching factor, $Q_{fac}(T_A)$ adequately described by the Lidhard theory [Lifhardt et al.(1963)]. More specifically in our estimate of $Qu(T_A)$ we assumed a quenching factor of the following empirical form [Lifhardt et al.(1963)], [Simon et al.(2003)]:

$$Q_{fac}(T_A) = r_1 \left[\frac{T_A}{1keV}\right]^{r_2}, \quad r_1 \simeq 0.256 \quad , \quad r_2 \simeq 0.153 \tag{6}$$

The quenching factor, exhibited in Fig. 3 for recoil energies appropriate for ^{131}Xe and ^{40}Ar, were obtained assuming a quenching of the form of Eq. (6). In the presence of the quenching factor as given by Eq.(6) the measured recoil energy is typically reduced by factors of about 3, when compared with the electron energy. With the above quenching factor this relationship is shown in Fig. 4. In other words a threshold energy of electrons of $E_{th} = 2$ keV becomes $E'_{th} = 6$ keV for nuclear recoils (see also Fig. 5).

The number of the observed events for each neutrino species is found to be:

$$N_{ev}(a, T) = \Phi_\nu(a, T)\sigma(a, T, E_{th})N_N(P, T_0, R) \tag{7}$$

$$N_N(P, T_0, R) = \frac{P}{kT_0}\frac{4}{3}\pi R^3 \tag{8}$$

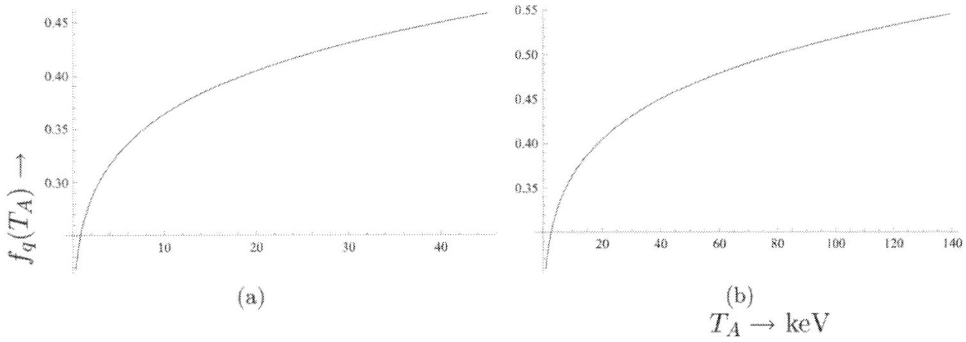

Figure 3: The quenching factor as a function of the recoil energy in the case of A=131 (a) and A=40 (b).

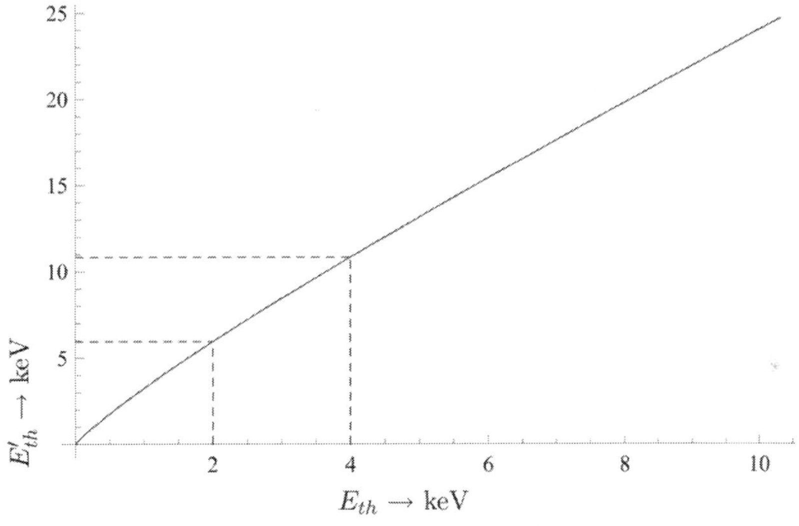

Figure 4: Due to quenching the threshold energy is shifted as shown in this figure, i.e. upwards from E_{th} to E'_{th}

where N_N is the number of nuclei in the target, which depends on the pressure, (P), the absolute temperature, (T_0) and the radius R of the detector. We find:

$$N_N(P, T_0, R) = 1.04 \times 10^{30} \frac{P}{10 \text{ Atm}} \frac{300\,^0\text{K}}{T_0} \left(\frac{R}{10\text{m}}\right)^3 \tag{9}$$

3.1. Results for the Xe target

The differential cross section for neutrino elastic scattering, obtained with the above neutrino spectrum, on the target $^{131}_{54}$Xe is shown in Fig. 5. Integrating the total cross section of Fig. 5 from $T_A = 0$ to infinity we obtain the total cross section given in table 4. The above results refer to an ideal detector operating down to zero energy threshold. In the case of non zero threshold the event rate is suppressed as shown in Fig. 6. The total cross section is, of course, also affected by the quenching factor. This is exhibited in Fig. 7. The effect of quenching is quite important.

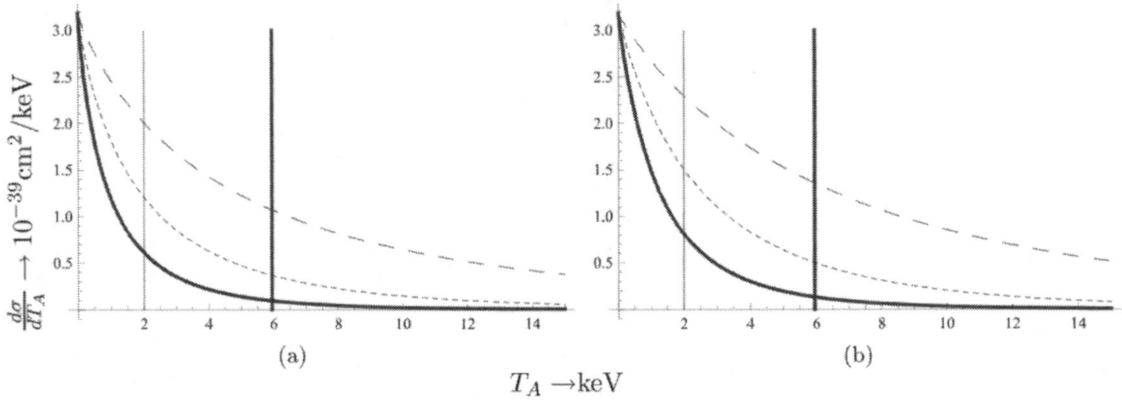

Figure 5: The differential cross section for elastic neutrino nucleus scattering in the case of the target $^{131}_{54}$Xe as a function of the recoil energy T_A in keV. In the case of a threshold energy of 2 keV, only the space on the right of the vertical bar is available. In the presence of quenching (see below) only the space on the right of the thick vertical bar is available. Otherwise the notation is the same as in Fig. 1

Table 4: The total neutrino nucleus cross section in the case of Xe target in units of 10^{-39}cm^2 assuming zero detector threshold.

a	$\sigma/10^{-39}\text{cm}^2$			
	ν_e T=3.5 MeV	$\tilde{\nu}_e$ T=5 MeV	$\sum_x \nu_x$ T=8 MeV	Total
0	4.117	8.312	19.764	32.194
0.75	4.361	8.815	20.921	34.097
1.50	4.749	9.608	22.727	37.083
2.00	5.104	10.330	24.346	39.780
3.00	6.074	12.288	28.621	46.083
4.00	7.408	14.966	34.147	56.521
5.00	9.118	18.364	40.546	68.028

Using Eq. 7 and the above total cross sections, after summing over all neutrino species (i.e. over all T), we obtain the number of events shown in Table 5. In the presence of a detector threshold of even 1 keV the above rates are reduced by about 20% (50%) in the absence (presence) of quenching.

3.2. The Ar target

The differential cross for neutrino elastic scattering on the target $^{40}_{18}$Ar is shown in Fig. 8. For comparison we are currently calculating the differential cross sections to the excited states of ^{40}Ar due to the neutral current. We also are going to calculate the charged current cross sections (ν_e, e^-) and $(\tilde{\nu}_e, e^+)$ on ^{40}Ar, which are of interest in the proposal GLACIER, one of the large detectors[3].

[3] V. Tsakstara, private communication

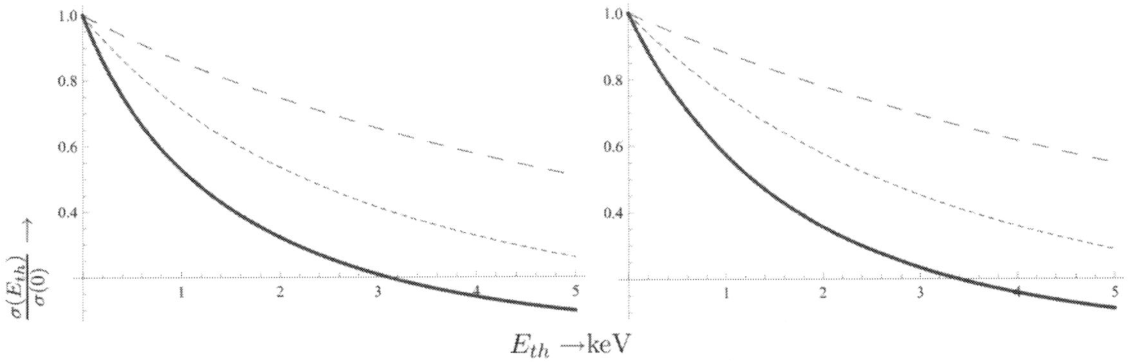

(a) (b)

Figure 6: The ratio of the cross section at threshold E_{th} divided by that at zero threshold as a function of the threshold energy in keV. Otherwise the notation is the same as in Fig. 1

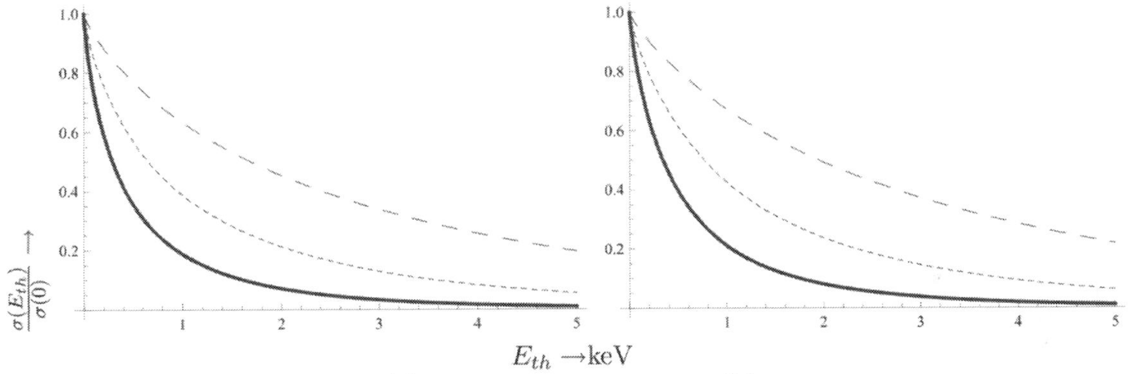

(a) (b)

Figure 7: The same as in Fig. 6
taking into account the effect of quenching.

In the presence of a detector threshold of even 1 keV the above rates are reduced by about 10% (30%) in the absence (presence) of quenching.

3.3. The Ne target
The differential cross for neutrino elastic scattering on the target $^{20}_{10}$Ne is shown in Fig. 11.

In the presence of a detector threshold of even 1 keV the above rates are reduced by about 5% (10%) in the absence (presence) of quenching.

4. Terrestrial sources with similar spectrum

The spherical TPC detector can become a dedicated SN neutrino spectrum. Untill such SN explosion takes place it can be tested using terrestrial neutrino sources with spectra similar to those of SN neutrinos. Two such possibilities come to mind:

Journal of Physics: Conference Series **309** (2011) 012031 doi:10.1088/1742-6596/309/1/012031

Table 5: The total event rate as a function of a in the case of a gaseous Xe target, under a pressure of 10 Atm and temperature 300 ^{0}K, summed over all neutrino species assuming zero detector threshold.

a	R=10m	R=4m
0	10872	696
0.75	11089	710
1.50	11427	731
2.00	11726	750
3.00	12483	799
4.00	13378	856
5.00	14288	914

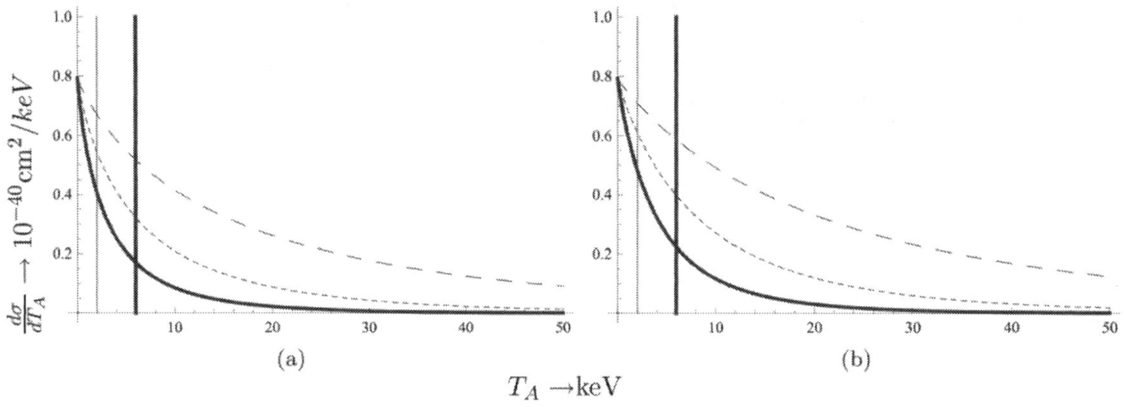

Figure 8: The same as in 5 in the case of the Ar target.

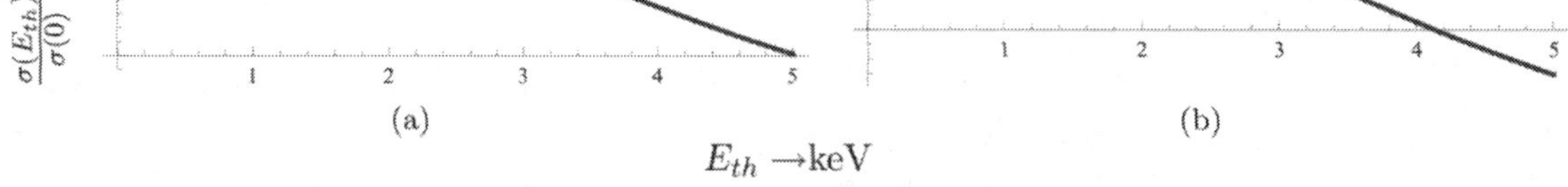

Figure 9: The same as in Fig. 6 in the case of the Ar target.

4.1. The Oak Ridge Spallation Neutron Source (SNS)

SNS is a prolific pulsed source of electron and muon neutrinos as well as muon antineutrinos [Avignone and Effremeko(2003)],[Vergados et al.(2009)Vergados, Avignone, and Giomataris].

Table 6: The total neutrino nucleus cross section in the case of Ar target in units of 10^{-40}cm^2 assuming zero detector threshold.

a	$\sigma/10^{-40}\text{cm}^2$			
	ν_e T=3.5 MeV	$\tilde{\nu}_e$ T=5 MeV	$\sum_x \nu_x$ T=8 MeV	Total
0	3.324	6.520	13.678	23.521
0.75	3.525	6.908	14.412	24.845
1.50	3.843	7.518	15.528	26.888
2.00	4.133	8.067	16.497	28.693
3.00	4.917	9.537	18.905	33.359
4.00	5.990	11.488	21.690	39.168
5.00	7.353	13.843	24.480	45.676

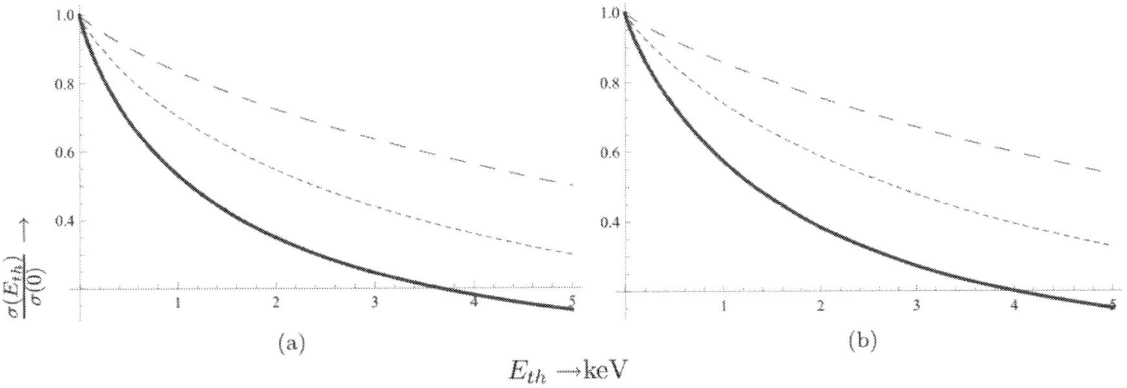

Figure 10: The same as in Fig. 7 in the case of the Ar target.

Table 7: The same as in Table 5 in the case of the target Ar.

a	R=10m 10 Atm	R=4m 10Atm	50 kton (coh)
0	193	12	2.8×10^4
0.75	197	13	2.8×10^4
1.50	203	13	2.9×10^4
2.00	209	13	3.0×10^4
3.00	223	14	3.2×10^4
4.00	242	15	3.5×10^4
5.00	262	17	3.8×10^4

The normalized neutrino spectra can be described very well by the shapes exhibited in Fig. 14 The differential cross sections obtained after folding with the neutrino spectra are shown in Fig. 15. Integrating these differential cross sections we obtain for A=131 at zero threshold the total

Journal of Physics: Conference Series **309** (2011) 012031 doi:10.1088/1742-6596/309/1/012031

Table 8: The same as in Table 5 in the case of the target Ar.

a	R=10m	R=4m
0	193	12
0.75	197	13
1.50	203	13
2.00	209	13
3.00	223	14
4.00	242	15
5.00	262	17

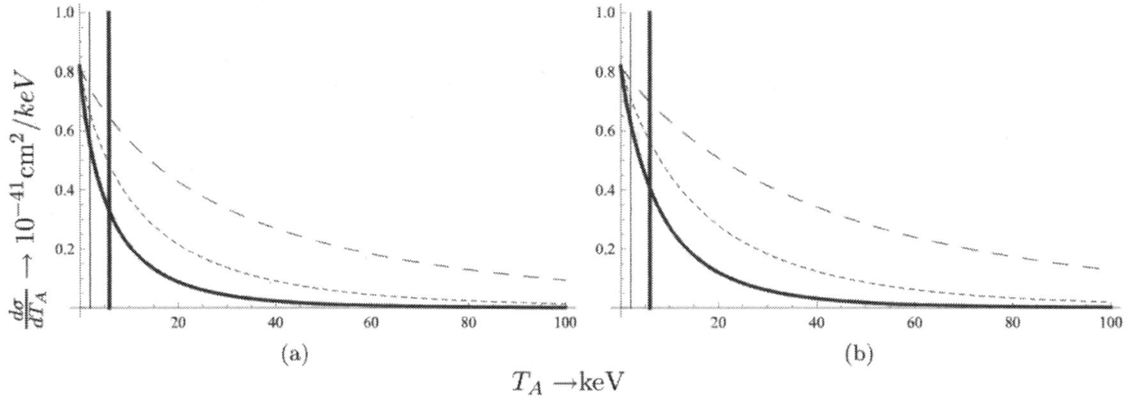

Figure 11: The same as in 5 in the case of the Ne target.

Table 9: The total neutrino nucleus cross section in the case of Ne target in units of 10^{-41}cm^2 assuming zero detector threshold.

a	$\sigma/10^{-41}\text{cm}^2$			
	ν_e T=3.5 MeV	$\tilde{\nu}_e$ T=5 MeV	$\sum_x \nu_x$ T=8 MeV	Total
0	6.861	13.456	28.232	48.548
0.75	7.277	14.258	29.747	51.281
1.50	7.934	15.515	32.049	55.497
2.00	8.531	16.649	34.049	59.229
3.00	10.150	19.683	39.021	68.854
4.00	12.364	23.708	44.772	80.844
5.00	15.176	28.568	50.537	94.281

cross sections:

$$(6.21, 19.1, 7.47) \times 10^{-38}\text{cm}^2$$

for ν_e, $\tilde{\nu}_e$ and ν_μ respectively. In the case of the A=40 we find

$$(5.33, 6.83, 4.58) \times 10^{-39}\text{cm}^2.$$

Journal of Physics: Conference Series **309** (2011) 012031 doi:10.1088/1742-6596/309/1/012031

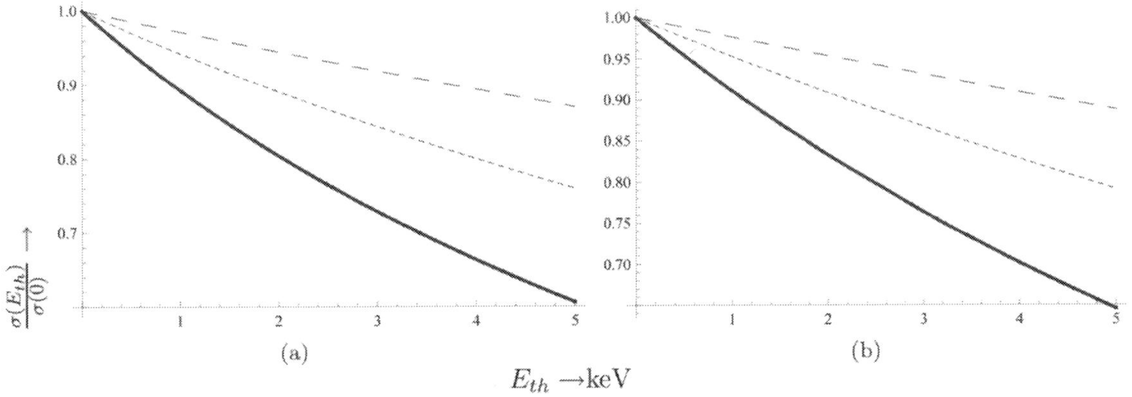

Figure 12: The same as in Fig. 6 in the case of the Ne target.

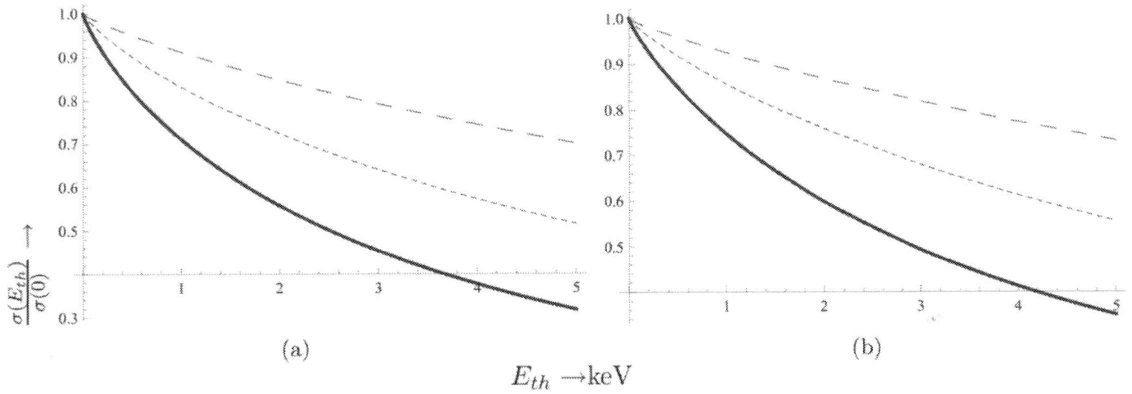

Figure 13: The same as in Fig. 7 in the case of the Ne target.

Table 10: The same as in Table 5 in the case of the target Ne.

a	R=10m	R=4m
0	160	10
0.75	163	10
1.50	167	11
2.00	170	11
3.00	177	11
4.00	185	12
5.00	190	12

the effect of quenching is similar to that of the SN neutrinos.

The SNS facility provides an excellent opportunity to employ and test the spherical TPC gaseous detector [Giomataris et al.(2008)]. There remain, however, some experimental issues that need be resolved first. These have do with the fact that the neutrino flux is changing from point to point inside the detector. The number of expected events for a vessel of volume V

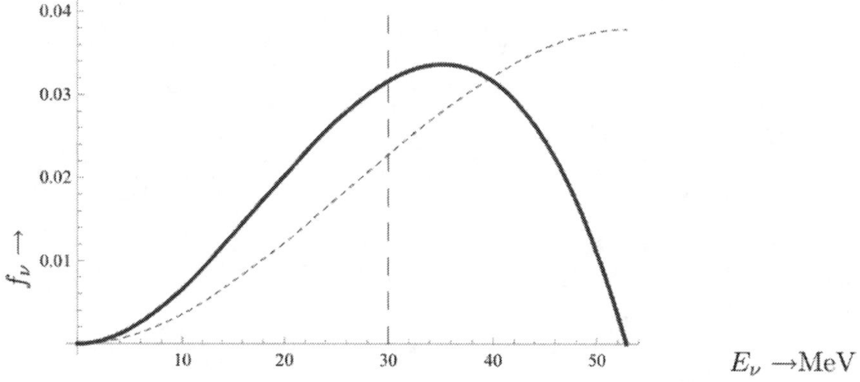

Figure 14: The neutrino spectrum from stopped pions. The normalized solid and dotted curves correspond to ν_e and $\tilde{\nu}_\mu$ respectively. Also shown is the discreet ν_μ spectrum (dashed vertical line).

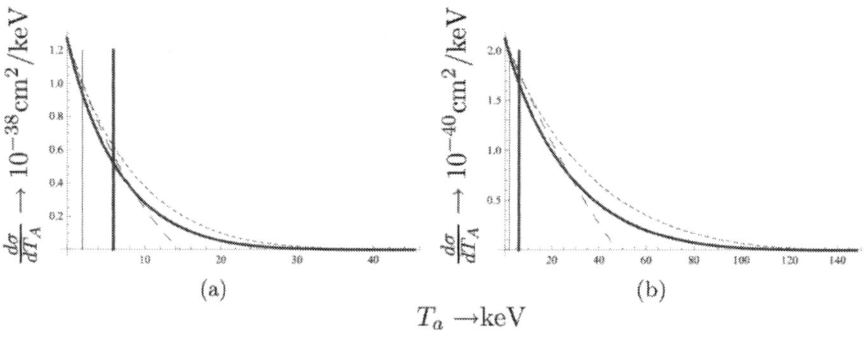

Figure 15: The differential neutrino nucleus cross section, in the case of SNS source, as a function of the recoil energy in keV for the A=131 target (a) and A=40 target (b). The notation is the same as that of the curves of Fig. 14. The fine vertical line corresponds to a threshold of 2 keV. The phase space on its left is unavailable. Due to the presence of quenching the excluded phase space becomes larger (on the left of the thick line), which leads to a smaller total rate.

under pressure P and temperature T takes the form:

$$\mathcal{R} = 3.156 \times 10^7 \frac{t}{1y} \Phi(\nu, L) \sigma(E_\nu, A, N) \frac{PV}{kT} s(V, L) \tag{10}$$

where the parameter $s(V, L)$ is a geometrical factor needed when a large detector is close to the source. It depends on the shape of the vessel and the distance L of its geometric center from the source. In the case of sphere of radius R with its center at a distance L from the source he function s(V,R) depends only on the ratio R/L, i.e. $s(V, L) \rightarrow s(R/L)$, and it is given by

$$s(R/L) = \frac{L^2}{(4/3)\pi R^3} 2\pi L \int_{1-R/L}^{1+R/L} x^2 dx \int_0^\pi d\theta \sin\theta \frac{1}{1 + x^2 + 2x\cos\theta}, \quad x = \frac{r}{L} \tag{11}$$

The above expression takes into account the fact that, if the sphere is close to the source, the neutrino flux is changing from point to point inside the sphere. Spherical coordinates (r, θ, ϕ),

Table 11: The number of events in a year originating from the Spallation Neutron Source (SNS) for nuclear recoils due to neutral current neutrino-nucleus scattering as a function of (P, R, L) where P is the pressure of the gas in the vessel, R is the vessel radius and L is the distance from the source. The detector is at temperature $T = 300\ ^0$K and is assumed to operate with zero threshold.

target	distance	P =5Atm			P =10Atm		
		R =10m	R=5m	R=1.3m	R =10m	R=5m	R=1.3m
$^{131}_{77}$Xe	L=50m	132164.	1241.77	0.133082	264329.	2483.54	0.266163
	L=100m	2483.54	22.7307	0.00234887	4967.08	45.4613	0.00469774
	L=150m	239.866	2.16741	0.000220058	479.733	4.33483	0.000440116
$^{40}_{18}$Ar	L =50m	2795.96	26.2698	0.00281536	5591.92	52.5397	0.00563072
	L=100m	52.5397	0.480871	0.0000496908	105.079	0.961742	0.0000993815
	L=150m	5.07441	0.045852	4.655×10^{-6}	10.1488	0.0917039	9.311×10^{-6}

are used to specify any point inside the sphere. The origin of coordinates was chosen at the center of the sphere with polar axis the straight line from the source to the center of the sphere. With the above choice the flux is independent of the angle ϕ. The function $s(R/L)$ is exhibited in Fig. 16.

Using

$$\Phi(\nu, L = 50\text{m}) = 1.95 \times 10^6 \text{cm}^{-2}\text{s}^{-1}$$

for each flavor we obtain the data of table 11.

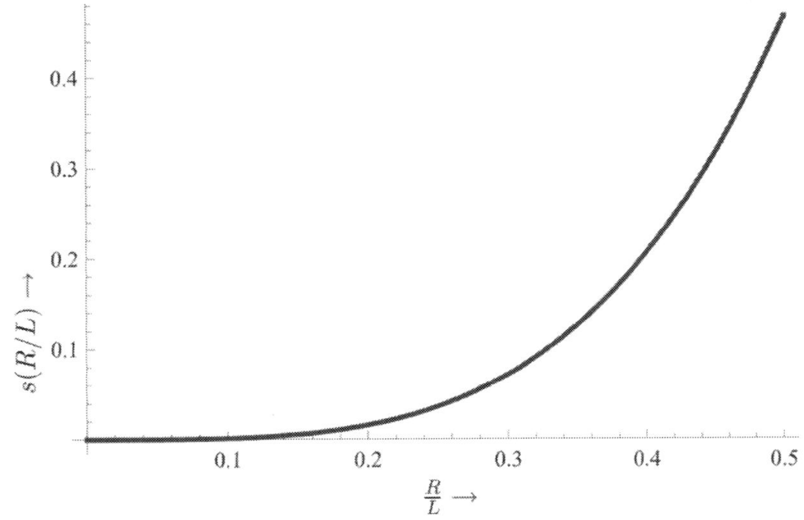

Figure 16: The parameter $s(R/L)$ (see text) in the case of a sphere of radius R whose center is at a distance L from the source.

4.2. Boosted Radioactive Sources

In recent years in the process of developing neutrino factories it has become feasible to obtain radioactive sources [Zucchelli(2002)] for both electron neutrinos and antineutrinos. There exist many candidate sources, which can be boosted up to high energies, with boosting factors $\gamma = E_b/M_b c^2$ ranging up to 20, yielding neutrino spectra not very different from those expected for SN neutrinos[Autin et al.(2003)]. These sources can live long enough ($\tau = \gamma\tau_0$) so that experiments testing the SN detectors can be performed. The spectra of two popular such sources, whose advantages have already been discussed [Autin et al.(2003)], are shown in Fig. 17.

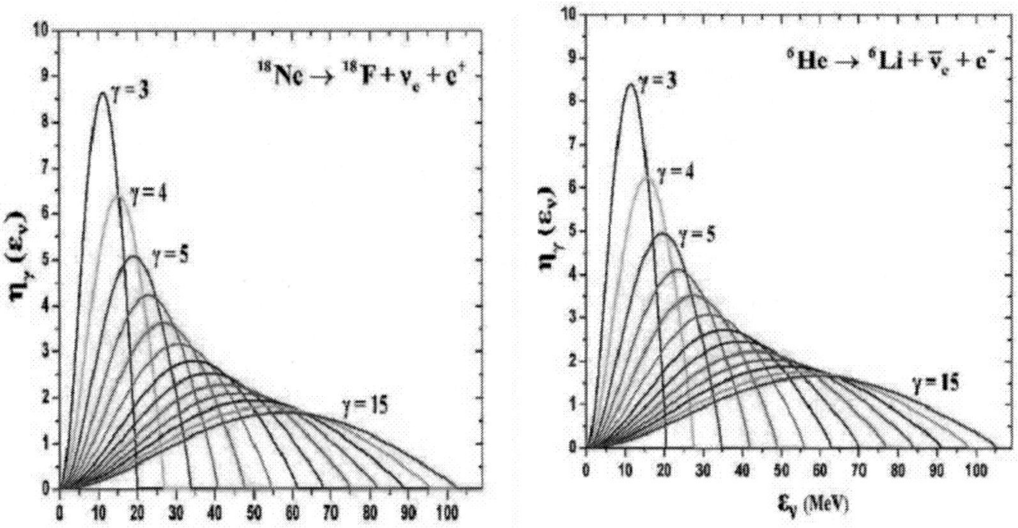

(a) (b)

Figure 17: The spectrum of a typical boosted neutrino source (a) and antineutrino (b) for various values of the boosting parameter γ.

5. Conclusions

From the above results one can clearly see the advantages of a gaseous spherical TPC detector dedicated for SN neutrino detection. The first idea is to employ a small size spherical TPC detector filled with a high pressure noble gas. An enhancement of the neutral current component is achieved via the coherent effect of all neutrons in the target. Thus employing, e.g., Xe at 10 Atm, with a feasible threshold energy of about 100 eV in the recoiling nuclei, one expects between 700 and 900 events for a sphere of radius 5m. Employing ^{40}Ar one expects between 200 and 300 events but with a vessel of larger radius (R=10m). If necessary, this detector can be tested with Earth neutrino sources, which have a neutrino spectrum analogous to that of a SN.

This detector, dedicated for SN detection, should not be viewed as a competitor of the huge constructions currently being envisioned for various purposes including supernova neutrino detection [Autiero et al.(2007)][4]. The information provided by the neutral current detectors, which are not sensitive to neutrino oscillation effects, will provide information about the primary supernova neutrino flux and aid the analysis of detectors exploiting the charged current interaction. It can also serve as an insurance, if anything goes wrong with the large detectors.

[4] See, e.g. in these proceedings: T. Patzak, Laguna: future Megaton detectors in Europe; A Curioni, The GLACIER project and related RD; K. Scholberg, "Supernova Neutrino Detection in Water Cherenkov Detectors"; A. Ianni, "Supernova Neutrino Detection With Liquid Scintillators" ; L. Koepke, Supernova Neutrino Detection with IceCube; I. Gil-Bottella, "Supernova neutrino detection in LAr TPCs".

The second idea is to build several such low cost and robust detectors and install them in several places over the world. First estimates show that the required background level is modest and therefore there is no need for a deep underground laboratory. A mere 100 meter water equivalent coverage seems to be sufficient to reduce the cosmic muon flux at the required level (in the case of many such detectors in coincidence, a modest shield is sufficient). The maintenance of such systems, quite simple and needed only once every few years, could be easily assured by universities or even by secondary schools, with only specific running programs. Admittedly such a detector scheme, measuring low energy nuclear recoils from neutrino nucleus elastic scattering, will not be able to determine the incident neutrino vector and, therefore, it is not possible to localize the supernova this way. This can be achieved by a cluster of such detectors in coincidence by a triangulation technique.

A network of such detectors in coincidence with a sub-keV threshold could also be used to observe unexpected low energy events. This low energy range has never been explored using massive detectors. A challenge of great importance will be the synchronization of such a detector cluster with the astronomical γ-ray burst telescopes to establish whether low energy recoils are emitted in coincidence with the mysterious γ bursts.

In summary: networks of such dedicated gaseous TPC detectors, made out of simple, robust and cheap technology, can be simply managed by an international scientific consortium and operated by students. This network comprises a system, which can be cheaply maintained for several decades (or even centuries). Obviously this is is a key point towards preparing to observe few galactic supernova explosions.

Acknowledgement

The author would like to express his appreciation to the organizers of the "FIFTH SYMPOSIUM ON LARGE TPCs FOR LOW ENERGY RARE EVENT DETECTION and workshop on neutrinos from Supernovae" for their financial support and hospitality.

References

[Tomas et al.(2004)Tomas, Kachellriess, G.G. Raffelt, Janka, and Schreck] R. Tomas, M. Kachellriess, A. D. G.G. Raffelt, A.-T. Janka, and L. Schreck, JCAP **0409**, 015 (2004).

[Janka et al.(2007)Janka, Langanke, Marek, Martinez-Pinedo, and Mueller] H. T. Janka, K. Langanke, A. Marek, G. Martinez-Pinedo, and B. Mueller, Phys. Rep. **442**, 38 (2007).

[Giomataris and Vergados(2006)] Y. Giomataris and J. Vergados, Phys. Lett. **B 634**, 23 (2006), hep-ex/0503029.

[Lifhardt et al.(1963)] J. Lifhardt et al., Mat. Phys. Medd Dan Vid. Selsk **33**, 1 (1963).

[Simon et al.(2003)] E. Simon et al., Nucl. Instr. Meth. **A 507**, 643 (2003).

[Avignone and Effremeko(2003)] F. A. Avignone and Y. V. Effremeko, J. Phys. G: Nuc.Part. Phys **29**, 2615 (2003).

[Vergados et al.(2009)Vergados, Avignone, and Giomataris] J. D. Vergados, F. T. Avignone, and I. Giomataris, Phys. Rev. D **79**, 113001 (2009), arXib:0902.1055 [hep-ph].

[Giomataris et al.(2008)] I. Giomataris et al., JINST **3**, P09007 (2008).

[Zucchelli(2002)] P. Zucchelli, Phys. Lett. B **532**, 166 (2002).

[Autin et al.(2003)] B. Autin et al., J. Phys. G **29**, 1785 (2003), arXiv:Physics/0306106.

[Autiero et al.(2007)] D. Autiero et al., JCAP **0711**, 011 (2007), arXiv:0705.0116 [hep-ph].